W9-BZG-425

BASIC DIEMAKING

Prepared under the direction of the Training Manual

Basic Diemaking

Committee of the National Tooling & Machining Association

by D. Eugene Ostergaard

GLENCOE
McGraw-Hill
New York, New York Columbus, Ohio Mission Hills, California Peoria, Illinois

OAKTON COMMUNITY COLLEGE
DES PLAINES CAMPUS
1600 EAST GOLF ROAD
DES PLAINES, IL 60016

BASIC DIEMAKING

Imprint 1997

Send all inquiries to:
Glencoe/McGraw-Hill
936 Eastwind Drive
Westerville, Ohio 43081

Library of Congress Catalog Card Number: 63-10842

ISBN: 0-07-046090-6

31 32 33 34 35 36 37 38 004 03 02 01 00 99 98 97

NATIONAL TOOLING & MACHINING ASSOCIATION
TRAINING MANUAL COMMITTEE

Rolf H. Berg (President, NTMA), Atols Tool & Mold Corp., Schiller Park, Illinois
Herbert Harig (Chairman), Harig Manufacturing Corp., Chicago, Illinois
William J. Fortin (Vice Chairman), Pamco Co., Fraser, Michigan
David J. Bathgate, Oval Tool & Die Corp., Detroit, Michigan
Donald W. Darrone, Allen Tool Corp., Syracuse, New York
John D. Dewhurst, Arrow Tool Co. Inc., Wethersfield, Connecticut
J. Kleinoder, Volkert Stampings, Inc., Queens Village, Long Island, New York
Archie McMillan, Helfrecht Machine Co., Saginaw, Michigan
Richard F. Moore, Moore Special Tool Co. Inc., Bridgeport, Connecticut
Jack H. Schron, Jergens Tool Specialty Co., Cleveland, Ohio
Donald E. Sutton, Northern Tool and Die Co., Chicago, Illinois
Edward J. Stanek, Stanek Tool Corp., Milwaukee, Wisconsin

ACKNOWLEDGMENT

The National Tooling & Machining Association decided in the mid-fifties to undertake the preparation and publication of texts devoted to apprentice training. In the succeeding years many members of the association have devoted time and effort to the project. None, however, has given of himself so continuously and unsparingly as Herbert Harig, a former president of the association, who has served as chairman of the Training Manual Committee from the very beginning of the project. He deserves the unstinting commendation of his fellow-members in the association and, indeed, of everyone interested in the training of skilled craftsmen.

ROLF H. BERG, *President*
National Tooling & Machining Association

PREFACE

When trouble is encountered either in building a die or in the way it functions after it is made, the trouble can all too often be traced to an inadequate grasp of the basics of die design and construction. It is the purpose of this book to present a practical nucleus of basic die-construction facts around which successful careers in the field of diemaking can be established. A bridge can, of course, be crossed from either end. For industrial training programs, this book explains the "whys" underlying applied diemaking and operation.

Essential facts of cutting and forming operations are explained and related to the manner in which the dies must function in order to achieve the desired results. Primary die components such as punches, punch plates, die blocks, strippers, etc. are discussed as individual entities in addition to their function as part of the complete die. Efficient utilization of the stock material is described. Methods of achieving efficient stock-strip layouts are explained, and the reasoning which determines the optimum choice of stock-strip configurations is discussed.

Each subject is begun at its necessary foundation level. It is then developed and expanded to the point where the diemaker can proceed on his own initiative, confident that he has a firm foundation supporting him.

Each subject is thoroughly illustrated, and the illustrations have been carefully selected in order to assure maximum pertinency. They have been kept as simple as possible in order to depict clearly the specific point of discussion.

It is obvious that the information presented in this book can be an asset not only to the potential diemaker, but to anyone concerned with tooling and/or production. Teachers have found that learning is facilitated when the study of a subject can be associated with practical application. It is an inherent feature of this book that, because of the subject matter and its presentation, it bridges what would otherwise be a gap between study and application. Therefore, it can be of considerable value in the education of technical students.

In addition to use in technical and vocational schools, this book has been designed as a text for apprentice-related training. We recommend that each apprentice have a copy, and that the instructor (journeyman diemaker, foreman, engineer, or any other qualified man) teach the apprentice in the classroom using "Basic Diemaking" as an integral part of the course of study.

We also recommend that the apprentice be introduced to this related training after he has learned the basic requirements of metal fabrication in the shop, as well as the use of measuring instruments and hand tools. It is advisable that his first classroom training be shop mathematics and blueprint reading before starting with this textbook.

The time it will take for an apprentice to get this preliminary training and experience will vary from shop to shop. It is our opinion that a logical point for an apprentice to start with this book would be approximately the beginning of his second year of apprenticeship.

It is intended that the instructor use the lecture approach in the classroom, with this textbook giving him his subject material. The student can study as homework a preassigned segment of the book and should be prepared to enter into class discussions and/or tests following the lecture.

This textbook is an ideal reference and refresher manual for the journeyman tool and diemaker. There are some fundamentals that are not recognized by many tool and diemakers which they could easily pick up by studying "Basic Diemaking." Therefore, it is our recommendation that diemakers purchase their own books or that employers have some copies available in a company library for employee use. Tool-and-die designers and engineers can also obtain much information applicable to their everyday work from this book.

It is hoped that college and university educators will accept "Basic Diemaking" as a part of the curriculum of mechanical engineering. The same principles recommended for teaching the apprentice in the classroom can be applied to the college student in either a two- or four-year program.

HERBERT HARIG, *Chairman*
Training Manual Committee
National Tooling & Machining Association

CONTENTS

CHAPTER 1 PRINCIPLES OF BLANKING AND/OR PIERCING DIES

INTRODUCTORY TERMINOLOGY

Before beginning the study of diemaking it is necessary to attain a clear understanding of the following terms:

piece part: A piece part is the product of a die. It may be a complete product in itself (the draftsman's erasing shield and a bottle cap opener are examples of this kind of product) or it may be only one component of a product consisting of many and different parts, such as an electrical terminal used in a television set. The die may or may not produce the piece part in a finished state.

stock material: General term for any of the various materials from which the piece part is made.

die: The word "die" has several definitions. This book utilizes two: (*a*) a complete production tool, the purpose of which is to produce piece parts consistently to required specifications, and (*b*) the female part of a complete die.

punch: A punch is a male member of a complete die which mates or acts in conjunction with the female die to produce a desired effect upon the material being worked. A die can be a simple tool composed of punch, die block, and stripper, or it can be an extremely complex mechanism which performs many and varied operations.

PIECE PARTS PRODUCED BY BLANKING AND/OR PIERCING

The piece part in Fig. 1·1 is produced by blanking. It is shown in relation to the stock strip from which it was blanked. The terms used to describe the various aspects of the work are indicated. The stock strip is stock material of the required thickness *T* which has been cut to strips of suitable width for the particular job. The stock strip in this instance is fed through the die from right to left, as indicated by the feed direction arrow. The advance is the distance the stock must be fed (advanced) to allow a clean blanking operation at each press stroke. The scrap bridge is the scrap remaining between the openings in the stock strip after blanking. The advance distance is equal to the scrap opening plus the scrap bridge. The back scrap is so called because it is located toward the back of the press (away from the operator). The front scrap is on the side toward the front of the press (toward the operator). The lead end of the stock strip is the end that is fed into the die first. The opposite end of the strip is called the tail.

The piece part shown in Fig. 1·1 would be called a blank if further work were required to complete it. This is illustrated in Fig. 1·2, where the piece part is the same as the first one except that there is an opening in it. When a die is used to make an opening in a blank or piece part, the operation is called piercing. The slug is the material that is cut out of the blank or piece part by the piercing operation.

To summarize, the piece part illustrated in Fig. 1·1 is produced by blanking. The piece part in

Figure 1·1 Relationship of piece part and stock strip (blanking operation).

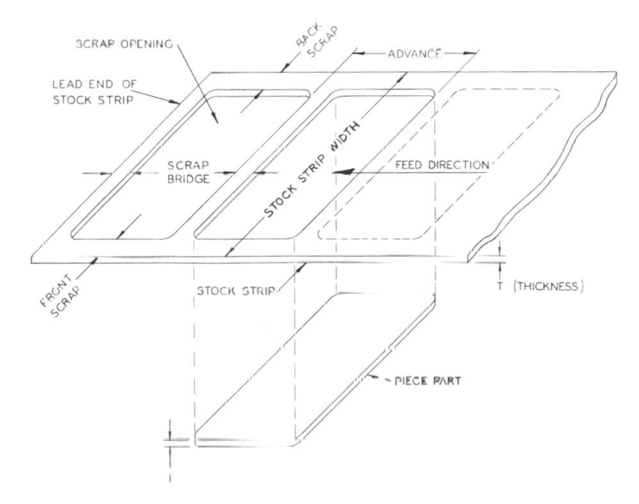

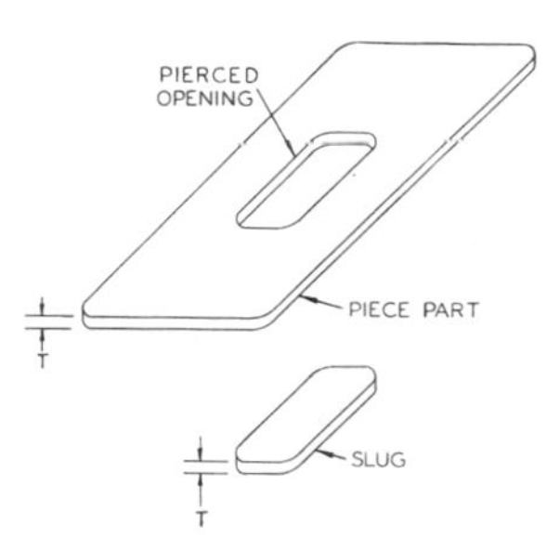

Figure 1·2 Relationship of piece part and slug (piercing operation).

Fig. 1·2 is made by blanking and piercing, and the slug is a result of the piercing operation.

An endless number and variety of stampings are produced by blanking and/or piercing. In addition, there is another endless variety of piece parts in which blanking and/or piercing complement other pressworking operations.

BASIC BLANKING OR PIERCING OPERATION

Whether the die shown in Fig. 1·3 is a blanking die or a piercing die depends solely upon its intended use. It is called a blanking die if it is meant to produce blanks *B* of a desired contour and size by cutting them out of the required type of material *A*. The blanks are the desired product (piece parts) made by the die. In most cases, the material remaining after the blanks have been cut out is considered scrap.

If the purpose of the die is to make openings of a desired contour and size in the required material *A*, it is called a piercing die. In this case *A* represents the piece part and *B* is called the slug. The slug is usually considered scrap.

Figure 1·3 Piercing or blanking die—closed position (bottom of press stroke).

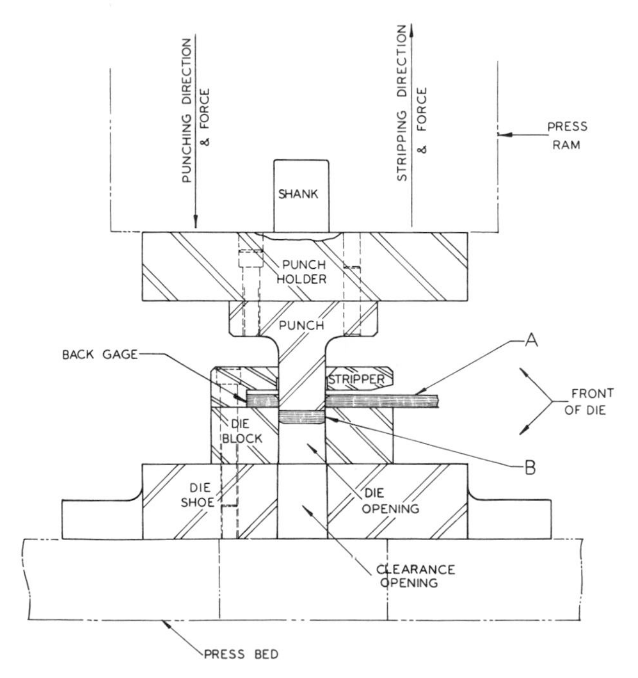

The basic elements of both dies are identical. The name of the die is derived from its intended use. The physical effects of the tool upon the stock material are the same whether it is punching out slugs in order to produce a desired opening in a piece part or whether it is punching out blanks which are the desired product of the die.

DESCRIPTION OF BLANKING OR PIERCING DIE

Figure 1·3 shows the basic elements of a blanking or piercing die. These elements are the die block in which the proper female die opening has been made, the punch, and the stripper. (The back gage which locates the back edge of the stock material is, in this instance, a part of the stripper.) They are mounted on a die set in order to achieve and retain proper matching of the punch and the die opening (see Chap. 16, *Die Sets*). The die block and stripper are secured to the die shoe by means of screws and dowels. The punch is screwed and doweled to the punch holder.

When the die is set up or mounted in the punch press, the punch holder is secured to the ram of the press and, of course, moves with the ram. The die shoe is secured to the press bed and remains stationary.

PRESS CYCLE WHEN BLANKING OR PIERCING

The "at rest" position of the press ram is at the top of the stroke. The ram is then at its greatest distance from the bed of the press. When the press is tripped, the clutch allows engagement of the crankshaft with the press flywheel. The crankshaft then rotates through 360° or one full turn. During the first half of the cycle, the ram is driven toward the press bed. During the last half of the cycle, the ram moves away from the press bed. The distance traveled by the ram during one half cycle (top of stroke to bottom of stroke) is called the press stroke.

ACTION OF BLANKING OR PIERCING DIE

Refer to Fig. 1·3. The stock material *A* is fed or loaded in the proper position on the top surface of the die block. When the press is tripped, the ram drives the punch through the stock material *A* into the die opening, thereby producing an opening in the stock material by cutting out the blank or slug *B*. This blank or slug remains in the die opening when the punch is withdrawn and is pushed through the die by the blanks or slugs produced subsequently.

Stripping. After the blanking or piercing has taken place, the punch is returned to the open position by the press ram as it completes its cycle. The stock material clings to the punch and will remain on the punch unless something is done to prevent it. This is the function of the stripper: it keeps the stock material from traveling with the punch on the return stroke. Because the stock material is held back by the stripper, the punch is withdrawn from the material on the return stroke.

REACTION OF THE STOCK MATERIAL: SHEARING ACTION

The result of the forces imposed on the stock material by the working of the blanking or piercing dies is a shearing action. This shearing action may be considered in three stages which are important to the diemaker because of their direct relationship to the dimensional qualities and appearance of piece parts. They are also related to the effective working and life of the die. These stages showing the reaction of the stock material are illustrated progressively in Fig. 1·4.

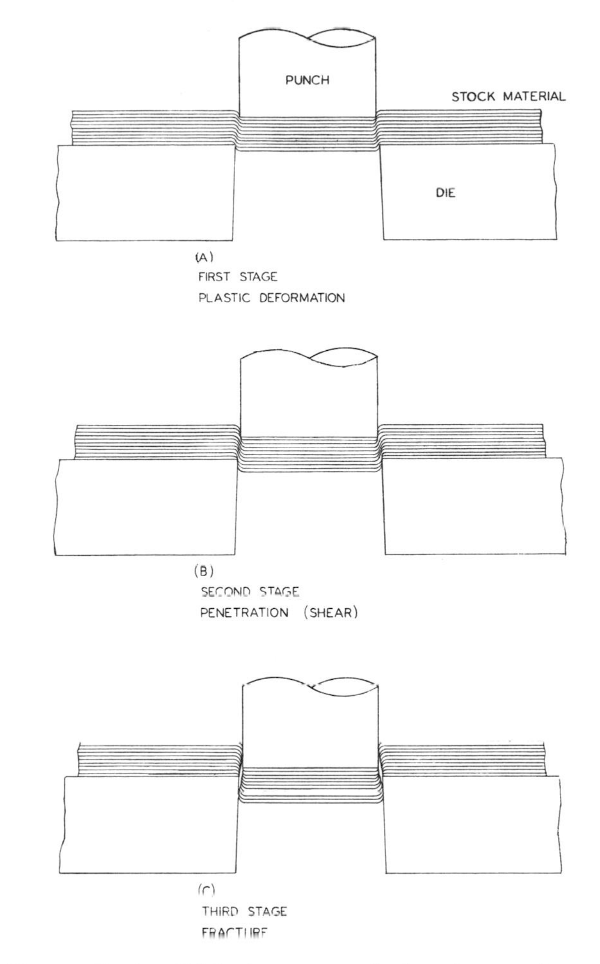

Figure 1·4 Critical stages of shearing action on metal.

Critical stages of shearing action on metal. *First Stage—Plastic Deformation, View A.* The stock material has been placed on the die, the press has been tripped, and the punch is being driven toward the die. The punch contacts the stock material and exerts pressure upon it. When the elastic limit of the stock material is exceeded, plastic deformation takes place.

Second Stage—Penetration, View B. As the driving force of the ram continues, the punch is forced to penetrate the stock material, and the blank or slug is displaced into the die opening a corresponding amount. This is the true shearing portion of the cutting cycle, from which the term "shearing action" is derived.

Third Stage—Fracture, View C. Further continuation of the punching pressure then causes fractures to start at the cutting edges of the punch and the die. These are the points of greatest stress concentration. Under proper cutting conditions, the fractures extend toward each other and meet. When this occurs, the fracture is complete and the blank or slug is separated from the original stock material. The punch then enters the die opening, pushing the blank or slug slightly below the die cutting edge.

These three stages of shear action are responsible for the characteristic appearance of piece parts produced by blanking and/or piercing.

Figure 1·5 Optimum cutting clearance.

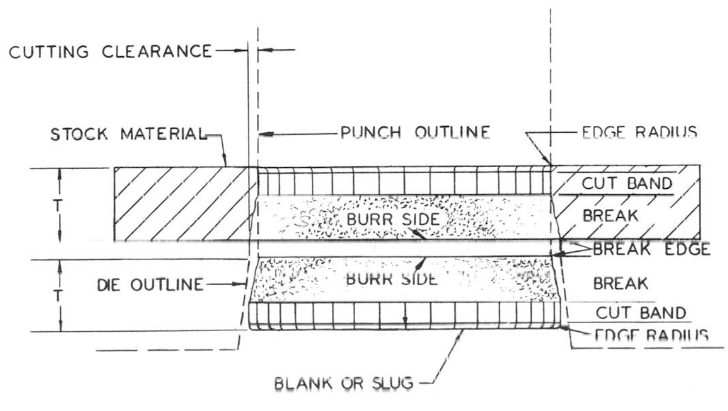

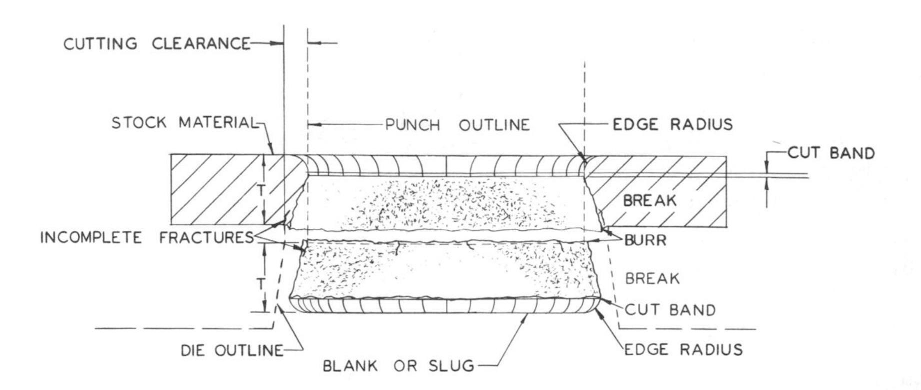

Figure 1·6 Excessive cutting clearance.

TYPICAL APPEARANCE CHARACTERISTICS

The appearance characteristics of material that has been blanked or pierced are factors of significance to the tool- and diemaker. A visual check or examination of these characteristics tells him whether or not the punch and die have the proper amount of clearance between them. He can also tell whether the punch is properly aligned with the die opening. The relationship of the appearance characteristics of piece parts to the cutting clearance is illustrated in Fig. 1·5.

Optimum cutting clearance. In Fig. 1·5 the blank or slug has been made under optimum cutting conditions. The edge radius is the result of the initial plastic deformation which occurred during the first stage of the shear action.

The highly burnished band, resulting from the second stage (penetration) of the shearing action, is the cut band. The width of the cut band is approximately ⅓ the thickness *T* of the stock material. The balance of the cut is the break, which results from the third stage (fracture) of the shearing action.

Cutting clearance is the space between a side of the punch and the corresponding side of the die opening at the cut edge when the punch is entered in the die opening. Therefore, the cutting clearance should always be thought of and expressed as the amount of clearance *per side.*

Burr side. The burr side is adjacent to the break. The burr side is so called because, if a noticeable burr condition develops, it will occur on this side. Burrs should be practically nonexistent if the cutting clearance between the punch and the die is correct and if the cutting edges are sharp. In fact, when a die is running in production, the degree of burr on the piece parts is an indication whether the die is ready for sharpening.

Refer to Fig. 1·5 and note that the characteristics of the blank or slug and the punched opening are inversely identical. The burr side of a blank or slug is *always* toward the punch. The burr side of a punched opening is *always* toward the die opening.

Excessive cutting clearance. Figure 1·6 illustrates the result of excessive cutting clearance. The comparatively large space between the punch and die cut edges allows the stock material to react to the initial punch pressure in a manner approaching that of forming rather than cutting.

The edge radius becomes larger and does not blend smoothly into the cut band. The cut band becomes smaller, sometimes degenerating to a mere line of demarcation between the break and the edge radius. The break shows greater irregularities due to tearing. These irregularities may extend into the cut band and occasionally into the edge radius. The stock material has been forced into the clearance space, and when the break occurs, large burrs are present at the break edge.

Residual incomplete fractures may sometimes be found in the finished piece parts. This is because the material starts to fracture in the normal way, beginning at the cut edges, but cannot complete a normal fracture due to the excessive clearance, which allows a new path of least resistance to be set up.

Insufficient cutting clearance. When the cutting clearance is slightly too small, the condition may be identified by a greater width and irregularity of the cut band. If, however, the proportional cutting clearance is further decreased, the stock material may react by showing two or more cut bands, as pictured in Fig. 1·7. This is a somewhat idealized concept of the actual condition, but it serves to typify the reaction of the stock material to insufficient cutting clearance. Because of the steeper angle between the punch and die cut edges, the resistance of the stock material to fracture is increased. The resulting accumulation of pressure may cause the initial fractures to originate in the clearance rather than at the cut edges.

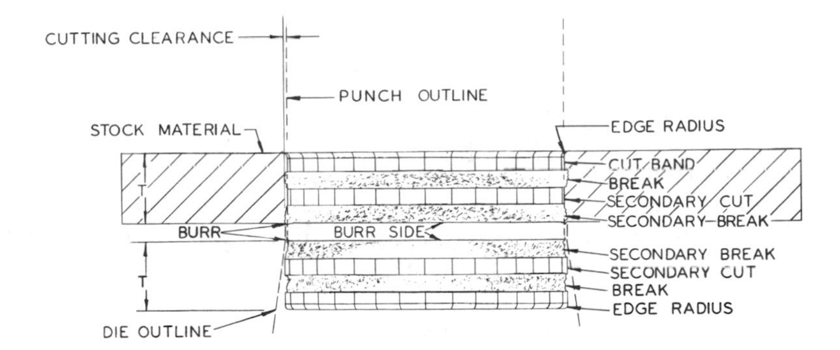

Figure 1 · 7 Insufficient cutting clearance.

If the fractures were extended, they would not meet and cause a clean and complete fracture. As a result, a partial break occurs. The pressure accumulation then causes a secondary cutting action to take place, resulting in a secondary break.

Objectionable burrs may appear on the piece parts if the cutting clearance is insufficient. This may seem paradoxical in view of the fact that excessive cutting clearance also causes burrs, but it is nevertheless true. In the case of excessive clearance the burr results from dragging of the material. With insufficient clearance the burr is caused by compressive forces.

Misalignment of punch and die. As shown in Fig. 1·8, the cutting characteristics indicate whether the punch and the die opening are in accurate alignment. This indication is of value to the tool- and diemaker when he is mounting his die. It also enables him to detect and correct misalignment conditions when they develop during the life of the die.

Importance of cutting clearance. Proper cutting clearance is necessary to the life of the die and the quality of the piece part. Excessive cutting clearance results in objectionable piece-part characteristics; insufficient cutting clearance causes undue stress and wear on the cutting members of the tool because of the greater punching effort required.

Exceptions. However, round cutting members can operate successfully with less than normal cutting clearance. Reduced cutting clearance is used in some cases to overcome the hazard of slug pulling. In other cases it is used where a wider cut band on the piece part is desirable.

In addition, fragile punches and die sections often benefit from using slightly greater than normal cutting clearance. Less punching force is required, and this, of course, reduces the stresses on the cutting members.

Determining cutting clearance. The physical properties and the thickness of the stock material are the factors that determine the amount of cutting clearance. The thickness is easily measured, but the physical properties in relation to cutting clearance are not. Therefore, the optimum clearance must often be determined by actual experiment. Always keep in mind that it is relatively simple to increase the amount of cutting clearance, but to decrease it may require the remaking of an entire punch or die block. When in doubt, the tool- and diemaker starts out with less cutting clearance than he estimates to be necessary. He then makes trial punchings in the proper stock material and "opens up" or increases the cutting clearance by removing stock from the punch or die opening, whichever is appropriate.

Cutting clearance should be expressed in terms of percentage of stock material thickness *per side.*

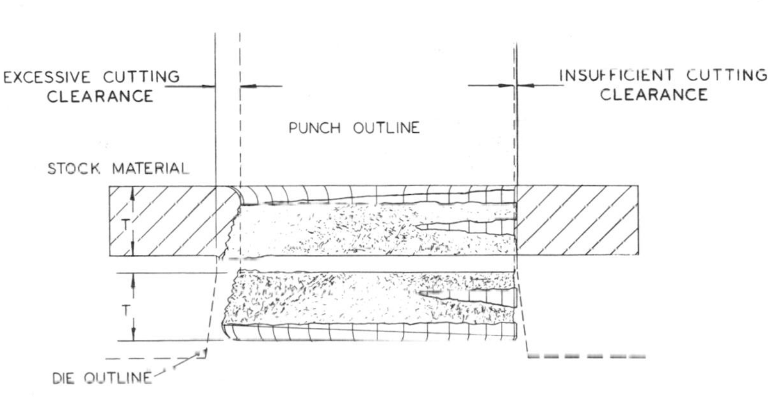

Figure 1 · 8 Misalignment of punch and die opening.

The percentage varies with the properties of the material. A suggested list of percentages for various materials is given in Fig. 1 • 19. Mica, fiber materials, plastics, etc., generally require less cutting clearance than any of the metals.

There is another result of the necessity for cutting clearance that must be studied and thoroughly understood by the tool- and diemaker. This is the effect of cutting clearance on the actual dimensions of the piece part as shown in Fig. 1 • 9.

RELATIONSHIP OF PIECE-PART SIZES TO PUNCH AND DIE SIZES

When pierced or blanked piece parts are measured, the measurement is made at the cut band. The blank or slug measures larger than the opening in the stock material from which it was punched. The reasons for this are obvious upon examination of Fig. 1 • 9. The actual cutting of the blank or slug is done by the cutting edge of the die opening. Therefore, the die opening determines the size of a blank or slug. The actual cutting of the opening in the stock material is done by the punch. Therefore, the size of a punched opening is determined by the punch. It is imperative that the tool- and diemaker have a clear understanding of this. It enables him to know which cutting members must be made to piece-part size and which ones should have the cutting clearance applied to them.

Figure 1 • 9 Piece-part sizes in relation to punch and die size.

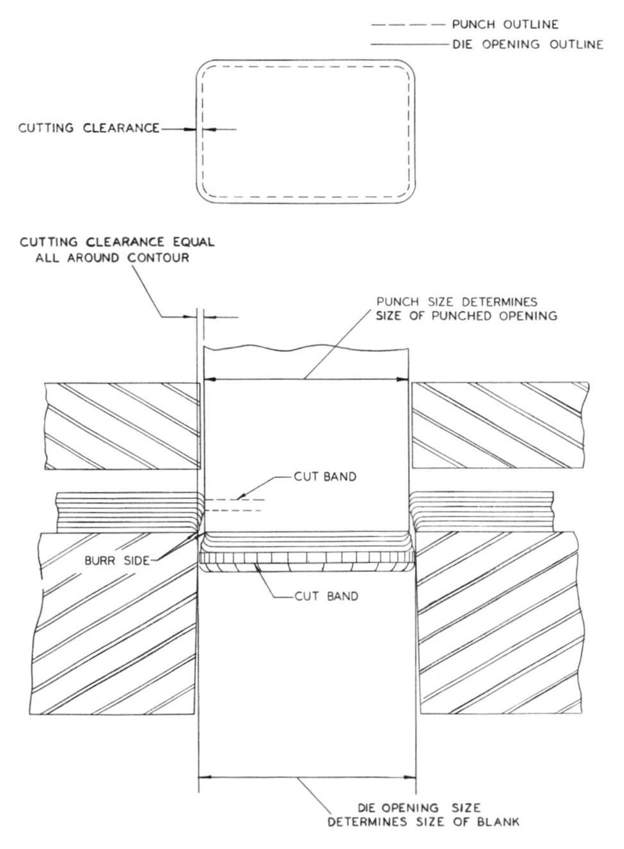

A careful dimensional check of a blank or slug will often reveal that its overall dimensions are slightly larger than the die opening that produced it. This is because the cutting action caused it to be compressed in the die opening. After it passed through the die the pressure was released; therefore, it expanded a slight amount. Conversely, a punched opening will often be slightly smaller than the punch which produced it. When the punch was withdrawn from the stock material during the stripping stage, the opening closed slightly. Most materials react this way. Satisfactory compensation for this condition is usually achieved by making the punch 0.0005 to 0.001 in. larger overall than the desired opening to be punched. If the blank is the desired product, the die opening is then made 0.0005 to 0.001 in. smaller overall than the desired blank.

ANGULAR CLEARANCE

Angular clearance is a draft or taper applied to the sidewalls of a die opening (Fig. 1 • 10) in order to relieve the internal pressure of the blank or slug as it passes through the opening.

Importance of angular clearance. In any blanking or piercing operation where the blank or slug passes through the die, it is imperative that angular clearance be provided on the sidewalls of the die opening. The importance of angular clearance cannot be overemphasized. The "blank-through" type of die will not run successfully unless there is relief for the pressure developed within the die opening by the blanks or slugs as they are forced through. This pressure can build up rapidly; failure to relieve it causes distortion and burring of the material passing through the die opening and culminates in serious damage to the die. Lack of angular clearance can result in pressure accumulations so great that punches break and die blocks burst. In some instances even the punch press has suffered severe damage because of blanks or slugs becoming jammed in the die opening.

Specifying angular clearance. Angular clearance should be expressed in terms of the amount of clearance *per side,* not as an overall or included angle figure. The optimum angular clearance for any given die opening is a variable. It depends upon the type of material to be run, the production requirements, and the method of die construction. The strength of the cutting members also influences the amount of angular clearance to be used.

Generally, soft materials require greater angular clearance than hard. Aluminum especially tends to jam or pack in die openings. Heavy gages of material require more angular clearance than thinner material of the same kind.

Maximums and minimums of angular clearance. With few exceptions, an angle of 2° per side can be considered a maximum desirable clearance angle. A practical minimum clearance angle is 0.002 in. per in. per side (⅛°). This minimum angle, however, is practical only for very-high-production dies having ground die openings. It is considered good general practice to use angles ranging from ¼ to ¾°, depending on the production requirements and the type of material to be run in the die.

Methods of providing angular clearance. Figure 1·10 illustrates the application of angular clearance to die openings. In view *A*, the opening is made with a cutting land contiguous to the cutting edge and extending into the die opening. The height of the cutting land should be equal to the thickness of the stock material but should not be less than 1/16 in. A height of ⅛ in. is considered good average practice for cutting material less than ⅛ in thick. This portion of the die opening must be made with maximum care. The cutting land should be perfectly straight and parallel to the punching direction or, especially in the case of filed die openings, should have a very slight angle to ensure against "bell mouthing." In any case, a bell-mouthed or opposite-taper condition should not be permitted to exist.

View *B* shows the angular clearance beginning at the cutting edge. This method is used when it is necessary or desirable to relieve the compression stored within the blanked part as soon as possible. Angular clearance beginning at the cutting edge is the preferred method for cutting materials that are abrasive. Silicon steels and stainless steels, for example, tend to bell-mouth the die opening very rapidly unless the taper begins at the cutting edge. As a general rule, soft materials, especially aluminum, will run better if the angular clearance begins at the cutting edge. Since in this kind of die opening the compression of the blank or slug begins to be relieved immediately after cutting, it is feasible to use a smaller clearance angle. A smaller clearance angle can compensate for the lack of a cutting land insofar as sharpening life is concerned.

When thin stock material is being cut, the angular clearance should be held to a minimum consistent with the production quality requirements. This will minimize the increase in cutting clearance which occurs when the die cutting face is ground away because of repeated sharpenings.

Angular clearance and sharpening. Sharpening is accomplished by grinding off the cutting face of the die block (Fig. 1·10). In view *A*, the size of the die opening remains relatively unchanged until the cutting face has been ground below the depth of the cutting land. Further sharpening causes the die opening to become larger. This enlargement due to angular clearance begins with the first sharpening when the die opening is made as shown in view *B*.

Enlargement of die opening due to sharpening. When sharpening extends below the start of the clearance angle, the amount of increase in die-opening size is directly proportional to the degree of angular clearance. This increase *per side* can be calculated as

$$A = B \tan a$$

where A = increase per side
B = amount removed by grinding cutting face
a = clearance angle, deg

Figure 1·10 Angular clearance.

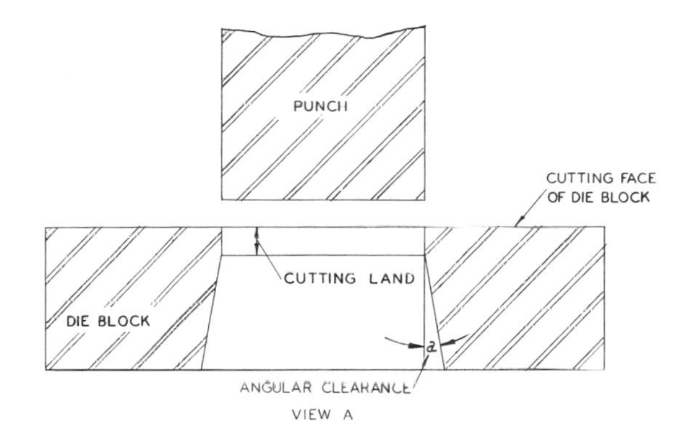

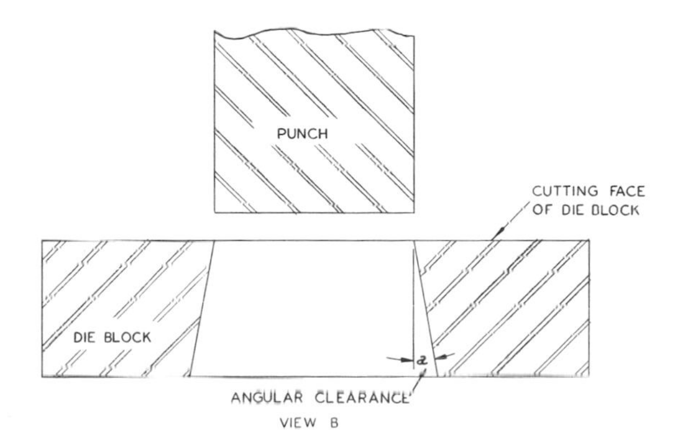

The resultant A added to the amount of cutting clearance before sharpening is the new cutting clearance after sharpening. If it is desired to know what the overall die-opening size will be after sharpening, add $2A$ to its overall dimensions before sharpening.

This formula may also be expressed as

$$\frac{A}{B} = \tan a$$

Expressed this way, the formula is used to aid in selecting the amount of angular clearance to apply to a die opening.

EXAMPLE:

Given a maximum permissible cutting clearance increase of 0.0025 in. and a desired die life of ¼ in. To satisfy these conditions, what degree of angular clearance should the diemaker use when making the die opening? Using the formula $A/B = \tan a$, by substitution we have $0.0025/0.25 = \tan a$, and performing the division gives us $0.01 = \tan a$. Referring to a table of natural trigonometric functions, we find the figure 0.01 lies between the tangent of 34′ and the tangent of 35′. Therefore, the angle which satisfies the given conditions lies between 34 and 35′. However, instead of using the exact angle, we round off to the nearest smaller nominal angle, in this case ½°. The rounding off is done for two reasons: (*a*) It is more convenient to work with a nominal figure; (*b*) using the slightly smaller angle has provided a margin of safety in the die-life to cutting-clearance ratio. The angle of ½° which resulted from the calculation is the maximum clearance angle to be used for the given conditions if the angular clearance begins at the cutting edge (Fig. 1 · 10, view *B*). If this die opening is made with a cutting land of ⅛ in. (see view *A*), the maximum clearance angle may then be 1°.

For convenience in estimating proportional die life, cutting clearance, and angular clearance, refer to Fig. 1 · 11. The table is set up in increments of 0.010 in. The 0.010-in. increment can be considered an average amount of stock removal from the cutting face of a die block per sharpening.

Angular clearance summarized. Angular clearance is of vital importance in any die where blanks or slugs pass through the die opening. Like cutting clearance, angular clearance is a "per side" measurement. A clearance angle of ¼° per side is suggested for die work of good quality when the stock material is less than 1/16 in. thick.

All die-opening walls should have smoothly finished surfaces throughout. When smaller degrees of angular clearance are used, the surface finish of the die-opening walls must be made correspondingly smoother.

As a rule, delicate die sections will be less apt to break if slightly more than normal angular clearance is used. Owing to the lessening of back pressure from blanks or slugs, small or delicate punches will also benefit from slightly increased angular clearance in the die opening.

When a die is being tried out, it is good pro-

a = CLEARANCE ANGLE, DEGREES
B = AMOUNT GROUND BELOW START OF TAPER, INCHES
A = INCREASE IN CUTTING CLEARANCE PER SIDE, INCHES

$A = B \tan a$

$\frac{A}{B} = \tan a$

B	CLEARANCE ANGLE				B	CLEARANCE ANGLE			
	0°15′	0°30′	1°	2°		0°15′	0°30′	1°	2°
0.010	0.00004	0.00008	0.00017	0.00034	0.260	0.00114	0.00226	0.00455	0.00907
0.020	0.00007	0.00017	0.00035	0.00069	0.270	0.00118	0.00235	0.00472	0.00942
0.030	0.00011	0.00026	0.00052	0.00104	0.280	0.00122	0.00244	0.00489	0.00977
0.040	0.00017	0.00035	0.0007	0.00139	0.290	0.00126	0.00253	0.00506	0.01012
0.050	0.00022	0.00044	0.00088	0.00174	0.300	0.00131	0.00262	0.00523	0.01048
0.060	0.00026	0.00052	0.00105	0.00209	0.310	0.00135	0.0027	0.0054	0.01082
0.070	0.0003	0.00061	0.00122	0.00244	0.320	0.00139	0.00279	0.00558	0.01117
0.080	0.00034	0.0069	0.0014	0.00279	0.330	0.00144	0.00287	0.00575	0.01152
0.090	0.00039	0.00078	0.00157	0.00314	0.340	0.00148	0.00296	0.00593	0.01187
0.100	0.00044	0.00087	0.00175	0.00349	0.350	0.00153	0.00305	0.00611	0.01222
0.110	0.00048	0.00095	0.00192	0.00383	0.360	0.00157	0.00313	0.00628	0.01257
0.120	0.00052	0.00104	0.00209	0.00418	0.370	0.00161	0.00322	0.00645	0.01292
0.130	0.00056	0.00113	0.00227	0.00453	0.380	0.00165	0.00331	0.00663	0.01327
0.140	0.0006	0.00122	0.00244	0.00488	0.390	0.00169	0.0034	0.0068	0.01362
0.150	0.00065	0.00131	0.00262	0.00523	0.400	0.00174	0.00349	0.00698	0.01397
0.160	0.00069	0.00139	0.00279	0.00558	0.410	0.00178	0.00357	0.00715	0.01431
0.170	0.00073	0.00148	0.00297	0.00593	0.420	0.00182	0.00366	0.00733	0.01466
0.180	0.00078	0.00157	0.00314	0.00628	0.430	0.00187	0.00375	0.0075	0.01501
0.190	0.00082	0.00166	0.00332	0.00663	0.440	0.00191	0.00384	0.00767	0.01536
0.200	0.00087	0.00175	0.0035	0.00098	0.450	0.00196	0.00393	0.00785	0.01571
0.210	0.00091	0.00184	0.00367	0.00732	0.460	0.002	0.00401	0.00802	0.01606
0.220	0.00096	0.00193	0.00385	0.00767	0.470	0.00204	0.0041	0.00819	0.01641
0.230	0.001	0.00201	0.00402	0.00802	0.480	0.00209	0.00418	0.00837	0.01676
0.240	0.00105	0.0021	0.0042	0.00837	0.490	0.00213	0.00427	0.00854	0.01711
0.250	0.0011	0.00218	0.00438	0.00872	0.500	0.00218	0.00436	0.00872	0.01746

Figure 1 · 11 Increase in die-opening size due to sharpening.

cedure to stop the die after a few parts have been run. The blanks and slugs remaining in the die openings should then be tapped through the die in order to be certain that there is no interference. This procedure should also be followed when a die is brought to the die shop for maintenance work, such as sharpening. In this case, the blanks or slugs should be tapped through before the maintenance work is begun. To tap them through, use a soft drift of suitable cross section placed against the blanks and tap the drift with a lightweight brass knocker. A series of light blows should clear a die opening of blanks or slugs. Caution! *Do not* use hardened drifts or hammers when performing this check.

BLANK AND SLUG OPENINGS THROUGH DIE SHOES

After a blank or slug has passed through its die opening, it falls through the clearance opening in the die shoe. Good judgment is required on the part of the diemaker when making these openings. The following requirements must be met: (*a*) The blanks or slugs must fall freely, without interference; (*b*) the contour of the clearance opening in the die shoe should be made as simple as possible; (*c*) the opening should not weaken the die shoe any more than necessary; (*d*) the contour of the die-shoe opening must be such that it provides adequate support for the die block.

Various typical die-shoe clearance openings are shown in Figs. 1 · 12 to 1 · 15.

Round slug holes for irregular die openings. The simplest type of die-shoe clearance is, obviously, a round hole. This fact can be put to advantage in many instances (see Fig. 1 · 12). When a round hole is used, it should be tapered clear through the die shoe. Its smallest diameter must be at the top of the die shoe next to the die block. The top

Figure 1 · 12 Round openings for blank or slug passage through die shoe.

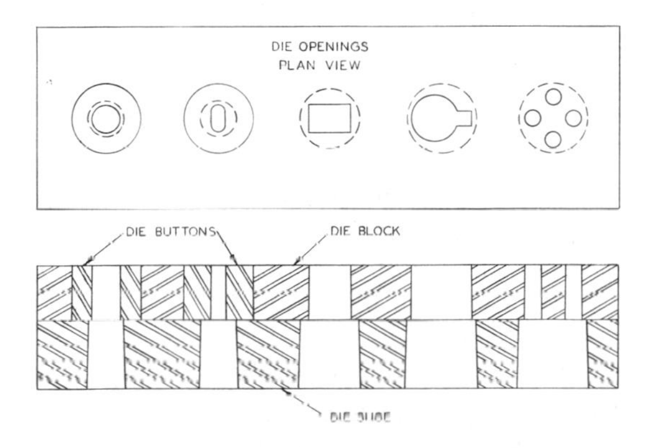

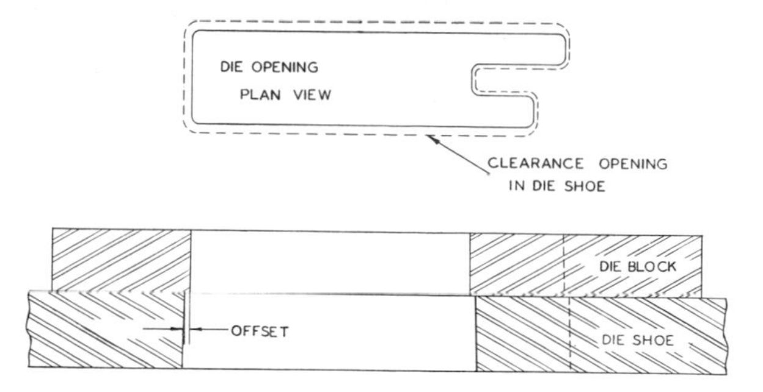

Figure 1 · 13 Die-shoe clearance opening contoured to provide support for die-opening contour.

(smallest) diameter of the die-shoe clearance hole must be larger than the largest dimension across the bottom of the die opening in the die block. The clearance holes must be located properly with respect to the die openings so as to achieve a maximum of support for the die block while avoiding any possibility of interference with the passage of the blank or slug.

The use of a round die-shoe clearance hole in conjunction with irregular die openings is good practice for small slugs and blanks. It can readily be seen, however, that this method is not practical for die openings that are larger in size. The relatively greater diameter required for the die-shoe opening would result in weakening the die shoe and would not provide adequate support for the die block.

Supporting weak die conditions. An example of a die opening that needs to be supported by the die shoe is shown in Fig. 1 · 13. This die opening has a comparatively long, narrow tongue projecting into it. The die-shoe opening is contoured with a matching tongue to support the projection in the die opening. If the projection is relatively narrow, less offset is used around this portion of the opening in order to provide maximum support for the projection in the die opening.

Figure 1 · 14 Straight-walled blank or slug clearance opening in die shoe.

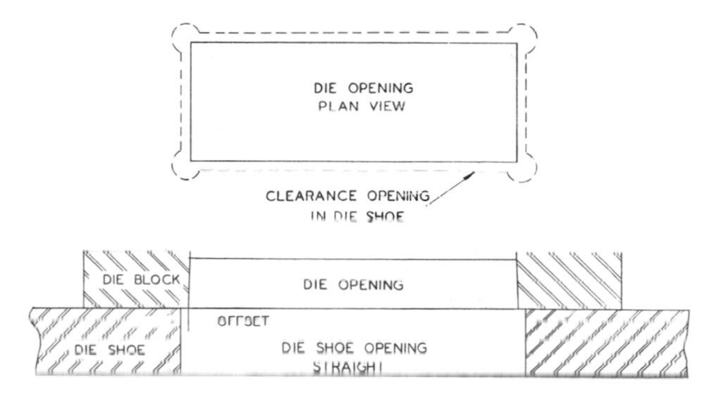

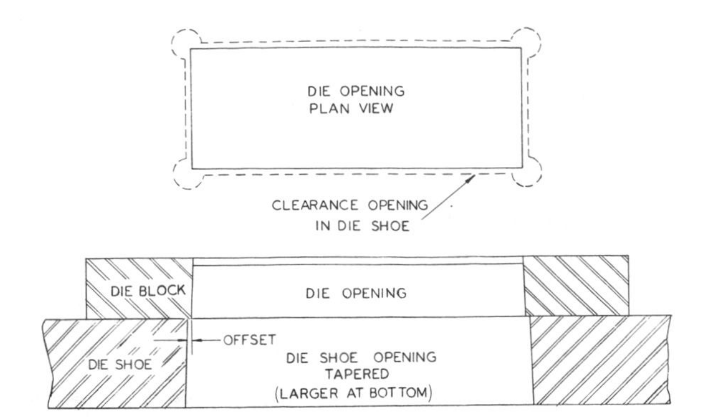

Figure 1 · 15 Tapered blank or slug clearance opening in die shoe.

Straight and tapered die-shoe openings. It is at times difficult to decide whether die-shoe clearance openings should be made straight or tapered (see Figs. 1 · 14 and 1 · 15). The tapered opening is unquestionably the safest. The straight opening, however, is often easier to make, especially when the die shoe is large and/or thick.

Fortunately, the larger blanks cause the least trouble. This usually allows the clearance openings for larger blanks to be made with straight sidewalls. Small slugs or blanks (less than 1 in. across) should have tapered clearance holes in the die shoe. This is true for both round and irregular shapes. The die-shoe openings should also be tapered if a die is to be run in an inclined press.

Amount of Taper. The amount of taper on the sidewalls of the drop-through opening in a die shoe is not critical. An angle of ½ to 2° will satisfy most cases. The angle must begin at the top of the die shoe (next to the die block) and extend clear through the die shoe. The degree of angle need not be constant all around the die-shoe opening. As long as there is some taper, the blanks will fall through because the taper causes the opening to be larger at the bottom of the die shoe.

Amount of offset. The top edge of the die-shoe opening is offset from the bottom edge of the die opening (Figs. 1 · 14 and 1 · 15). The amount of offset is not critical; it may vary from a step of 1/64 in. to a step of ⅛ in. or more at times. Straight openings are usually offset more than tapered openings.

Extra clearance for sharp corners. Since sharp corners on blanks or slugs can dig into the sidewalls of the die-shoe clearance opening, it is a good idea to provide extra clearance for sharp corners. Refer to the plan views in Figs. 1 · 14 and 1 · 15 and note the extra clearance shown at the corners. In these instances the extra clearance is provided by drilling holes at the corners far enough out that the sharp corners on the blanks cannot touch the walls of the die-shoe opening. These holes also admit the saw blade and make it easier to saw out the opening.

Die-shoe openings summarized. Generally, die-shoe clearance openings for larger blanks may be made with straight walls. Small blanks or slugs, whether round or irregular in contour, should have tapered die-shoe openings.

The clearance opening in the die shoe must be larger than the die opening it serves (see offsets, Figs. 1 · 14 and 1 · 15). However, the offset must be held to a minimum when the die shoe is needed for support. In tapered die-shoe openings, the angle on the walls is in the same direction as the angular clearance in the die opening—that is, the opening is larger at the bottom. The sidewalls of die-shoe clearance openings must be smooth enough to prevent any interference with the progress of the blanks or slugs as they drop through.

When the diemaker is trying out the die, he should check to make certain that all blanks and slugs fall freely. When a blanking and/or piercing die is installed in a press for production, the same check should be made immediately. If all blanks and slugs are not falling through freely, production must be stopped until the condition is corrected. A die that is successfully running should also be observed from time to time to be certain the blanks and slugs continue to drop through without interference.

CUTTING FORCE

The manner in which a blanking or piercing die functions requires that the punch or punches be driven through the stock material. This action was described earlier (see "Action of Blanking or Piercing Die").

The force required to drive a punch through the stock material is the cutting force for that particular punch. If a die has more than one punch acting simultaneously, the cutting force for that die is the sum of the forces required to drive each punch through the stock material.

Importance of cutting force. For any cutting die, the cutting-force requirement is the major factor used to select a punch press of proper tonnage rating for the job. Because of this, the cutting force should be determined before building the die. A diemaker should understand and be able to calculate cutting forces.

Shear strength of materials. When a cutting punch is driven through the stock material, shear action takes place. The shear resistance of the stock material is overcome by the punching force. Because of this, the shear strength of the stock material must be known in order to calculate cutting force.

The shear strength of a material is the force necessary to sever, by direct shear action, one square inch of that material. In Fig. 1·16, the average ultimate shear strength of various materials is given in tons per square inch. The table is expressed in tons because of the association of cutting force with press tonnages. This simplifies calculation for the diemaker.

Computing cutting force. The force required for direct-shear cutting (blanking, piercing, etc.) is the product of the shear strength of the stock material and the cut-edge area. The cut-edge area is the product of the length of cut and the thickness of the material being cut.

A formula for computing direct-shear cutting force may be expressed as

$$C = LTS \qquad \text{tons}$$

where C = cutting force, tons
L = length of cutting edge, in.
T = thickness of stock material, in.
S = shear strength of stock material, tons

EXAMPLE:

How much cutting force is required to blank out a piece part from 0.05-in.-thick mild steel if the periphery of the part is 30 in.? In Fig. 1·16, the shear strength of mild steel is given as 25 tons. By substituting values, the formula becomes $C = 30 \times 0.05 \times 25$. Then, by multiplying, $C = 37.5$ tons.

Cutting force for round work. The formula $C = LTS$ is used to determine the cutting-force requirements for any contour. In the case of round blanks or holes, $L = \pi D$, where D is the diameter of the blank or pierced hole. Therefore, in the special case of round work, the cutting-force formula may be expressed as

$$C = \pi DTS$$

Cutting force for multiples of punches. If a die has more than one punch cutting simultaneously, the cutting force for that die is the sum of the cutting forces required for each punch. The total cutting force required for a die can be obtained by calculating the force required for each punching and then adding them together. Another and often simpler procedure is to add all the cut-edge lengths first. The total cut-edge length is then substituted for L in the cutting-force formula.

Cutting force related to shearing action. As described earlier, there are three critical stages of shearing action: plastic deformation, penetration, and fracture. These stages are related to cutting force in the manner shown in Fig. 1·17. This figure represents a typical load curve of the cutting force as a blanking or piercing punch is being driven through the stock material.

Resistance begins when the punch contacts the stock material. The load builds up rapidly during the plastic-deformation stage and continues to in crease while penetration is taking place. The accumulated load is suddenly released when fracture occurs. If a proper cutting clearance condition exists between the punch and the die, fracture will occur when the cutting force equals the shear strength of the stock material.

The curve then tends to level off near the bottom. This last portion of the load curve represents frictional resistance as the punch travels on through the stock material. It includes the resistance of the blank or slug, which is being forced deeper into the die opening. The load curve does not level off completely but continues a very gradual descent until the bottom of the cutting cycle is reached. This continuing load reduction occurs because the blanks or slugs move more freely as the angular clearance in the die opening takes effect.

STRIPPING FORCE FOR BLANKING AND PIERCING

Reasonably accurate calculations of stripping-force requirements can be made. However, it is not usually practical to lay down general rules. There are so many variable factors involved that an accurate calculation must be a highly specialized computation for a specific job. It will be valid for that job only.

Important factors which affect stripping force

1. Stock Material. Materials which have a high friction value and materials which tend to cling are more difficult to strip.

2. Condition of Cutting Edges. When the cutting edges are sharp, less stripping effort is required.

3. Surface Condition of Sidewalls. A punch which has a smooth finish on its sidewalls strips more easily than a punch which is not as smooth.

4. Distance between Punches. More effort is required to strip punches that are close together.

5. Area of Stock Material to be Stripped. Figure

AVERAGE ULTIMATE STRENGTH OF MATERIALS TONS PER SQUARE INCH		
MATERIAL	SHEAR	TENSILE
ALUMINUM, 2S-O	4.8	6.5
2S-½H	5.5	8.5
2S-H	6.5	12.
24S-O	9.	13.5
24S-T	20.5	34.
52S-O	9.	14.5
52S-½H	10.5	18.5
52S-H	12.	20.5
ASBESTOS, SHEET	2.	
BRASS, CARTRIDGE, Cu 70%, Zn 30%, SOFT	16.	23.5
ROLLED STRIP & SHEET, Cu 65%, Zn 35%, SOFT	16.	23.
½ HARD	22.	32.5
HARD	25.	39.
SPRING	28.5	47.5
BRONZE, PHOSPHOR, GRADE A SHEET, ANNEALED	20.	27.5
SPRING TEMPER	32.5	52.5
COPPER, SOFT	12.	16.
HARD	18.5	27.5
FIBRE, HARD	12.	
GOLD, 14 CARAT, SOFT	21.	29.
LEATHER, CHROME	3.5	5.
RAWHIDE	6.5	
NICKEL (GERMAN) SILVER, Ni 18%, Cu 65%, Zn 17%, SOFT	17.5	29.
PAPER, HOLLOW CUTTER	6.	
SILVER, STERLING, SOFT	13.5	18.5
STEEL, MILD	25.	30.
BOILER PLATE	27.5	35.
SILICON	30.	36.
STAINLESS, 18-8, ANNEALED	37.5	47.5
CUTLERY, KNIFE 410, FULL HARD	72.5	
CUTLERY, FORK & SPOON 420 ¼ HARD	32.5	
SOFT, DEEP DRAWING	21.	26.5
¼ HARD	22.5	30.
½ HARD	25.	36.
HARD	30.5	46.
S.A.E. 1010, COLD ROLLED	21.	28.
1020 COLD ROLLED	26.	34.5
1030 COLD ROLLED	31.5	42.5
1050 COLD ROLLED	41.	55.

Figure 1·16 Strength of materials.

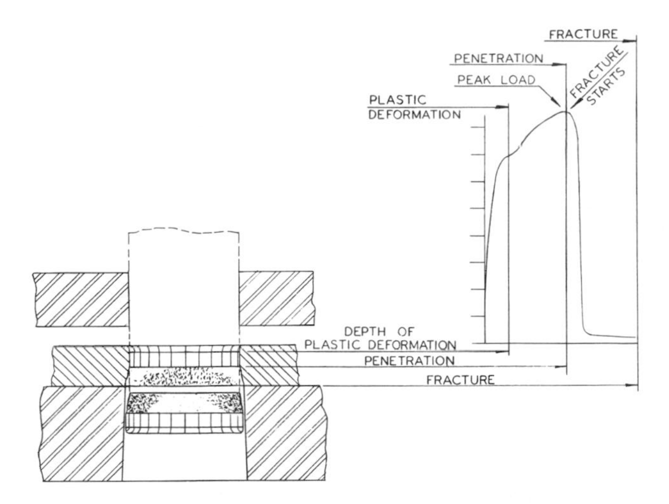

Figure 1·17 Cutting force related to stages of shearing action.

1·18 shows two piece parts, one larger than the other. The thickness and the type of stock material are identical. The pierced opening is the same size in both parts; in fact, both parts are pierced by the same punch and die. The cutting force is, of course, the same for both parts. However, the larger piece part requires the greater stripping effort. The larger area of stock material surrounding the punch is stronger and causes the material to cling more tightly to the punch.

These factors are some of the reasons why it is in general difficult to calculate stripping-force requirements accurately.

Formula for stripping force. Representative metal fabricators have reported that stripping-force requirements range from 5 to 20 per cent of the cutting-force requirements. For estimating stripping force, it is suggested that the following formula be used:

$$K = \frac{LTS}{5} \quad \text{tons}$$

Figure 1·18 Differential in opening size and area of stock material to be stripped.

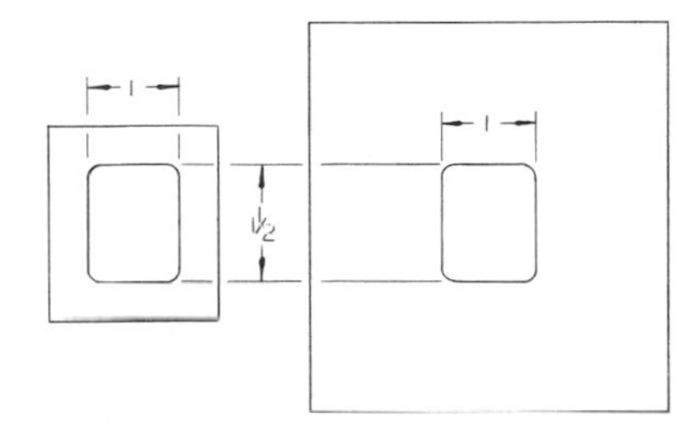

where K = stripping force, tons
L = length of cutting edge, in.
T = thickness of stock material, in.
S = shear strength of stock material, tons (from table in Fig. 1·16)

This formula will result in a figure equal to 20 per cent of the cutting force required for the blanking or piercing operation.

If a die has more than one punch, the stripping force for that die is the sum of the stripping forces required for each punch. This is true whether the punches cut simultaneously or not.

Stripping force and the diemaker. It is obvious that in a plain blanking die where the scrap is weak, considerably less than 20 per cent of the cutting force will be adequate for the stripping force. In the course of time, the diemaker develops judgment in this. He can judge reasonably well how strong a stripper plate should be or how much spring pressure will be required in the case of a spring stripper. This is one of the skills of diemaking. A good craftsman develops a reliable sense of mechanical proportion which he uses to advantage in these matters.

Inadequate stripping. The consequences of underestimating stripping-force requirements can be serious indeed. When in doubt, the diemaker should "play it safe" and figure that the job requires more stripping force rather than less.

Figure 1·19 Cutting clearance.

CUTTING CLEARANCE PER SIDE (PERCENTAGE OF STOCK THICKNESS)

MATERIAL	IRREGULAR CONTOURS	ROUND
ALUMINUM		
SOFT, LESS THAN 3/64 THICK	3%	2%
SOFT, MORE THAN 3/64 THICK	5%	3%
HARD	5%–8%	4%–6%
BRASS & COPPER		
SOFT	3%	2%
1/2 HARD	4%	3%
HARD	5%–6%	4%
STEEL		
LOW CARBON SOFT	3%	2%
1/2 HARD	4%	2%
HARD	5%	3%
SILICON STEEL	4%–5%	3%
STAINLESS STEEL	5%–8%	4%–6%

HEAVIER GAGES GENERALLY REQUIRE HIGHER PERCENTAGES

CHAPTER 2

ELEMENTARY BLANK DIES AND PIERCE DIES

BLANK-THROUGH DIE (EXAMPLE)

Figure 2 · 1 is an assembly drawing of the die which produces the blank for the piece part illustrated. This die does the blanking operation only. The holes in the piece part are pierced in a subsequent piercing die.

In Fig. 2 · 1, each component of the die is identified by detail number and name. The number required of each component or detail is also shown. Socket head cap screws are indicated by the abbreviation S.H.C.S.

Operation of die. Refer to Fig. 1 · 1 for terminology. The stock material is furnished in strip form 2¹¹⁄₁₆ in. wide. It is fed across the die face from right to left until the lead end contacts the stop. The press is then tripped, causing the first piece to be blanked out. After the blanking, the scrap bridge is lifted over the stop. As soon as the bridge passes over the stop, the strip is dropped back to the die face. The head of the stop is now within the previously blanked-out opening. The stock strip is then advanced until the edge of the blanked-out opening contacts the stop. This process is repeated at every press stroke until each strip is used up.

Fixed stop. If a plain-pin (no head) type of fixed stop were used, it would have to be installed at a safe distance from the cut edge in order to avoid weakening the die block. As a result, the scrap bridge would be wider than necessary, which would mean, of course, that stock material was being wasted. In order to minimize the amount of

Figure **2 · 1** Blank-through die.

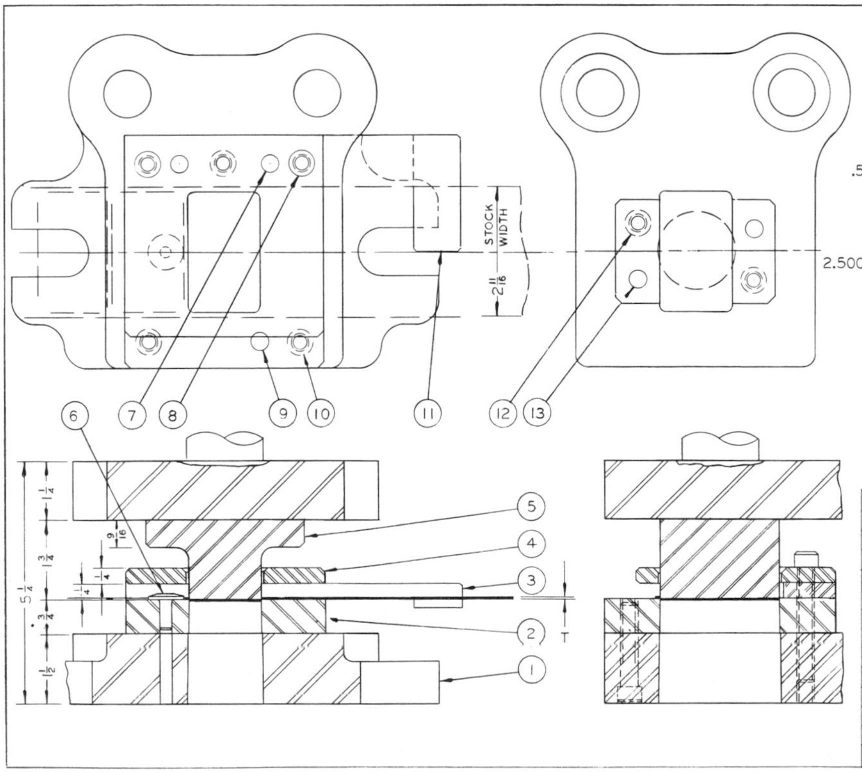

DET.	REQ.	DESCRIPTION
13	2	DOWEL
12	2	S.H.C.S.
11	1	STOCK REST
10	3	S.H.C.S.
9	1	DOWEL
8	2	S.H.C.S.
7	2	DOWEL
6	1	STOP
5	1	PUNCH
4	1	STRIPPER
3	1	BACK RAIL
2	1	DIE BLOCK
1	1	DIE SET

scrap while still maintaining adequate die-block strength, the die shown (Fig. 2 · 1) has a headed fixed-pin stop. The diameter of the head is made rather large in proportion to the shank diameter, which enables the mounting hole to be located a safe distance from the edge of the die opening. The thickness or height of the stop head should be held to a practical minimum in order to make the feeding of the stock strip as easy as possible. The stop is shown detailed in Fig. 2 · 2.

Stripping. A gap-type fixed stripper is used. The gap is made relatively high for ease of feeding. The height of the gap and the open front provide visibility and accessibility for the press operator.

Since the scrap bridge must be lifted over the fixed stop each time the stock strip is advanced, the stripper gap must be high enough to allow free passage of the stock strip during feeding. In this die (Fig. 2 · 1), the head of the stop is 3/16 in. high. This combination allows a space of 5/16 in. between the stop and the stripping surface. The stock material thickness T is, in this case, 0.050 in. The feeding clearance between stop and stripper is slightly more than $6T$, which is ample for ease of feeding.

Cutting clearance. (See Chap. 1, "Determining cutting clearance.") Before the punch and die can be fitted, it is necessary to know how much cutting clearance should be used. The amount of cutting clearance required for any die depends upon the thickness and type of stock material being cut. The piece part (see Fig. 2 · 1) specifies 0.050-in.-thick half-hard brass. To determine the required cutting clearance, refer to Fig. 1 · 19. In this table the recommended clearance for half-hard brass is 4 per cent of the thickness T per side. Taking 4 per cent of 0.050 in. yields a cutting clearance of 0.002 in. per side for this particular job.

Because the die (Fig. 2 · 1) is a blanking die, the die-opening sizes are derived directly from the piece-part dimensions. The punch dimensions are determined by deducting the cutting clearance on each and every side from the die-opening dimensions. Keep in mind that, with blanking dies, the cutting clearance exists all around the cutting periphery. (See Fig. 1 · 9 and "Relationship of Piece-part Sizes to Punch and Die Sizes," Chap. 1.)

DETAILS OF BLANK-THROUGH DIE

The die details are shown in Fig. 2 · 2. For the sake of clarity, only the overall dimensions and the dimensions pertinent to producing the piece part are shown. A brief explanation of why the various pertinent dimensions were chosen is included. Abbreviations used in Fig. 2 · 2 are O.H.T.S. for oil-hardening tool steel, C.R.S. for cold-rolled steel, and R. C 60–62 to indicate the hardness of the hardened tool-steel components as measured on the C scale of a Rockwell tester.

Die block. This die block (Fig. 2 · 2, detail 2) is for a blanking die. Therefore, the piece parts produced will be sized by the die opening. The piece part specifies a width of $1.500 \begin{smallmatrix} +0.000 \\ -0.005 \end{smallmatrix}$ in. The tolerance is negative (minus). This gives the width dimension a range of 1.495 to 1.500 in. The mean dimension is 1.4975 in. With this in mind, a figure of 1.497 in. was chosen for the die-opening width.

The length of the piece part is $2.500 \begin{smallmatrix} +0.005 \\ -0.000 \end{smallmatrix}$ in. This tolerance is positive (plus). The range is 2.500 to 2.505 in., and the mean dimension is, of course, 2.5025 in. The die-opening figure selected for this dimension is 2.502 in.

A tangent radius of 1/8 in. is specified on all four corners of the piece part. The die-block detail shows a radius of 0.125 in. at these corners. This figure is based on using a standard 0.250-in.-diam reamer or end mill to produce these corners.

The figures chosen give a die-opening size of 1.497 by 2.502 in. These figures are the optimum die-opening sizes for the given conditions. A die opening made to these figures will yield a blank (or piece part) that will measure very nearly the exact mean of the dimensions specified on the piece-part drawing.

Tolerances are specified on the piece-part drawing. The purpose of the tolerances is to indicate a range of acceptable sizes: All piece parts produced within the range are acceptable. This does *not* mean that the die opening can be made to the same tolerance range as the piece part.

The diemaker should use the optimum dimensions as his target. However, because of the tolerances shown on the piece-part drawing, it is not necessary to spend an excessive amount of time attempting to achieve a finished die opening that measures the exact optimum. Instead, he should decide on a maximum and minimum acceptable figure for each dimension and produce a die opening within that range. A good choice for minimum acceptable die-opening size is 1.4955 by 2.5005 in. The maximum acceptable die opening should be 1.498 by 2.503 in. Because this is a blank-through die, the die-opening tolerance is

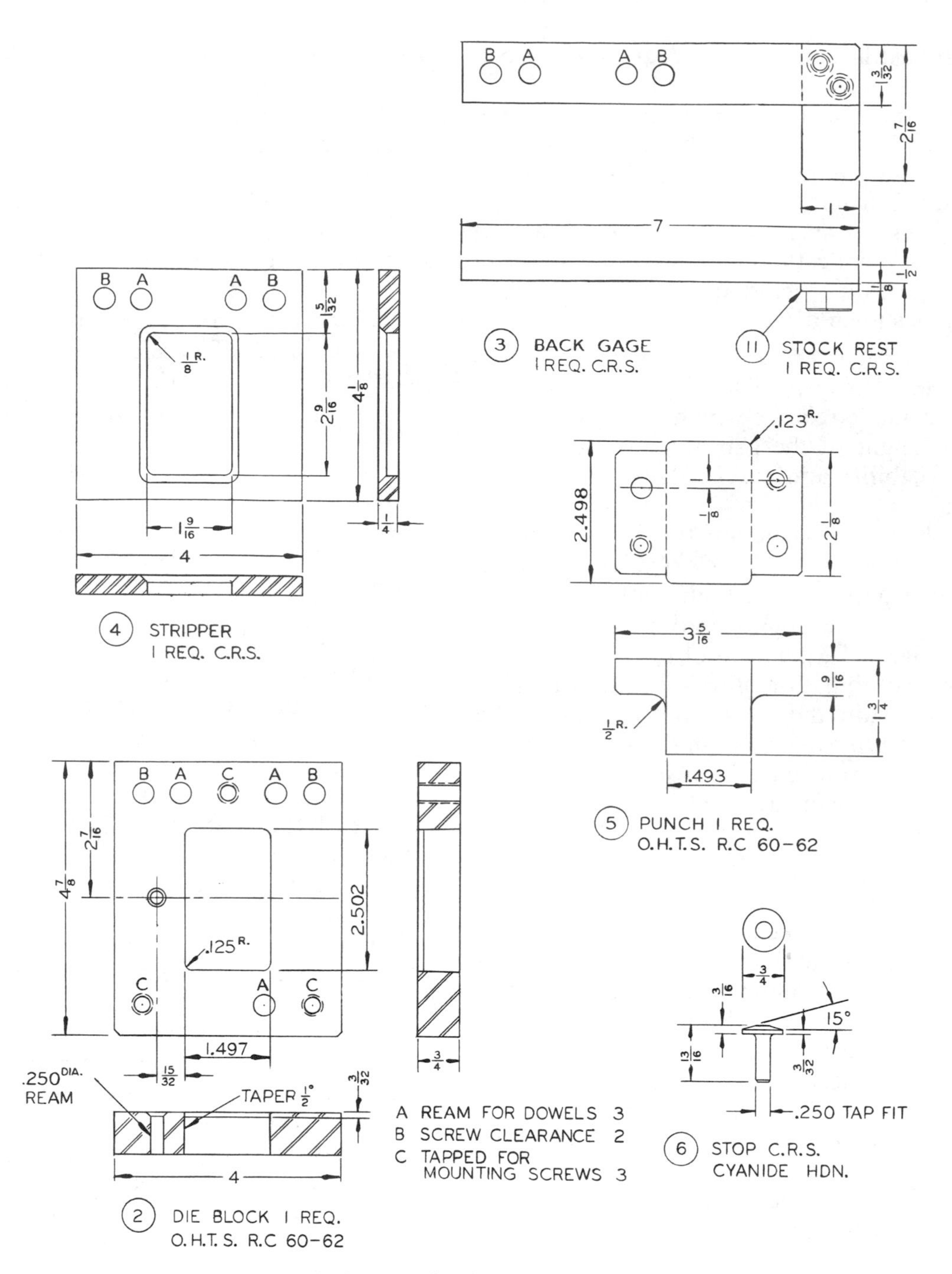

Figure 2·2 Details of blank-through die.

set farther from the piece-part high limit than from the low limit. This is done because of the eventual die-opening size increase that will occur after repeated sharpening has caused the cut edges to be affected by the angular clearance (see Chap. 1, "Angular clearance and sharpening").

To take advantage of the tolerances as described is one reason why the die opening should be entirely finished before the punch is completed to its final size. This should not be construed to mean that the entire making of the punch should wait until the die block has been finished. The punch can be made complete except for a final finishing allowance of 0.005 to 0.010 in. per side on the cut-edge dimensions. Thus, most of the work on the punch can be done at any convenient time, which aids in efficient work procedure and organization.

Punch. The punch sizes are shown (Fig. 2·2, detail 5) as 1.493 by 2.498 in. with a corner radius of 0.123 in. These figures are the result of deducting the cutting clearance from the optimum die-opening dimensions. It is standard die design procedure to show punch and die dimensions in this manner.

The die opening should have been completely finished first. If the finished die-opening size happens to be the optimum figures shown, the punch is then made to its optimum dimensions as specified on the drawing. If it is found, however, that the die opening is some size other than the optimum (within the acceptable range), the diemaker proceeds to alter the indicated punch dimensions in order to maintain the proper cutting clearance.

Note that one dowel is offset 1/8 in. This is important. Once the punch has been properly doweled in alignment with the die opening, the offset dowel ensures that the punch will always be remounted properly if for any reason it has to be removed during the life of the die. Of course, a symmetrical screw and dowel arrangement could be made so precisely related to the cutting contour of the punch that the punch would be reversible. However, such precision is time-consuming and absolutely unnecessary for the purpose. Except for relatively rare instances where punches or die sections must be made reversible, the mounting holes should always be foolproofed.

Back gage. The back gage (Fig. 2·2, detail 3) is 1 3/32 in. wide. This dimension is determined as follows: The stock-material width is specified as 2 11/16 in. One-half of this stock width is 1 11/32 in. The distance from the center line of the die opening to the back of the die block is 2 7/16 in. Subtracting 1 11/32 from 2 7/16 leaves a difference of 1 3/32 in. from the edge of the stock strip to the back of the die block. Using this 1 3/32-in. dimension for the width of the back gage aids the mounting of the gage to the die block. All that is necessary is to clamp the gage to the die block with the back edges flush and with the left end of the back gage flush with the left side of the die block. The dowel holes *A* and the screw clearance holes *B* are then transferred from the hardened die block to the back gage. The stock rest (detail 11) may be attached to the back gage at any convenient time. The mounting holes in the back gage should be transferred from the die block after the die block has been heat-treated. This is also true for the stripper.

Stripper. The stripper opening (Fig. 2·2, detail 4) measures 1 9/16 by 2 9/16 in. These dimensions give a clearance space of 1/32 in. on each side between the punch and the stripper. The purpose of the clearance is ease of fabrication and assembly. The 1 5/32-in. dimension from the opening to the back edge allows the stripper, like the back gage, to be lined up flush with the back edge of the die block in order to expedite the mounting procedure.

PIERCE DIE (EXAMPLE)

Figure 2·3 is an assembly drawing of a die to pierce two holes per piece part illustrated. This die performs the piercing operation only; the blank was produced by the blank-through die shown in Fig. 2·1. For convenience, the piece part has been shown on each die assembly drawing.

The pierce-die drawing (Fig. 2·3) is set up in the same manner as the blank-through die drawing (Fig. 2·2). The components are identified and the number of each required is stated. Socket head cap screws are abbreviated S.H.C.S. Socket head stripper screws are indicated by the abbreviation S.H.S.S.

Spring stripper. This die has a moving stripper actuated by springs. The designer elected to use this type of stripper for reasons of accessibility for the press operator. Using a spring stripper for this die also ensures a flatter piece part after piercing, because the stripper pressure holds the part against the face of the die block while the cutting action is taking place.

Instruction plate. Note that the drawing calls for an instruction plate. The plate, which reads "load part burr side down," must be clearly lettered and located at the front of the die where it is easily seen by the operator. The reason for the plate is that the piece-part drawing specifies a burr side, and because the blank is symmetrical, there is no way to foolproof the loading of the blank in the die. As a result, the press operator must be instructed to feed the part into the die in such a manner that the burr side of the blank and the burr side of the pierced holes will agree, as specified.

The burr side of a punched opening always faces away from the punch and toward the die opening (see Chap. 1, "Burr side"). Therefore in this instance the blank must be placed in the die with the burr side facing downward, toward the die openings.

Instruction plates, such as this one, are used when there is no practical positive way to achieve proper loading or feeding. Naturally, positive or foolproofed loading is always preferred if it can be achieved.

Locating and feeding. Proper location of the pierced holes in the piece part is achieved by a

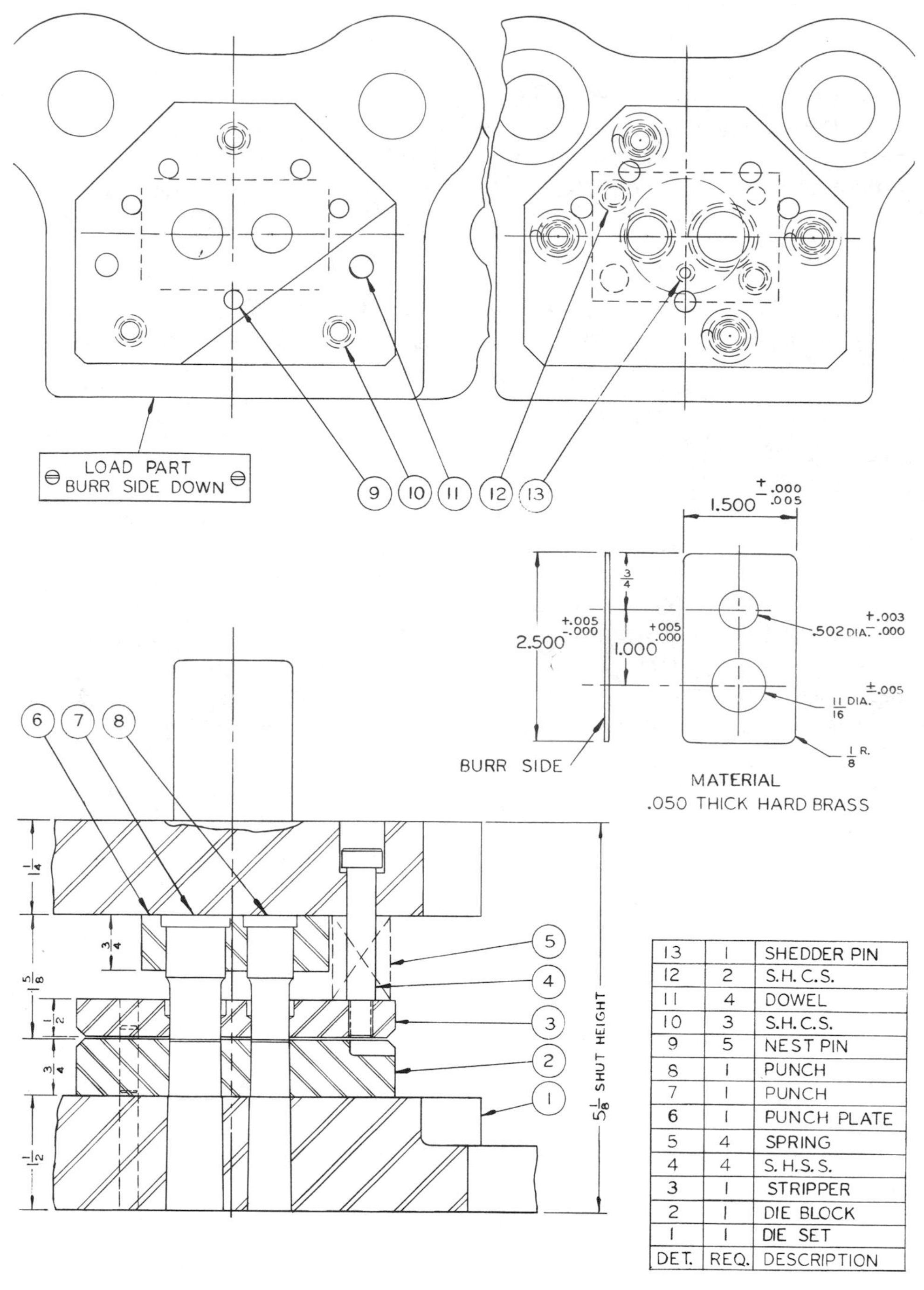

DET.	REQ.	DESCRIPTION
13	1	SHEDDER PIN
12	2	S.H.C.S.
11	4	DOWEL
10	3	S.H.C.S.
9	5	NEST PIN
8	1	PUNCH
7	1	PUNCH
6	1	PUNCH PLATE
5	4	SPRING
4	4	S.H.S.S.
3	1	STRIPPER
2	1	DIE BLOCK
1	1	DIE SET

Figure 2·3 Pierce die.

nest area created by the nest pins (detail 9). When the blank is placed in the nest area, it is confined between the nest pins, which are correctly located with respect to the die openings and punches.

In this die, the operator uses a large pair of tweezers to load and unload the piece parts. The step shown on the die block provides clearance for the nose of the tweezers in order to facilitate loading and unloading.

Cutting clearance. (See Chap. 1, "Exceptions.") When the cutting clearance for the blank-through die was determined, the figure was found to be 0.002 in. per side. However, for the pierce die, owing to the fact that the piercings are round, a cutting clearance figure of 0.0015 in. per side was chosen.

DETAILS OF PIERCE DIE

Figure 2·4 is the detail sheet for the pierce die. Major overall sizes and dimensions associated directly with the piece-part requirements are shown.

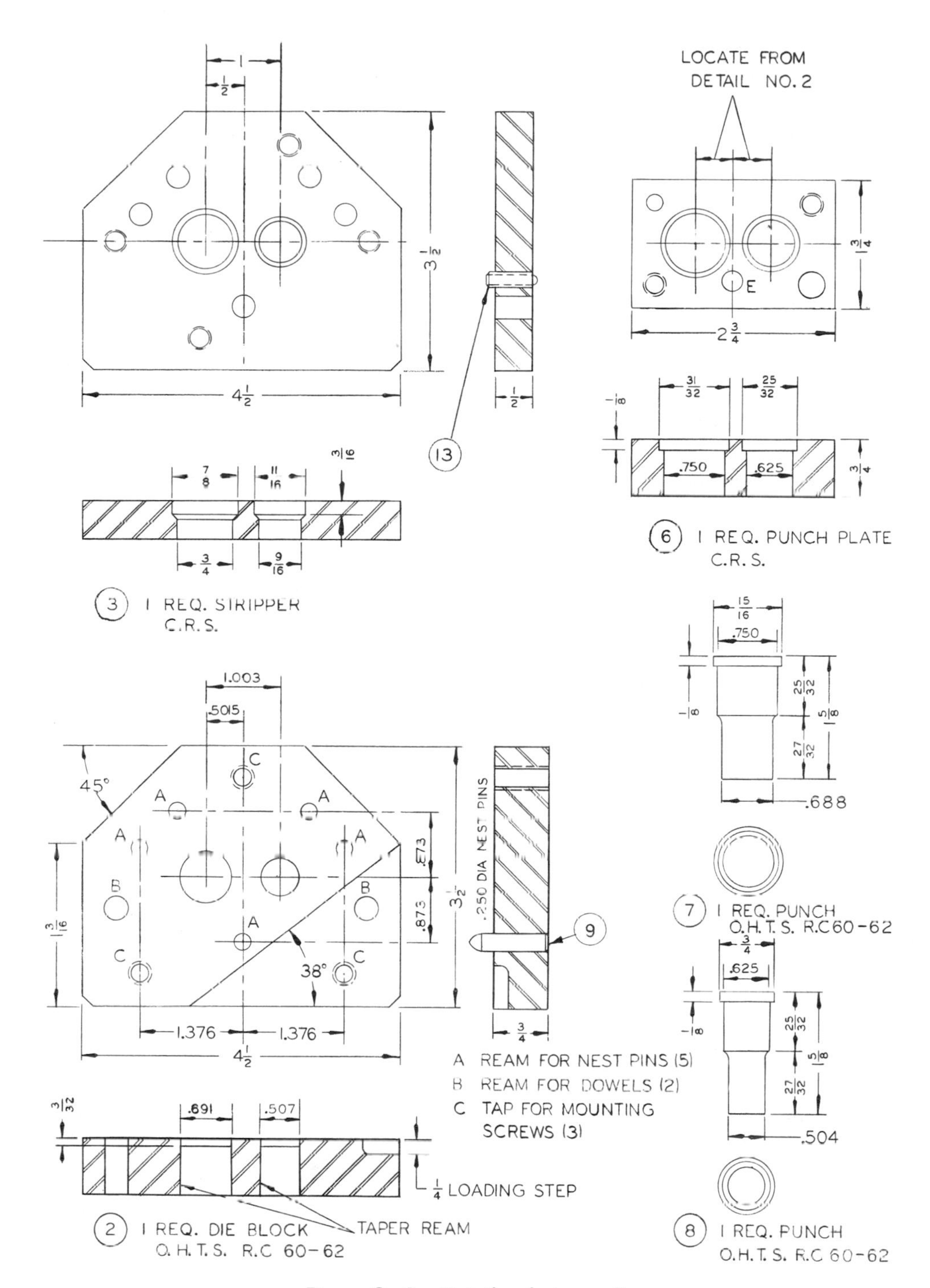

Figure **2·4** Details of pierce die.

Die block. As this die block (Fig. 2·4, detail 2) is for a piercing die, the size of the die openings is derived from the punches. The punch diameters are 0.688 and 0.504 in. A cutting clearance of 0.0015 in. per side is used. The die-opening diameters are, therefore, 0.691 and 0.507 in.

Center-to-center distance between the die openings is taken directly from the piece-part specifications, which give $1.000 \begin{smallmatrix} +0.005 \\ -0.000 \end{smallmatrix}$ in. Thus the dimensional range is 1.000 to 1.005 in. For this range the exact mean dimension is 1.0025 in., which is rounded off for convenience to a full-thousandth figure of 1.003. As far as the idea of rounding off is concerned, the figure could just as well have been 1.002. However, it was felt that there was a possibility of a slight amount of shrinkage during the heat treatment of the die block. For this reason the center distance was rounded off to the next higher thousandth of an inch instead of the next lower.

After the optimum die-opening size and location is decided, the next consideration is to assure proper location of the pierced openings in relation to the contour of the piece part. This is the function of the nest pins (detail 9). In order to serve their purpose, the nest pins must be accurately located. The nest must fit the blank closely enough to assure the required accuracy of location. Some clearance, however, should exist between the nest pins and the blank to permit ease of loading and unloading. The amount of clearance to allow, of course, depends upon the accuracy requirements of the piece part. In this particular case a clearance of 0.0005 to 0.001 in. between the edges of the blank and the nest pins is satisfactory.

Strictly speaking, to achieve the above conditions the diemaker would be required to have an actual blank produced by the blanking die in order to determine the exact distance required between the nest pins. This would mean that the blanking die would have to be completed before the nest-pin mounting holes could be located. It is not necessary, however, to know the exact blank size; all that needs to be done is to proceed as this die was made. The nest pins are located so that they will adequately gage the smallest acceptable blank size. Then, whenever blanks are available, the gaging portion of the nest pins can be polished down to suit.

Punch plate. Refer to Fig. 2·4, detail 6. The 0.750- and 0.625-in. diameters are chosen because they are the next-larger standard diameter for the respective required punch sizes. This allows the use of standard reamers for sizing these holes when jig-boring.

The center distance is "picked up" or transferred from the hardened die block. This procedure is followed to eliminate the possibility of minor alignment discrepancies which might result from slight dimensional changes caused by the die-block heat treatment.

The counterbores which receive the heads of the punches are made larger than the punch heads. Clearing for the heads in this way eliminates the possibility of misalignment resulting from contact between the head of the punch and the sidewall of the counterbore. The bottom of the counterbores must be flat and square with the axes of the punches in order to insure perpendicularity.

Foolproofing is accomplished by using dowels of differing diameters.

A spring pin (detail 13) is provided in the stripper. It is a standard purchased item, the length of which cannot be altered. It protrudes from the back of the stripper and might eventually strike against the punch plate, which would result in damage to the tool. To preclude this possibility, a clearance hole or pocket E of sufficient depth is provided in the punch plate, thus allowing the full life potential of the punches to be realized without interference from the shedder spring pin.

Punches. Refer to Fig. 2·4, details 6 and 7. One of the openings in the piece part is dimensioned $0.502\,{}^{+0.003}_{-0.000}$ in. diam. For this opening, a punch diameter of 0.504 in. was decided upon. A punch which has a point or working diameter of 0.504 in. will produce an opening in the piece part of 0.5035 to 0.504 in., which is optimum for this particular opening.

The other opening is specified as $^{11}/_{16} \pm 0.005$ in. diam. Since the mean dimension for this figure is 0.6875 in., a punch diameter of 0.688 in. is used.

It is an accepted practice to make the shank diameters of round punches in $^1/_{16}$-in. increments up to ½ in. diam and in ⅛-in. increments for diameters larger than ½ in. Using diameters of these size increments means that the punch-plate openings can be sized by reamers that are common toolroom equipment. Following this practice led to a choice of 0.625- and 0.750-in. diameters for the respective punch shanks.

The punch shanks are fitted to the punch plate. The proper fit is a light tap fit, not a press fit.

Stripper. The diameters of the stripper openings (Fig. 2·4, detail 3) are $^9/_{16}$ and ¾ in., respectively. These figures give a clearance of $^1/_{32}$ in. per side between the punches and the stripper. So much clearance is permissible because of the stock material thickness and is desirable from the diemaker's viewpoint because it eliminates the need for precise location and sizing of the stripper openings.

Spring strippers, such as this one, must have shedder pins to insure against the possibility of a piece part adhering to the stripper. In this stripper, the shedder pin (detail 13) is a standard purchased item. The spring pin is contained within a threaded cylinder. A suitable tapped hole is provided in the stripper to receive this unit, which is shown in assembly with the stripper (Fig. 2·4, details 3 and 13).

BLANK-THROUGH DIE AND PIERCE DIE: PROFICIENCY EXERCISE

Figures 2·5 and 2·6 are assembly drawings of the blanking die and the piercing die used to produce the piece part shown. For convenience, the piece part is illustrated on each die drawing.

The student is to make detail drawings for these

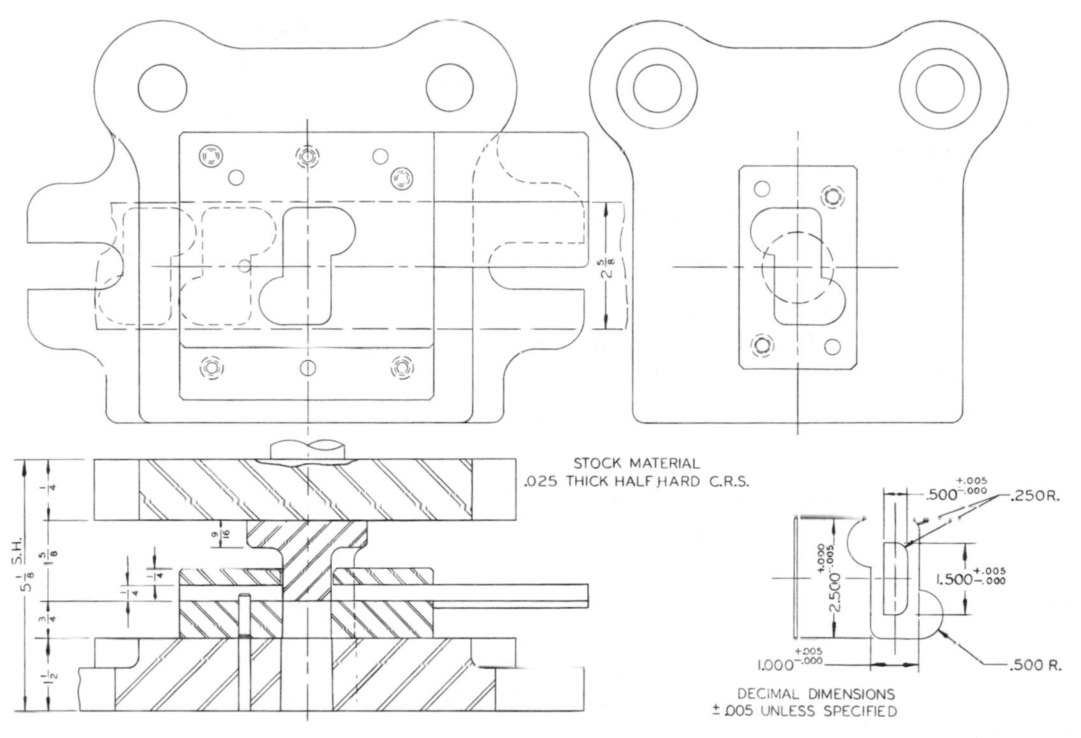

Figure **2·5** Blank-through die (problem).

Figure **2·6** Pierce die (problem).

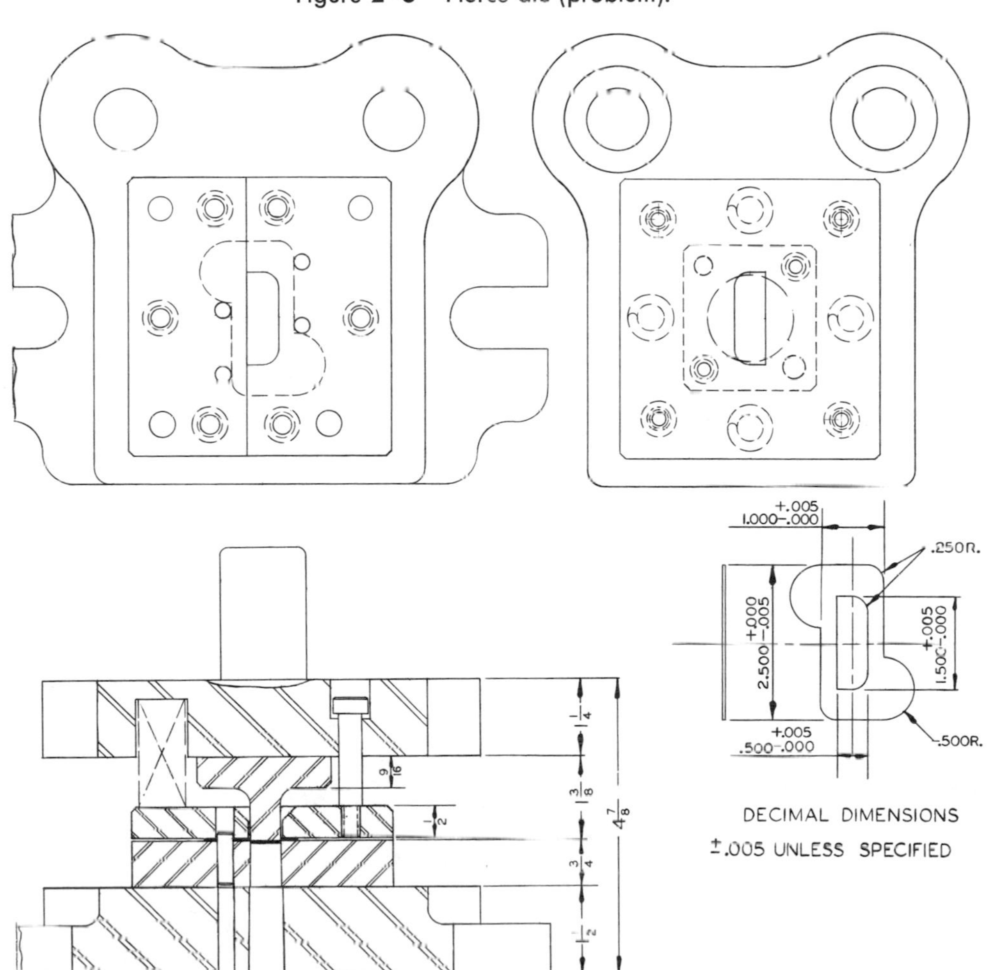

dies in a manner similar to that used in the examples (Figs. 2·2 and 2·4), with this exception: The student's detail drawings, taken from Figs. 2·5 and 2·6, are to be completely dimensioned down to the last screw and dowel.

It would be ideal if at this stage of study the apprentice (or student) were actually to build the two dies shown. Doing this would enable him to associate the information he has studied with the actual construction of a simple blank-through die and a simple pierce die. However, it is not practical to do this, nor is it absolutely necessary. Much the same result can be attained by proceeding as follows:

To avoid confusion, do one die at a time. However, keep in mind that these are associated dies, both of which must be used in order to produce the piece part.

Study each die thoroughly. Attain a clear understanding of the function of each part of the die and its relationship to the piece part.

Then, in lieu of actually making each detail, make a drawing of each. The detail drawings should be completely dimensioned—as would be necessary, after all, if the dies were actually being built. In order for the details to be considered complete, the following information must be clearly shown:

1. Overall dimensions, to show consideration of die proportions for strength and economy of construction
2. All punch sizes and die-opening sizes, these to be properly related to piece-part sizes and to each other
3. Size and location of stops, gages, and/or nest pins
4. Size and location of stripper openings
5. Size and location of screws and dowels
6. Whether details are hardened or left soft

For each detail include a brief explanation of why the various dimensions are chosen. If it seems necessary or desirable to make certain details before making others, state the sequence and explain the reasons for it.

CHAPTER 3

BENDING

BASICS OF BENDING

When used in reference to stamping dies, the term bending is restricted to a definite type of forming operation. In bending operations, the material is formed around a straight axis which extends completely across the material at the bend lines. A bending operation produces a plane surface which is at an angle to the original plane of the workpiece (or stock strip). The piece parts shown in Fig. 3·1 are basic examples of piece parts produced by bending.

The illustration serves also to depict fundamental elements of bends. Shown here are the bend radius, bend lines, and bend axis. Refer to view *A* and note the following relationships:

1. The bend radius is tangent to the inner plane surfaces of the piece part.
2. The bend lines occur at the tangency of the bend radius with the inner plane surfaces.
3. At the bend lines, the bend radius is perpendicular to the inner plane surfaces.
4. The bend axis is located at the center of the bend radius.

In view *B*, the inside corner of the bend is sharp. Therefore, the bend radius is equal to zero. When this is the case, the axis of bend and the bend lines coincide and are located at the intersection of the inner plane surfaces of the piece part.

Figure 3·1 Basic bends and bend elements.

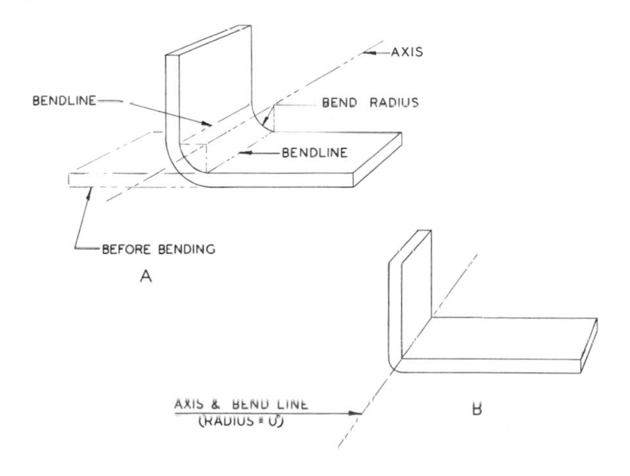

Bending stresses. Figure 3·2 represents the typical stress pattern which is set up within a metal bar when the bar is subjected to a deflecting force. Application of the force causes deflection of the bar, creating an inner surface and an outer surface. The inner surface tends to become shorter as force is applied, and the outer surface tends to lengthen. Somewhere between the outer surface and the inner surface, there exists a neutral plane which does not change in length. The resulting stress pattern is indicated by the directional arrows: Material inside the neutral plane is in compression; material outside the neutral plane is in tension. If the material is uniform in section, and if its elastic limit is not exceeded, the neutral plane will coincide with the center line of the material, as shown.

If the elastic limit of the material is not exceeded, plastic flow does not occur, and the material will spring back to its original shape when the force is removed. If, instead of being removed, the force is increased, the material will bend (Fig. 3·3). Whenever the applied force becomes strong enough, it causes the stresses (particularly toward the inner and outer surfaces) to increase beyond the elastic limit of the material. Exceeding the elastic limit forces the material

Figure 3·2 Bending stresses.

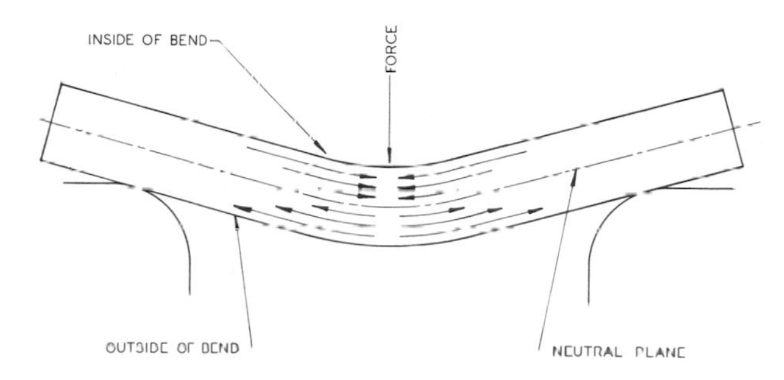

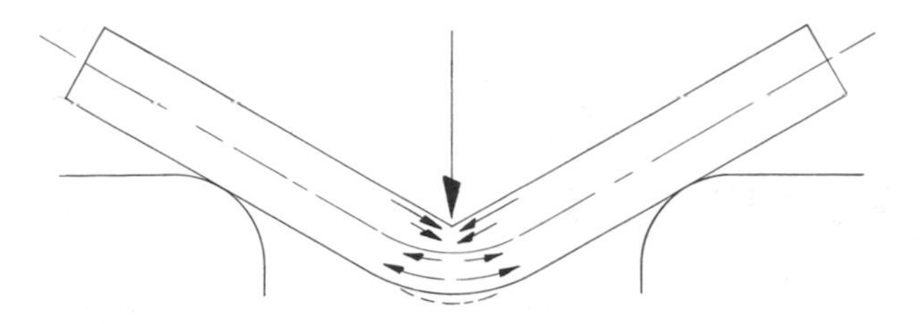

Figure 3·3 Bending.

into a state of plastic flow within the bend area, resulting in permanent deformation of the material.

Plastic deformation due to bending. As far as the pressworking of metals is concerned, the shape of the material can be changed. The volume of the material, however, is considered constant. (Volumetric change on the order of a small fraction of 1 per cent may occur, but this is ignored.)

Bending operations change the shape of the material in a specific manner, which is described at the beginning of this section. Because the material changes in shape but not in volume, the act of bending causes that portion of the material which is within the area of the bend to become distorted.

Actually, some distortion, proportionate to the ductility of the material and the severity of the bend, exists beyond the bend area. However, for discussion purposes it should be disregarded. Since bending is a specific type of forming operation, all bends are distorted in a similar manner. This distortion is called plastic deformation because it is associated with the characteristic plastic flow of the material within the bend area.

Figure 3·4 is a convenient and appropriate method of visualizing the plastic deformation which occurs in bending operations. For illustrative purposes, the crystalline structure of the metal is represented by cubical units equal in size and shape. This is the condition of the material before bending (views *A* and *B*).

After bending, the units are displaced and deformed in the manner indicated in views *C* and *D*. The pattern of distortion is directly associated with the type of strains induced by bending: Material toward the inner bend surface is under compression; material toward the outer surface is in tension. Therefore, on the outside of the neutral plane the units are stretched longitudinally (view *C*), which causes a corresponding reduction in their cross-section area (view *D*). The units inside the neutral plane are compressed longitudinally, which increases their cross-section area. This combination of strains, which is typical of all bending operations, deforms the material in the manner illustrated. In the compressed part of the bend, the material bulges wider than its original width *W* (view *D*). On the tension side of the bend, the material is reduced in both width *W* and thickness *T*, as indicated by the dotted lines in views *C* and *D*.

Thinning of material on the tension side of the bend is more pronounced than the bulging on the compression side. This is because the resistance of the material in compression builds up more rapidly than the resistance of the material in tension. Thus, relatively, the resistance of the material in tension decreases. Therefore, the greater portion of internal movement (plastic flow) takes place on the tension side of the bend. This also accounts for the fact that the neutral plane tends to shift toward the compression side of the bend.

For any given type of material, the amount of plastic deformation varies directly with the severity of the bend. That is, the amount of plastic deformation increases as the bend becomes more severe. The major factors which determine the severity of a bend are stock material thickness, size of the bend angle, and size of the bend radius. These factors, singly or in combination, influence bend severity as follows:

1. Increasing the stock material thickness increases the severity.
2. Increasing the bend angle increases the severity.
3. Decreasing the bend radius increases the severity.

The type of material influences the amount of plastic deformation which can occur in a bend. In general, softer (more ductile) material does not offer as much resistance to plastic flow as harder

Figure 3·4 Plastic deformation due to bending.

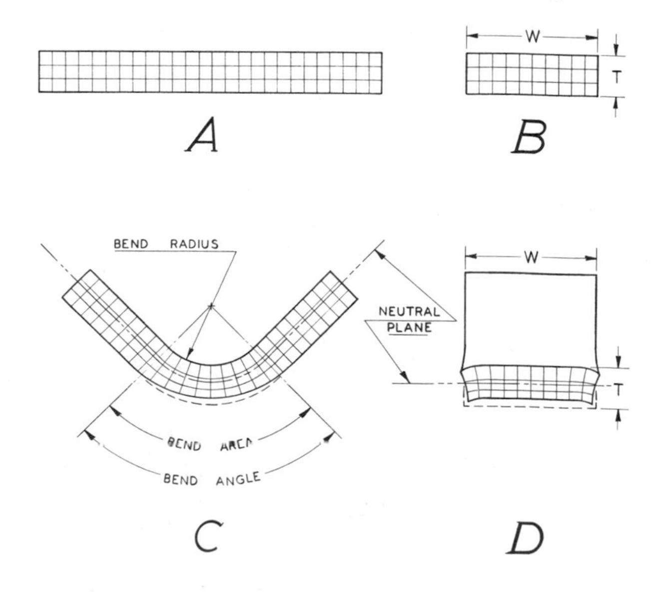

(less ductile) material. Therefore, the more ductile materials can be plastically deformed to a greater degree before their ultimate strength is reached. Because of this the more ductile materials can be bent more radically than the less ductile.

The compression bulge on the inside of the bend causes a slight, localized increase in the width of the piece part at the bend (view *D*). Occasionally, this condition may affect the utility of the piece part. If the localized increased width is not acceptable, it may be necessary to decrease the width of the flat blank. If this in turn is not acceptable, it may be necessary to undercut the sides of the blank at the bend area. If either of these procedures is required, it will affect the preceding operation. Therefore, the acceptability of the normal bulged width should be considered and verified when necessary.

The bend-allowance curve. The curved neutral plane of the bend area is called the bend allowance. It is shown in relation to the other bend elements in Fig. 3 • 5. Since the bend-allowance curve is subtended by the bend angle, the length of the curve is directly in proportion to the size of the angle. Precisely speaking, the bend-allowance curve is not a true arc described about the bend axis. It approximates such a true arc so closely, however, that for all bending calculations it is considered to be one.

The other bend elements, shown in this illustration, have been discussed in the preceding part of this section. It is important for the diemaker to be thoroughly conversant with the elements of bends and how they are interrelated.

Bend elements summarized. The bend axis is a straight-line axis around which the material is formed at the desired bend radius. The bend axis is the natural center of the bend radius. Length of the bend axis is the same as the width of the workpiece at the bend area.

Figure **3 • 5** Bend elements (any bend angle).

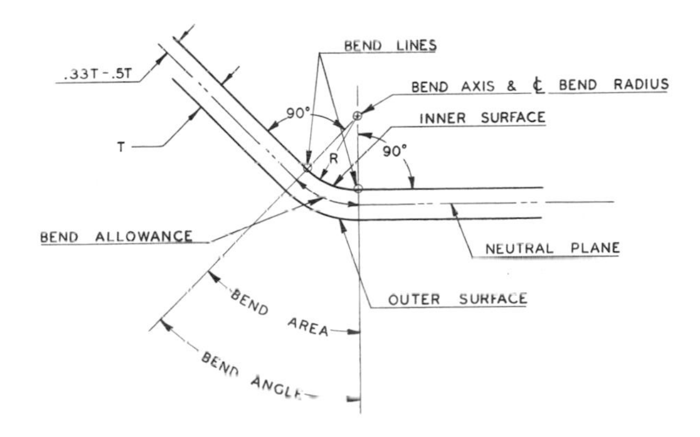

The bend radius is the radius of the inner arc of the bend. The extreme positions of the bend radius are tangent to the inner flat plane surfaces of the workpiece. Therefore, the extreme positions of the bend radius are perpendicular to the inner flat plane surfaces, as illustrated.

Bend lines are imaginary lines created by the tangency of the bend radius with the inner flat plane surfaces. There are two bend lines associated with each bend.

The bend angle is the angle included between the two extreme positions of the bend radius. It originates at the bend axis. The bend angle is *not* the included angle of the piece part, which originates at the intersection of the flat plane surfaces of the piece part. In the special case of right angled bends, both the bend angle and the included angle of the piece part are right angles. In this special case, both angles happen to be equal in magnitude. However, as described above, each angle still has its own separate origin.

Specifically, the bend area is the area included within the bend angle. It is not actually a bend element. The term "bend area" is used in a definitive sense in association with the elements of bends. It is also applied in the same manner to the stock material, plastic deformation, and other circumstances associated with bends and bending.

The inner and outer surfaces of a bend are self-evident. The inner surface is on the compression side of the bend and is squeezed shorter as a result of bending. The outer surface is on the tension side of the bend and is stretched longer as a result of bending.

The neutral plane is a theoretical plane originated by the inherent bending stresses. It delineates an imaginary separation between the material in compression and the material in tension. The neutral plane occurs at a distance of $0.33T$ to $0.5T$ from the inner surface, as illustrated. The exact distance varies according to the type of material and the circumstances under which bending takes place. The length of the neutral plane does not change as a result of bending.

It is essential for the diemaker to think of bends and be able to discuss them in a logically organized manner. As an aid in this direction, key bend terms are associated by illustration in Fig. 3 • 6. As a further convenience, these terms are shown in relation to different degrees of bending.

ESTIMATING FLAT-BLANK LENGTHS

When a given piece part is produced by bending, the length of its required workpiece "in the flat" is the same as the length of its neutral plane. Estimating the required flat workpiece length is a

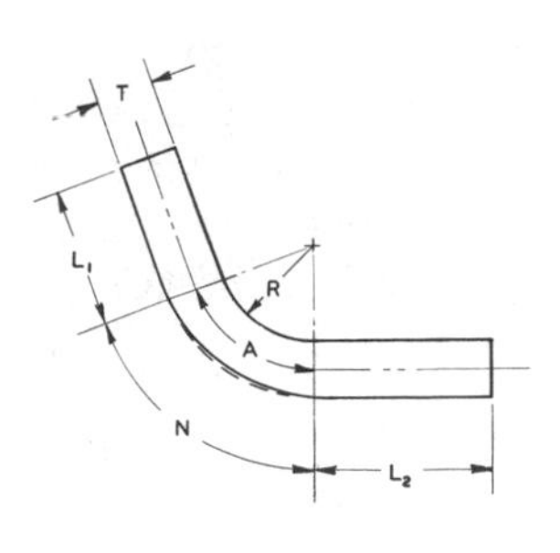

LOW-ANGLE BEND

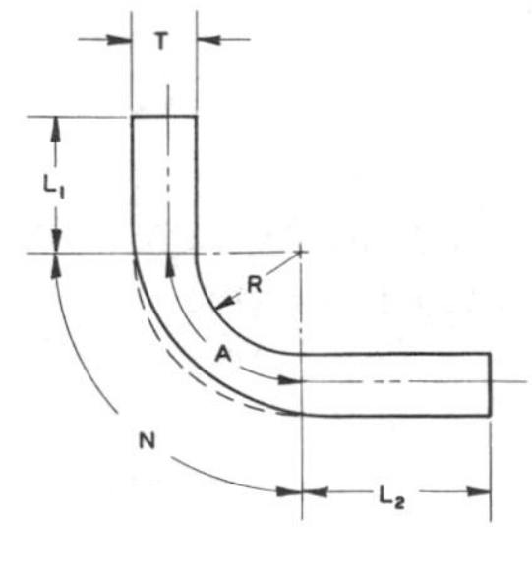

RIGHT-ANGLE BEND
90° BEND

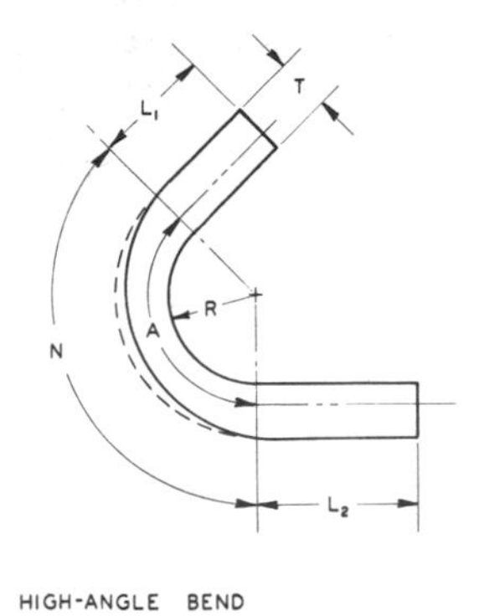

HIGH-ANGLE BEND

T = MATERIAL THICKNESS
R = BEND RADIUS
N = BEND ANGLE (DEGREES)
L = LENGTH OF FLAT PLANE (LEGS)
A = BEND ALLOWANCE
C = DISTANCE: INNER SURFACE TO NEUTRAL PLANE

Figure 3·6 Bend terms.

matter of determining the theoretical length of the neutral plane. To determine the theoretical length, it is necessary to calculate the bend allowance A for each bend. The estimated blank length will then be the sum of the lengths L of the bend legs and the allowances. Then, where B is the length of the flat blank,

$$B = L_1 + A + L_2 \qquad \text{etc.}$$

To find A:

$$A = (R + C)\frac{2\pi N}{360} \quad \text{or} \quad A = (R + C)\,0.01745N$$

Distance C from the inner surface to the neutral plane is a variable factor depending largely upon the ratio of the stock thickness T to the bend radius R. Optimum value for C may be considered to be

where $R < 2T$	$C = 0.33T$
$R = 2T$ to $4T$	$C = 0.4T$
$R > 4T$	$C = 0.5T$

Thus, for the bend of any angle

where

$R < 2T$	$A = (R + 0.33T)\,0.01745N$
$R = 2T$ to $4T$	$A = (R + 0.4T)\,0.01745N$
$R > 4T$	$A = (R + 0.5T)\,0.01745N$

For 90° bends

$$A = (R + C)\frac{2\pi 90}{360}$$
$$= (R + C)\frac{\pi}{2}$$
$$= (R + C)\,1.5708$$

Therefore, for 90° bends

where $R < 2T$	$A = (R + 0.33T)\,1.5708$
$R = 2T$ to $4T$	$A = (R + 0.4T)\,1.5708$
$R > 4T$	$A = (R + 0.5T)\,1.5708$

EXAMPLE 1:

In Fig. 3·7 assume (dimensions are inches)

$T = 0.030$	$R_1 = 0.18$
$L_1 = 1$	$N_1 = 90°$
$L_2 = 2$	$R_2 = 0.09$
$L_3 = 1.25$	$N_2 = 45°$

Then

$$A_1 = (0.18 + 0.015)\,1.5708$$
$$= (0.195)\,1.5708$$
$$= 0.306$$
$$A_2 = (0.09 + 0.012)\,45(0.01745)$$
$$= (0.102)\,0.785$$
$$= 0.080$$

Since $B = L_1 + A_1 + L_2 + A_2 + L_3$
$= 1 + 0.306 + 2 + 0.080 + 1.25$
$= 4.636$

EXAMPLE 2:

$T = 0.060$	$R_1 = 0.09$
$L_1 = 1$	$N_1 = 90°$
$L_2 = 2$	$R_2 = 0.25$
$L_3 = 1.25$	$N_2 = 40°$

Then

$$A_1 = (0.09 + 0.02)\,1.5708$$
$$= (0.11)\,1.5708$$
$$= 0.173$$

$$A_2 = (0.25 + 0.03)\,40\,(0.01745)$$
$$= (0.28)\,0.698$$
$$= 0.195$$

Since $B = 1 + 0.173 + 2 + 0.195 + 1.25$
$= 4.618$

Sharp-corner bends are special cases. Calculation can be simplified as follows:

For 90° bends where $R = 0$,

$$A = 0.5T$$

(Bend angles, other than 90°, are proportionate.)
Truly sharp-corner bends are seldom encoun-

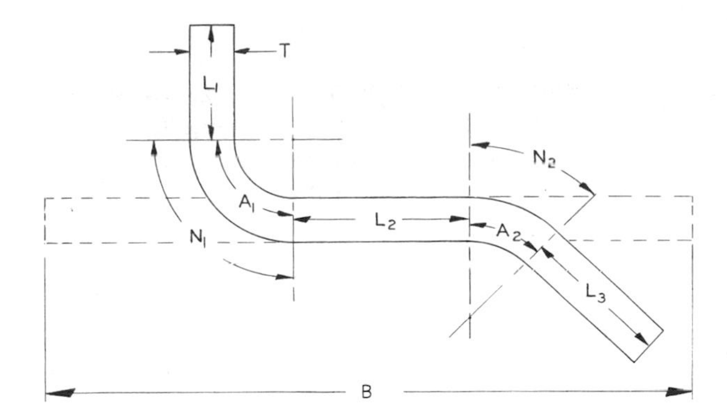

Figure 3·7 $B = L_1 + A_1 + L_2 + A_2 + L_3$.

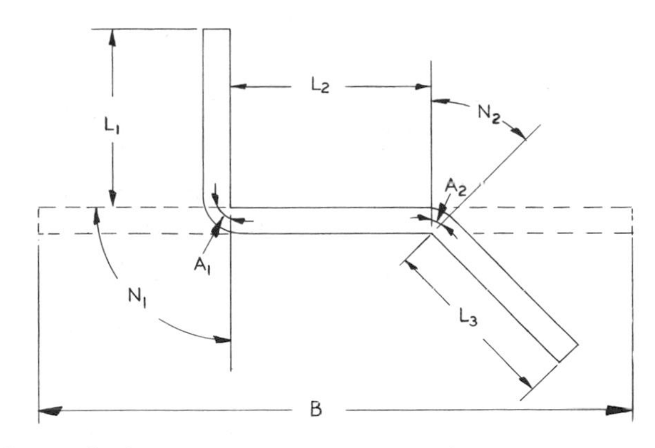

Figure 3·8 $B = L_1 + A_1 + L_2 + A_2 + L_3$.

tered. Generally a so-called sharp-corner bend will have a bend radius of 0.005 to 1/64 in.

In this case, for 90° bends

where $T <$ 1/16 in. $A = 0.4T$
$T =$ 1/16 in. to 1/8 in. $A = 0.45T$
$T >$ 1/8 in. $A = 0.5T$

(Bend angles other than 90° are proportionate.)

EXAMPLE:

In Fig. 3·8, assume

$T = 0.040$ $L_3 = 1.5$
$L_1 = 1.5$ $N_1 = 90°$
$L_2 = 1.75$ $N_2 = 45°$

Then $A_1 = 0.4T = 0.4(0.04) = 0.016$

Since $N_2 = \frac{N_1}{2}, A_2 = 0.008$

Since $B = L_1 + A_1 + L_2 + A_2 + L_3$
$B = 1.5 + 0.016 + 1.75 + 0.008 + 1.5$
$= 4.774$

NOTE:

For precision work, all estimated sizes should be confirmed by experiment. Corner setting affects estimated sizes and must be allowed for. Pressure-pad dies generally require slightly less bend allowance than V dies.

V-BENDING DIES IN PRESS BRAKES

V-bending dies are so called because of their characteristic cross-section shape, which resembles the letter V. A bending operation as produced in a V die also resembles the letter V. Dies of this type are widely used in conjunction with press brakes. Some generalized examples of these dies are shown in Fig. 3·9.

View *A* shows a 90° V-bend die, which is one of the most commonly used press-brake dies. The die in view *B* is called an acute-angle V die, because its shape in cross section is an acute angle. Both dies (views *A* and *B*) can be used to produce a wide range of angular bends simply by the adjustment of the press ram to secure different punch entry depths as required. The acute-angle die will, of course, have the wider range. This type of bending procedure where the punch does not hit home is called "bending in air." Naturally, piece parts produced by bending in air will not be as consistently accurate as they would be if the punch bottomed (hit home). Note that when the punch is bottomed in an acute-angle die, the resulting bend is a high-angle bend (see Fig. 3·6).

The die in view *C* is a V die for offsetting. Its cross-section shape is a double V with one V in-

Figure 3·9 V dies (press-brake).

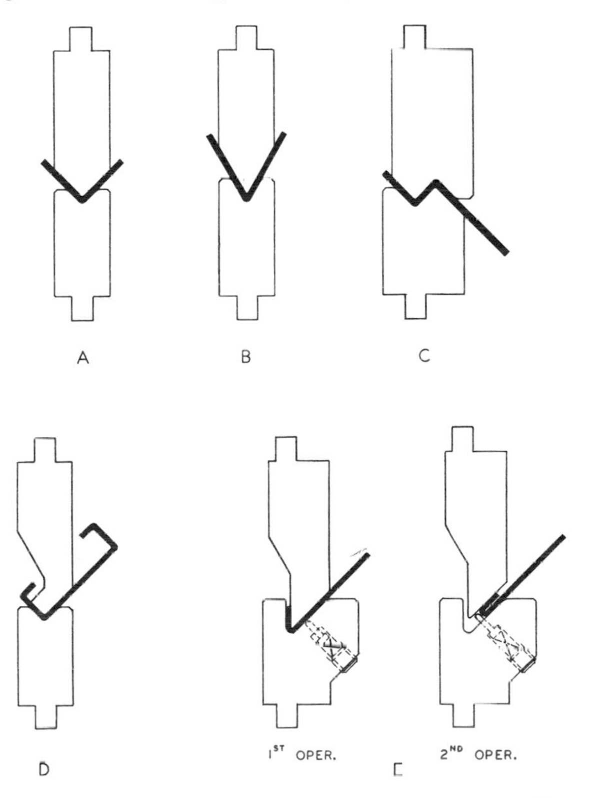

verted. View *D* is a gooseneck die for return-bending operations. It is essentially a simple V-bend die with clearance provided for return bends.

View *E* depicts one type of hemming die. It is simply an acute-angle V die with suitable spring pins provided for gaging the second operation. The required number of workpieces are bent to the acute angle in the first operation. In the case of the die shown, the punch entry depth is then adjusted to suit the required hem, and the workpieces are run through the second operation.

The scope of press-brake dies is varied and extensive, including cutting, bending, and forming operations of all sorts. To work with these dies, it will be necessary either to secure the required press-brake specifications from the press-brake manufacturer or to have access to the brake in which the dies will be run.

SIMPLE V DIE FOR PUNCH PRESS OPERATION

A typical V die and the piece part it produces are illustrated in Fig. 3·10. The piece part is a plain angle bracket. The stock material is ⅛-in.-thick hot-rolled mild steel.

Figure **3·10** V-form bending die for right- and left-hand angle bracket.

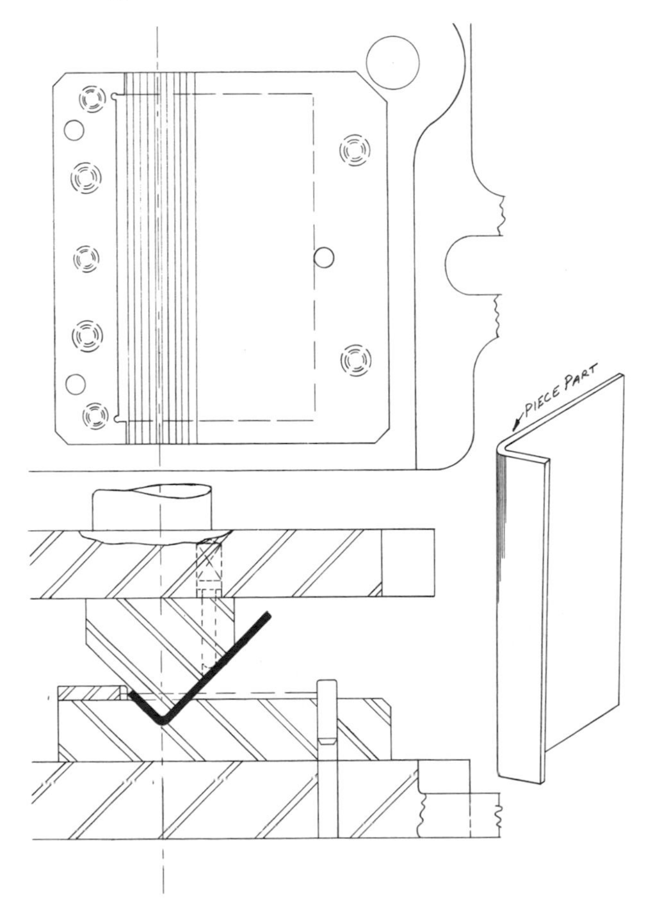

In both design and construction, the illustrated die is the essence of simplicity. It is composed of V-type punch and die members of basic design and suitable dimensions, together with a nest arrangement. The entire assembly is appropriately mounted on a standard die set which is equipped with guideposts. The die set provides accurate alignment, which facilitates setting up the die in the punch press. These dies are sometimes mounted on open, or plain, die sets (no guideposts). However, this practice is usually limited to low-production dies or to instances where it is not anticipated that the die may have to be run from time to time.

The illustrated die is quite typical of simple V dies which are designed and built to produce specific piece parts. It is not intended to be a general-use die, even though it is obvious that, with changes in the nest arrangements, the basic die could be used to produce a variety of similar piece parts.

In operation, the flat blank (workpiece) is loaded into the nest. When the press ram descends, the punch forces the material downward into the die opening. The contours of the die channel cause the legs of the bend to swing upward to the desired angles, as shown. A spring pin (or pins, if necessary) is incorporated in the punch to assure that the piece part does not adhere to the punch on the up stroke. When the die opens, the piece part remains on the lower die surface, to be removed (in this case) by the operator as he loads the next workpiece into the nest.

Although the design and construction of this type of die are obvious in their simplicity, there are at least two major considerations which the diemaker should double-check before proceeding with construction. These are punch height and die-block thickness. They are mentioned here because they are all too often overlooked, especially in cases where V bending is incorporated, along with other operations, in progressive dies.

Punch height. The punch must be high enough to permit the ends of the bend legs to be in the clear when at their maximum formed-up position (die closed). For larger piece parts, where the legs swing higher, it is often economically desirable to install a soft cold-rolled steel spacer block behind the punch to effect savings in tool-steel and heat-treat costs. At times it may be necessary to provide clearance pockets, or notches, in the punch holder. However, clearance in the punch holder should not be provided indiscriminately but with due regard for structural soundness and operator safety.

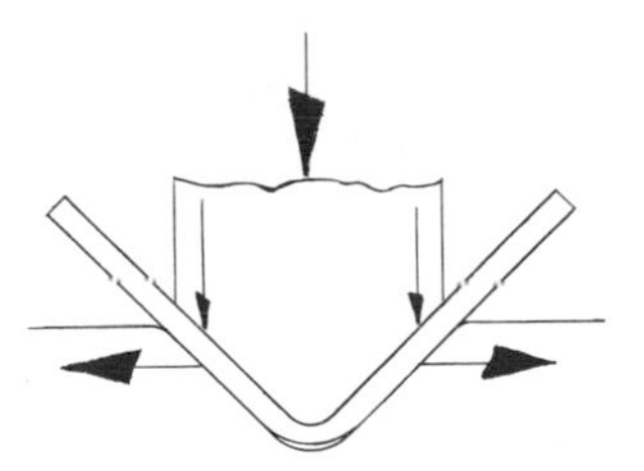

Figure 3·11 Wedge action in V dies.

WEDGE ACTION IN V DIES

Because of its V shape, the punch is a potential wedge which, if circumstances permit, can split the die block. This wedge action is pictured in Fig. 3·11. When the descending punch drives the stock material against the walls of the die opening, lateral force is transmitted to the die block. The lateral forces are directed outward from the center of the V, as indicated by the horizontal arrows. The die block must be strong enough to resist these forces or it will split. Consequently, the thickness of the die block must be adequate.

It is true that, for a given type of stock material, light-gage work requires less effort than heavy-gage work. However, once the bending operation has reached the point where the material is sandwiched between the punch and die walls, the lateral forces transmitted to the die block will *not* be less for light-gage stock material.

BENDING ACTION IN V DIES

The distinctive manner in which bending takes place in V dies is illustrated in Fig. 3·12. Here, a typical V-bending sequence is shown in three stages in order to slow down the action pictorially for analysis.

Radius R on the nose of the punch is, of course, the bend radius, which will be reproduced in the piece part. Radii indicated at R_2 are the bending radii. The bending radii are necessary to the performance of the bending act; they are not reproduced on the piece part. Do not confuse bend radii with bending radii. As an aid to differentiation, associate them in this way:

Bend radii are produced as a result of bending.
Bending radii are required to perform the act of bending.

View A depicts the early stages of bending action as it occurs in a V die. Here, the punch has descended until its nose radius R is tangent to the top plane of the die block. At the center, the punch has depressed the workpiece a distance equal to the stock material thickness T. The extremities of the bend legs have reacted by swinging upward, rocking a slight distance around the arcs of the bending radii R_2.

The motion of the material at R_2 is not quite a true rocking action. It is accompanied by a slight drag, or skidding, between the material and the arc surfaces of the bending radii. The amount of drag is negligible during the early stages of bending and becomes more pronounced as the bending action progresses. This is evident by examination of view B. Here, the punch has entered deeper into the die-opening channel. Reaction of the workpiece has advanced accordingly: It has been pulled deeper into the center and swung higher at its extremities. This action continues until the punch bottoms (view C).

V-bending action is sometimes described as a free, or unrestrained, bending action. This is true in a relative sense when it is compared to pad-type bending operations. However, it is not true in the absolute sense. From the time the punch begins exerting force upon the workpiece, the piece is constrained by its own resistance to the forces acting upon it. From views A and B it is evident that, during the bending action, the workpiece is restrained between the punch radius R and the die radii R_2.

Figure 3·12 Bending action in V dies.

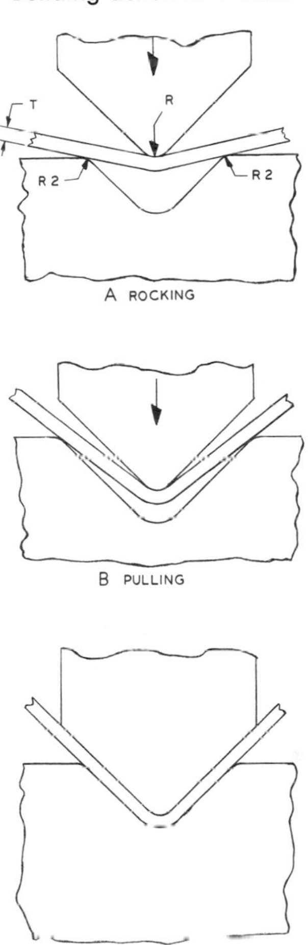

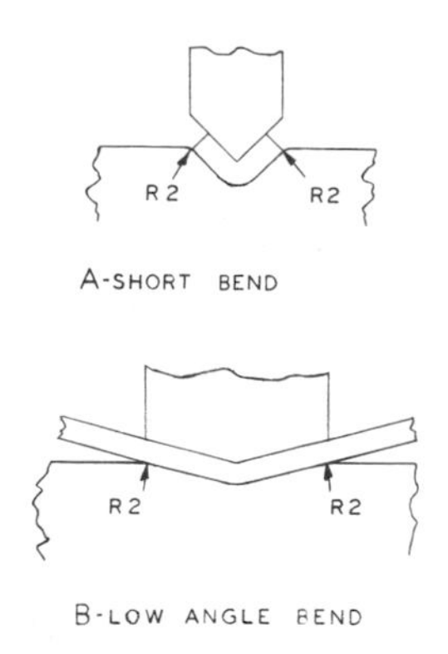

Figure **3 · 13** Bending radii.

BENDING RADII FOR V DIES

Optimum sizes for bending radii (Fig. 3 · 12, view *A*, R_2) depend upon circumstances peculiar to the individual job. For average conditions, bend radii of ½*T* to *T* are generally practical. As a rule, heavier stock thicknesses *T* require proportionately larger bending radii than light-gage materials. Softer materials also tend to require proportionately larger bending radii than harder materials. Very heavy material may require bending radii of 2*T* or more.

Short bends (Fig. 3 · 13, view *A*) and low-angle bends (view *B*) require the size of the bending radii to be quite small; in many instances, the radii may be less than ½*T*. In some extreme cases, the bending radii are made by just the very slightest rounding with a hand stone. In cases where the bend legs are relatively long, on the other hand, larger-than-average bending radii may be required. Proportionately larger bending radii are also required for high-angle bends which are made in acute-angle dies.

During the act of bending, relatively smaller bending radii tend to have a better gripping effect upon the stock material and thereby tend to be more efficient for controlling the reaction of the workpiece. Thus, smaller bending radii tend to produce bends which are more accurate than those produced with larger bending radii. In view of this tendency, it is good practice to make the bending radii slightly smaller than is deemed optimum. The size can then be increased, if necessary, at tryout. There is another very important reason for this procedure: It is much easier to increase the size of the bending radius, if required, than to decrease it if it is found to be too large.

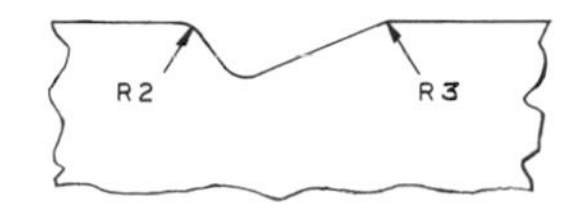

Figure **3 · 14** Bending radii, asymmetrical V die.

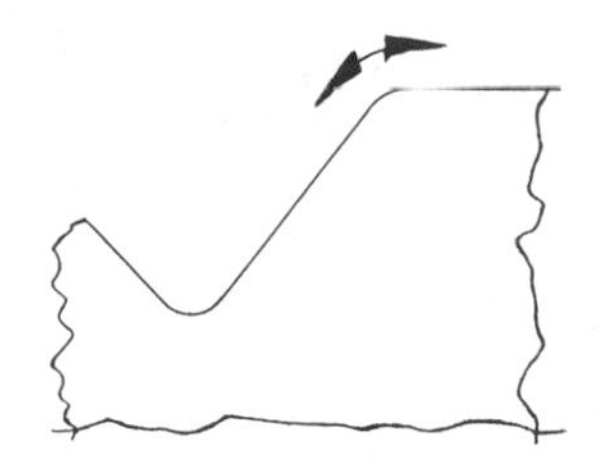
Figure **3 · 15** Lay of finish on bending radii.

For symmetrical V-die channels, such as the one pictured in Fig. 3 · 12, the size of both bending radii should be identical. If the bending radii differ in size, the bending action is thrown out of balance and the workpiece tends to shift in the direction of the smaller radius. Quality of surface finish too should be the same on both bending radii.

For asymmetrical V-die openings (Fig. 3 · 14), the bending radii should differ in the manner shown. Radius R_2 on the high-angle side of the bend should be larger than radius R_3, which is on the low-angle side. This is a compensatory procedure. Making R_2 larger permits the metal on that side to slide more freely than otherwise, and this tends to equalize the bending action.

Quality of surface finish on bending radii is of paramount importance. The arc surfaces of all bending radii should be finished to a high polish. The lay, or directional qualities of the finish, should be parallel to the pulling motion of the stock material during bending, as shown in Fig. 3 · 15.

SIZE OF DIE CHANNEL

Suitable die-opening proportions (Fig. 3 · 16) vary according to the requirements of the specific bending operation. For the average class of medium-sized work, the size of a V-die channel is not critical. A given piece part can often be adequately produced throughout a relatively large range of overall die-opening sizes. This does not mean that these dies can be made carelessly in re-

Figure **3 · 16** V-die channel proportions.

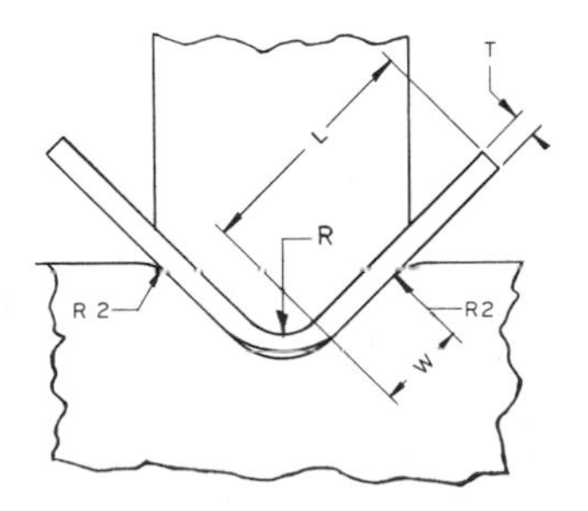

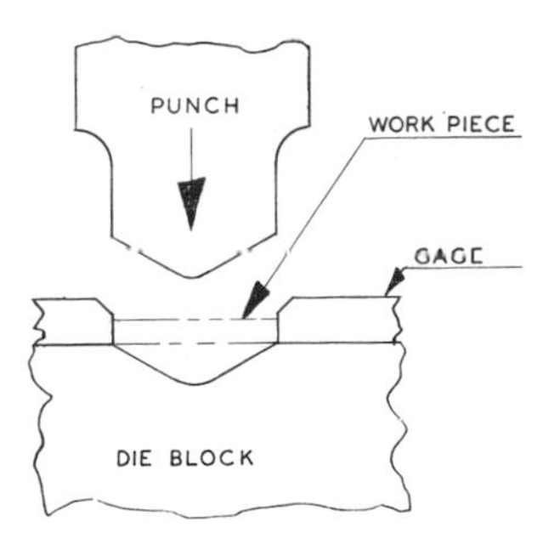

Figure 3·17 Short bend.

gard to appropriate die-opening sizes; it means, simply, that this phase of diemaking is not as exacting as many others. It is obvious that a die opening can be made either too large or too small for a given operation. It follows also that for a specific practical bending operation, there is an optimum die-opening size.

Because of factors which vary according to the specific bending operation, it is not practical to assign a simple rule or formula which will precisely determine optimum die channel sizes for all V-bending operations. However, for average conditions the following proportions may be used as a guide to determining suitable channel sizes:

Assuming T is from $\frac{1}{32}$ to $\frac{1}{4}$ in., then where L is $5T$ to $50T$, W may be $4T$ to $10T$ in proportion to L.

The above applies to bend angles up to 90°. High-angle bends (acute-angle dies) may require longer die wall straights W. Where T is less than $\frac{1}{32}$ in., W may have to be larger in proportion to L.

Harder and/or high-tensile-strength materials tend to require wider die openings than softer materials. Longer bend legs L require wider V openings.

For short bends, the proportion W/L must be larger than for long bends. That is, as length L decreases, W must increase in proportion to L. In extreme situations, dimension W may be longer than the bend leg L. A situation of this kind, where $W/L > 1$, is illustrated in Fig. 3·17. Note that bending radii are not required in this case.

When doubt exists as to the appropriate channel size, it is usually better to make the opening slightly larger than what is estimated to be optimum. The die channel can then be reduced in size if this is proved necessary at tryout. As a rule, it is simpler to reduce the size of the die opening than to increase it. Do not make this a standard order of procedure, however. Analyze each specific situation in advance to decide whether or not it will be advantageous to proceed in this manner.

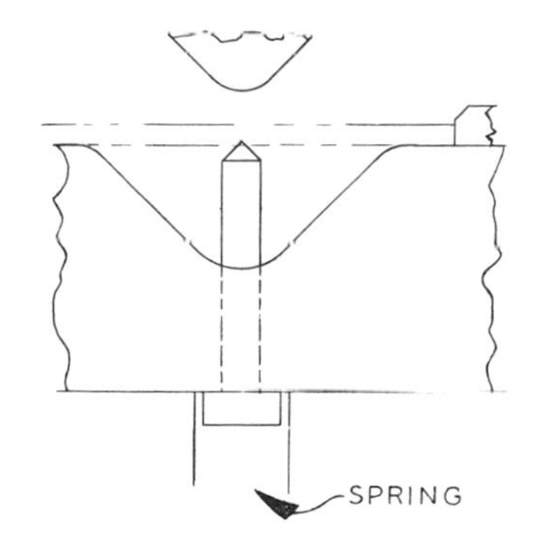

Figure 3·18 Pointed antislip pin.

Piece parts which have relatively long bend legs and/or large bend radii will often require V-die channels which are disproportionately large in relation to their stock material thickness. This condition aggravates the tendency of the workpiece to slip and shift laterally during bending. As a means of alleviating this tendency, pointed spring pins are often installed, as shown in Fig. 3·18. The number of pins required varies according to the width of the workpiece. A variation of this practice is to install spring pins which have a knurled face instead of the point.

The above practice is not necessarily limited to disproportionately wide die channels. Because V bending is inherently somewhat lacking in precision, antislip pins may be installed in dies of optimum proportions in order to produce more accurate piece parts.

V-DIE ANGLES

Nominally, the included angle M (Fig. 3·19) of a V-die assembly is the supplement of the required bend angle N of the piece part. That is,

$$180° - N = M$$

One exception to this is the practice of air bending, where the punch does not bottom the workpiece in the die opening. For bending in the air, angle M may be any practical angle smaller than the supplement of N. Another exception is the

Figure 3·19 Angular relationship.

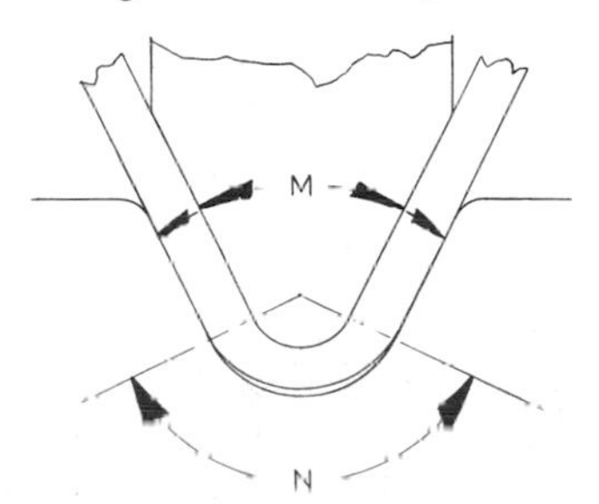

practice of overbending to compensate for springback.

SPRINGBACK

In bending operations, the elastic limit of the metal in process is exceeded, but its ultimate strength is not. Therefore, some of the original elasticity of the stock material is still present at the peak of the bending cycle, which occurs at the bottom of the press stroke. Because of this, when the force (punch) is withdrawn, the material on the compression side of the bend tends to expand slightly and the material on the tension side tends to contract. The combined result is that the workpiece tends to resume its original shape, causing the bend to spring open some small amount. This reaction of the material is called springback. Springback varies according to the thickness, type, and condition (temper, etc.) of the stock material. It also varies directly in proportion to the size of the bend radius: The larger the bend radius, the greater the springback.

Overbending in V dies. In concept, overbending is the simplest way to correct for springback. In V dies, overbending is accomplished by making angle *M* smaller by the required amount. For soft steel, brass, aluminum, or copper, springback may be 0 to 1°. For ¼ to ½ hard material, springback may be 1 to 5°. For harder materials, it may be 12 to 15° or more. These figures can only be considered approximations because of the many variables which influence springback. The only practical method of determining springback is by experiment. It is, of course, obvious that the correct stock material must be used in order to develop springback adequately by experiment. Even with the most careful development, there will be slight variations due to nonuniformity of the stock material. Because of this lack of constancy, overbending, while simple in concept, is not necessarily simple in practice.

Cornersetting in V dies. Setting is, generally, the most effective method of dealing with the problem of springback. Practically speaking, the idea of setting the bend is more one of eliminating springback than of making a compensating allowance. In some instances, it may be necessary or desirable to combine setting with overbending. This may be done because the springback is extreme, or it may be done to avoid excessive straining of the material. It may also be the case that the design of the die lends itself naturally to a combination of setting and overbending.

Setting is accomplished by coining (squeezing) the stock material at bend area. The coining effect causes additional compressive strains within the material. The extra compression strains overcome the characteristic springback tendencies of a normal bend. Because of the manner in which springback is overcome within the stock material, setting is more accurate and consistent than overbending. Since the set is made in the bend area, the practice of setting in bending operations is specifically referred to as corner setting.

Developed set. One way of setting a bend is illustrated in Fig. 3·20, where the setting condition is shown considerably exaggerated for purposes of illustration. With this method, the die-opening fillet *RR* is made to squeeze the stock material in the bend area. A very slight squeeze is enough; compression ranging from 2 to 5 per cent of *T* will be generally satisfactory. Excessive coining of the material must be avoided, as it will weaken the bend.

Because of thinning, the outer surface of a bend is not a true radial arc described from the bend axis. The outer surface does, however, very closely approximate a larger true arc similar to *RR* in the illustration. The die-opening fillet radius *RR* should be a very slight amount larger than the radius of the outer bend surface in order to coin the stock material. Required size of the fillet radius can be determined by experiment, or it can be estimated by comparison with similar bending operations which have been made previously.

One method of developing the size of the die fillet is to finish the punch first. Then, use the punch to either air-bend or hand-bend a piece of the proper stock material. Springback is corrected by pinching the bend legs closer together after bending. The sides of the piece should be machined after bending, to remove deformed edge

Figure 3·20 Developed die radius for setting to overcome springback.

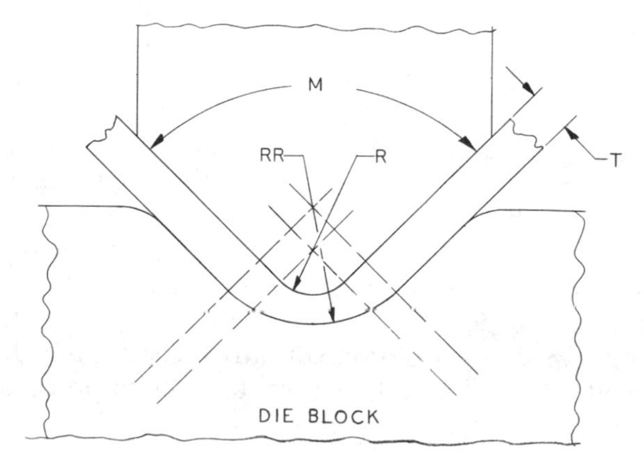

material (see Fig. 3 · 4, view D). The piece can then be used as a template for finishing the die opening.

Another method is to make the fillet radius equal to $R + \frac{5}{4}T$. The radius is then reworked at tryout to suit. This allowance assumes a 90° bend in soft steel where the bend radius R is equal to $T/2$.

Keep in mind that the material in the bend area becomes thinner as the bend angle becomes greater. Keep in mind also that the thinning becomes more pronounced as the size of the bend radius R decreases. For 90° bends in soft steel, thickness of the material cross section at the center of the bend area will be approximately

$$96\%\ T \text{ where } R = 2T$$
$$94\%\ T \text{ where } R = T$$
$$92\%\ T \text{ where } R = \frac{T}{2}$$

Advantages of This Method

Does not weaken piece part, assuming, of course, that coining is not overdone

Appearance of piece part not affected

Disadvantage

Time-consuming: therefore the most costly method

Offset-punch method. The face of the punch in Fig. 3 · 21 is offset in order to achieve a coining penetration in the bend area. Offset dimension S should not be made unnecessarily deep, as this can weaken the piece part. An offset depth of 5 per cent T is a good average figure for most work.

In the case illustrated in view A, the set area is equal to the inside bend area. For larger bend radii, the set area may have to be reduced. One method of reducing the set area is to relieve the nose of the punch, as shown in view B. Depth of this relief may be equal to S. Where the bend radius is less than T, the length of land K (view C) may be made equal to T, with a minimum of $\frac{1}{64}$ in. for thinner stock material. It should not be necessary to match die-opening fillet RR to the outer surface of the bend. Radius RR may be made equal to $R + T$. If $R + T$ equals an off-standard dimension, radius RR can usually be made the next-closest nominal size.

Advantages of This Method

Simple

Easy to modify, if required

Low in cost

Disadvantages

Indentation formation at inside corner of piece part

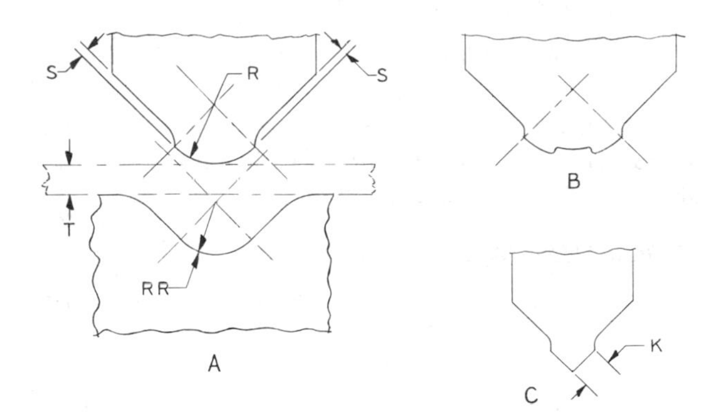

Figure 3 · 21 Offset punch for corner-setting.

Tendency to weaken piece part—although this can normally be disregarded if set depth is not overdone

Angular punch relief. Figure 3 · 22 illustrates another effective method of counteracting spring-back. Here, an angular differential is provided between the included angle M of the punch and the included angle of the die opening. The amount of differential will, of course, vary in accordance with the specific job conditions. However, because of the squeezing action, the required differential is relatively small. The reduced punch angle permits the workpiece to overbend while the setting action takes place, which accounts for the effectiveness of this method. It is not necessary to match the die-opening fillet RR to the outer bend surface. Radius RR may be made equal to $R + T$.

Advantages of This Method

Simple

Highly effective

Easy to modify, if necessary

Low in cost

Does not appreciably weaken piece part

Does not affect appearance of piece part

Setting is, in effect, a coining operation. To attempt to coin large areas of the workpiece would require excessive punching force. This condition, combined with the fact that setting is effective only in the general area of bend, is the reason for reducing the set area to the smallest practical size.

Because setting increases the compressive

Figure 3 · 22 Angular punch relief for corner-setting.

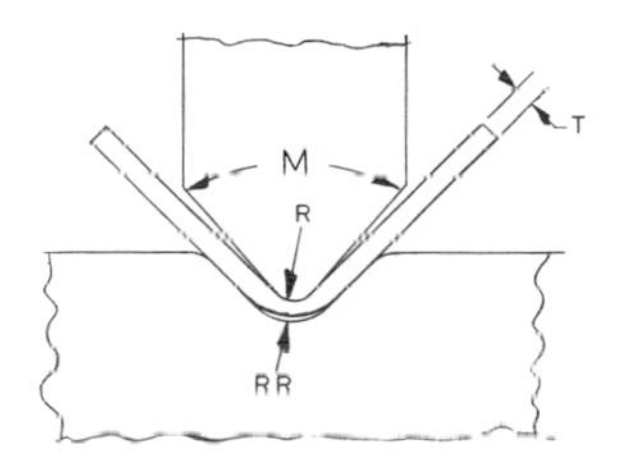

strains in the bend, the compression bulge on the inside of the bend (refer to Fig. 3·4, view *D*) will be exaggerated somewhat. The diemaker must keep this condition in mind and make proper allowance for it where necessary.

V DIES SUMMARIZED

Generally speaking, V dies are not considered as accurate as pad-type bending dies.

The most common errors in V-die construction are that the die opening and/or the bending radii are made too large. As a rule, both die opening and bending radii should be held to the smallest practical size.

The V-bending principle lends itself well to the making of very short bends as well as long ones.

For average work, V dies which bend upward (that is, in which the extremities of the bend legs swing upward) are prevalent. However, some applications may require inverted V dies. This is especially true when V dies are used for large work, where the bend legs are quite long.

PRESSURE-PAD BENDING DIES

The dies in this group are so designated because, as a basic design principle, they are equipped with pressure pads. Pressure pads are commonly actuated by springs or rubber cylinders. Where more pressure is necessary, the die is usually installed in a press which is equipped with an air cushion. Hydraulic cylinders also can be used to actuate pads when strong pressures are required.

In bending dies, pressure pads perform the following functions, which are listed in order of operational sequence:

1. They hold the workpiece during bending.
2. They serve as bottoming blocks for setting up the bend (or bends).
3. They act as strippers or shedders to aid in removing the piece part from the die.

U-BENDING DIES

These dies are called U-bending dies because the bending operations produced in them bear a resemblance to the letter U. They are often simply referred to as U dies. Typically, they are equipped with pressure pads, which usually perform all three of the functions listed above.

Bending action in a U die is typical for bending dies equipped with pressure pads. This action is depicted in Fig. 3·23. Four phases of the bending action are shown in views *A*, *B*, *C*, and *D*.

In view *A*, the descending punch has contacted the workpiece, forcing it against the pressure pad. Since the punch force is opposed by the pad pressure, the workpiece is gripped between the punch and the pad. The gripping force is equal to the pressure exerted by the pad. If the pad is spring- (or rubber-) actuated, the pad pressure will increase as the pad is forced down. If the pad is

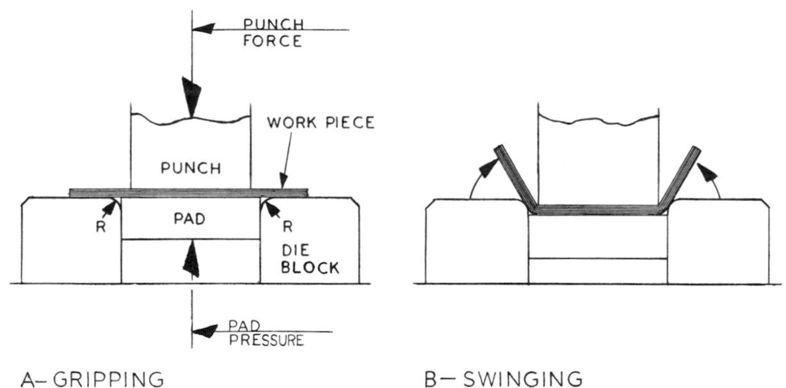

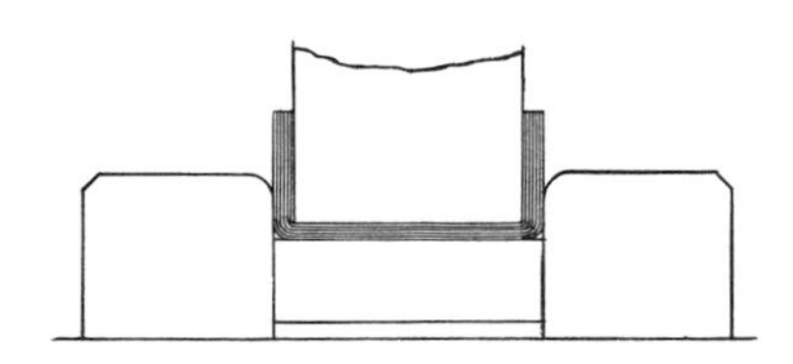

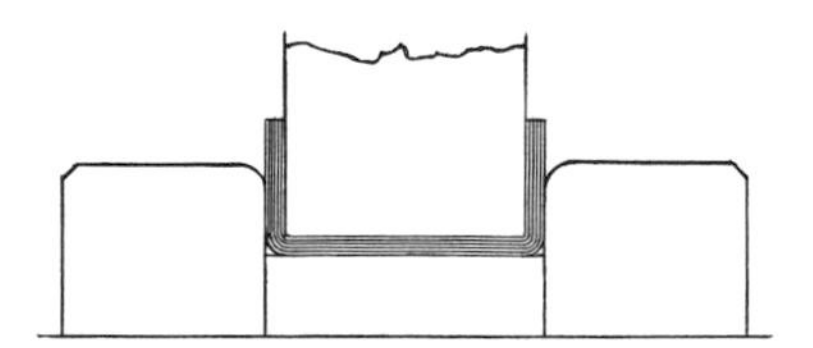

Figure 3·23 Bending action, pad-type die.

air- (or hydraulically) actuated, the pressure normally remains constant throughout the pad travel.

As the punch descends farther, the bend legs swing upward as shown in view *B*. This stage combines a rocking and pulling action similar to that described in Fig. 3·12, views *A* and *B*.

In view *C* (Fig. 3·23), the bend legs have completed their swing, and the stock material is being "wiped" between the vertical surfaces of the punch and the die opening. Wiping continues until bottoming occurs (view *D*). The wiping action should not be confused with "ironing." In a true ironing operation, the material is deliberately thinned and stretched. Ironing is not required and can be objectionable in normal bending operations.

PAD CONSTRUCTIONS

Figure 3·24 represents a U die equipped with a shouldered pressure pad. Pad actuation is achieved by springs, which are installed in spring pockets drilled into the die shoe. Occasionally, because of space restrictions, it may be necessary to provide spring pockets in the pad also, as indicated by the dotted outline.

When the die is open, the pad face should project above the die face a slight distance *d*, to assure that the workpiece will be securely held before bending begins. A *d* dimension of as little as 0.001 in. can be effective. However, for reasons of facility, *d* is usually made 0.005 to 0.010 in., which has proved to be satisfactory for the average application.

The sidewalls of the pad should be a close sliding fit to the die walls. This is to prevent lateral shifting of the pad during the bending cycle. A close fit also prevents the intrusion of foreign particles between the pad and the die block, which is an important consideration. For the great majority of applications a clearance *C* should be provided between the pad shoulder and the recessed wall of the die block. This obviates the need for unnecessary exacting fits and smooth finishes.

The corner formed by the pad shoulder should have a small fillet radius r_1. This fillet may have to be quite small, but it must be present. The inside corner of the die recess should also be provided with a fillet r_2. Clearance for these fillets is provided by beveling the mating members as shown.

When dies are equipped with internal springs, the pad shoulders seat against the top of the die recess at *S* as the die opens. Because of this action, the shoulders must be strong enough to resist the impact shock at *S*, which occurs on the upstroke.

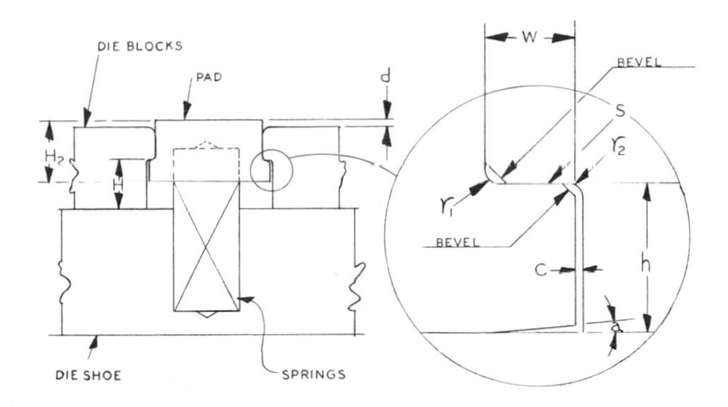

Figure **3·24** U die with internal springs and shouldered pressure pad.

Assuming that the peripheral length of the shoulder is adequate, the following minimum shoulder heights *h* are suggested:

For light to medium work

$$h = 1\tfrac{1}{2}w$$

For heavier work

$$h = 2w$$

Correct heat treatment is an important factor in pressure-pad construction. The pad proper may be hardened as required; generally a hardness of Rockwell C 58 to 60 is satisfactory. The pad shoulder, however, should be softer, to lessen the possibility of breakage. As a rule, the shouldered portion of the pad should be tempered down to Rockwell C 44 to 48.

For severe bottoming conditions, it is a good idea to relieve the back of the pad shoulders. One method of accomplishing this is to provide a small angular relief *a*, as shown.

Throughout its entire travel, the pad should be engaged in the die opening. To accomplish this, overall pad height H_2 must be greater than recess height *H*. The following is suggested as a minimum for H_2:

$$H_2 = H + \tfrac{1}{8} \text{ in.}$$

Any pad which bottoms must be completely enclosed. This is absolutely necessary in order to prevent ingress of foreign material of any kind. If a foreign particle of any consequence enters the pad cavity, it can be a source of malfunction and damage to the die.

Advantages of Internal Springs

The setup is simple; a minimum of work is required to mount the die in the press.

In some instances, less work is required to make the die, although it should be kept in mind that this is not always true.

Disadvantages

Pad shoulders strike against the top of the retaining recess on the upstroke. This must be considered a disadvantage.

Pad pressure is limited in accordance with the size and number of springs practical to install within the available space.

There is a lack of adjustment for pad pressure.

It is possible that spring breakage can cause damage to the die.

Instead of having shoulders, the pad pictured in Fig. 3·25 is retained by stripper bolts (shoulder screws). With this construction, the stripper-bolt heads hit home against the counterbored shoulder *S* when the die opens. Because of this, it is necessary to be certain of the adequacy of the stripper bolts: They must be large enough and not too few in number. The most common error is using stripper bolts which are too small. In order to use this method of construction, the pad area must be large enough to accommodate the required springs easily, and the stripper bolts as well. This must be accomplished without crowding.

Distance *E*, a very important factor with this type of construction, must be greater than the pad travel *e* to eliminate the possibility that the stripper bolts may impact against the bolster plate (or other obstruction). To determine minimum *E* dimension, the following is suggested whenever possible:

$$E = e + 1/8 \text{ in., min.}$$

When a pad is retained with standard stripper bolts, the stripper bolts should be checked and selected to assure uniform length. Stripper-bolt holes should have the same clearance as a standard cap screw.

In Fig. 3·26, the pad is actuated by an externally mounted spring. Rubber cylinders (rubber bumpers) are often used instead of the conventional spring shown. In either case, pad function and the general die construction will be the same.

Figure **3·25** Die with internal springs, pad retained with stripper bolts.

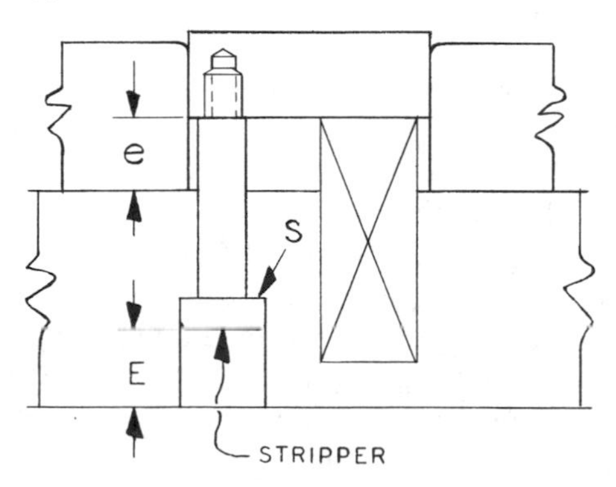

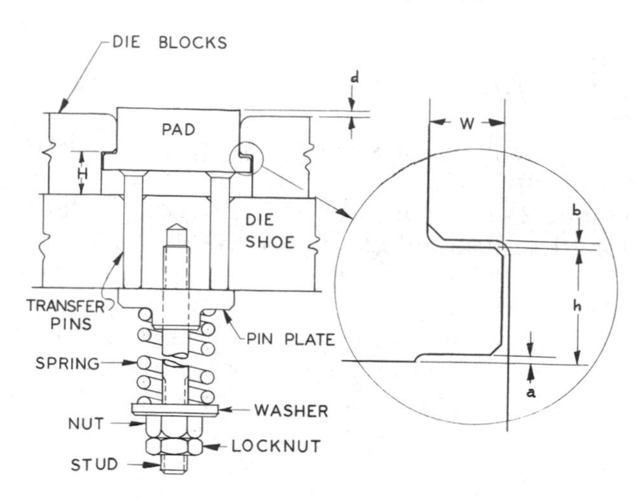

Figure **3·26** U die with external spring and shouldered pressure pad.

An important feature of any shouldered pad actuated by external means is that the pad shoulders are *not* driven against the recess shoulder. Recess height *H* is made to provide a clearance *b* when the pad is at its highest operating position. A clearance *b* of 1/32 in. is convenient for average die work. More clearance may sometimes be desirable where die construction is large and massive. Since the pad shoulders do not impact against the retaining surface, they do not need to be as rugged as when they are used with internal springs. For this reason, minimum shoulder height *h* may be reduced somewhat, if desirable for construction purposes. However, as a minimum shoulder height, the following proportion should be maintained:

$$h = w$$

In this illustration (Fig. 3·26) another method of providing bottoming relief for the pad shoulders is shown at *a*. An offset *a* of 0.004 to 0.005 in. will be satisfactory as an average figure.

For dies constructed as shown, overall length of the transfer pins is made equal to the die-shoe thickness plus pad travel. For dies used in conjunction with a press-mounted cushion, transfer-pin length is made to suit the top position of the cushion. In either case, the clearance *b* is provided in recess height *H*.

The inside diameter of the actuating spring is often considerably larger than the stud diameter. When this is the case, some means of assuring concentricity of the spring to the stud should be provided. One method is that shown: A boss is turned on the pin plate to act as a spring pilot.

The following pointers are offered as additional guides to this type of die construction:

Transfer pins should generally be retained in the die shoe or in the pin plate to facilitate mounting the die in the press.

Transfer pins should slide freely in the holes provided for them.

The hole in the pin plate should permit the plate to slide freely on the stud.

Some provision should be made to prevent rotation of irregularly shaped pin plates.

Top end of the stud should be threaded so as to provide a jam fit in the tapped hole in the die shoe.

Threaded length at bottom end of stud should provide ample spring adjustment.

Stud should be provided with wrenching flats for assembly into die shoe.

For most applications transfer pins, pin plates, and studs do not require heat treatment. For very severe production conditions, transfer pins may be made from a steel which can be heat-treated. A hardness range of Rockwell C 47 to 54 is optimum for such applications. If it is necessary to heat-treat a pin plate, the same range applies. Another procedure which can be followed is the use of pretreated machinable alloy steel (Rockwell C 28 to 34). When this is done, the material is used as supplied.

Punches and those die blocks which perform the bending operation should generally be hardened to the highest practical degree consistent with the type of material from which they are made.

Pad retainer blocks, etc., require hardening also but do not as a rule need to be as hard as the components which perform the bending operation.

Die shoes should be made of steel (not cast iron), in order to better withstand the force of bottoming. This holds true even when hardened thrust plates are installed to resist the impact of the pad when it bottoms.

Advantages of External Spring Construction

Shedder shoulders do not impact when die opens.

Die-shoe strength (no spring pockets) is maximum.

Longer pad travels are feasible.

Much greater spring pressure may be applied within a smaller area.

Spring pressure is adjustable.

Spring breakage is not likely; if it does occur, die is not apt to suffer damage.

If the pad is retained by stripper bolts instead of shoulders, pad action will still be the same, in principle. The pin plate will hit home against the bottom of the die shoe. Pad travel will be determined by the length of the transfer pins. Stripper-bolt heads will not strike against the counterbore shoulders when the die opens. If the

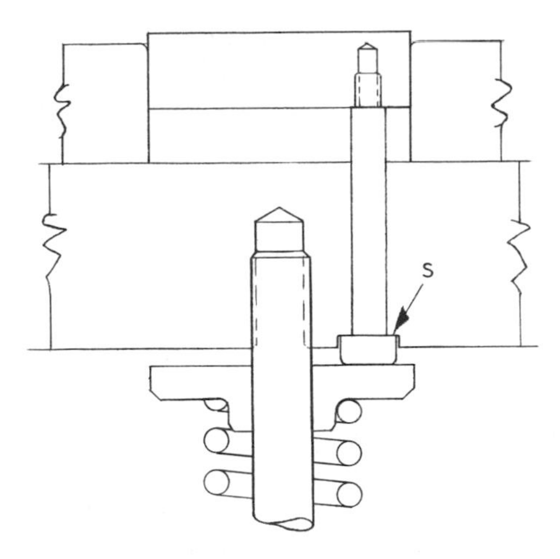

Figure 3·27 Stripper bolt as transfer pin.

stripper bolts are located within the area covered by the pin plate, they may be permitted to extend below the die shoe at the bottom of the stroke. However, if the stripper bolts are located outside the area of the pin plate, the situation is the same as that shown at *E*, Fig. 3·25.

In the construction pictured in Fig. 3·27, the pad is retained by stripper bolts. The stripper bolts serve also as transfer pins. In this case the pin plate does not necessarily hit home against the die shoe. The stripper bolts may hit home against the counterbore shoulders *S*. Do not attempt to make the stripper-bolt holes a slip fit for the stripper bolts. Drill them oversize for clearance, in the same manner as for a standard cap screw. Incidentally, this practice should be followed for the great majority of stripper-bolt applications.

FUNCTIONAL PROPORTIONS

For pad-type bending dies, size of the bending radius *R* (Fig. 3·28) will normally be from $2T$ to $4T$. All bending radii should have a highly polished surface. The lay of the surface finish should be parallel to the motion of the stock material as it slides around the arc surfaces (see Fig. 3·15). In the case of U dies, the opposing bend-

Figure 3·28 Proportions, bending radius and die wall.

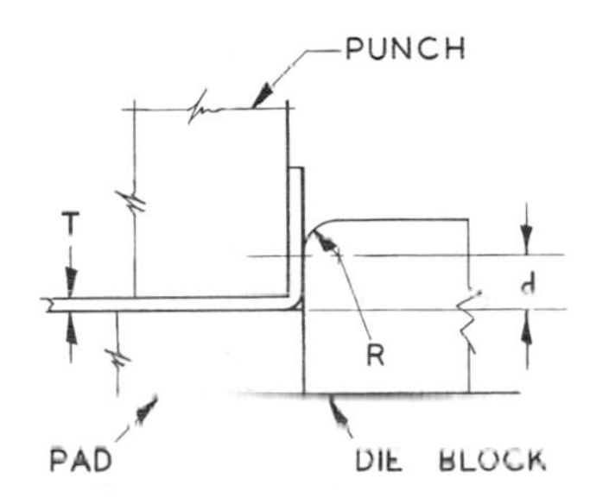

ing radii should be identical in all respects. If these radii are not alike, the workpiece during bending will tend to shift laterally in the direction of the smaller and/or rougher radius. If there is any doubt as to the required size of the bending radii, it is better to make the radii slightly smaller than is estimated to be optimum. It is then a simple matter to increase the size of the radii, if required, at tryout.

As a rule, when the stock material is more than 1/16 in. thick, dimension *d* should be $2\frac{1}{2}T$, min. For stock material less than 1/16 in. thick, *d* dimension should generally be 3/16 in. as a minimum. These proportions for *d* are subject to variation, but they will satisfy many situations and may be used as a guide in most instances.

LATERAL FORCES

Strong lateral forces are imposed on the die structure during bending. These are represented in Fig. 3·29. The punch tends to move inward (direction *P*). Die blocks tend to be displaced outward (direction *D*).

In the case of U dies, lateral forces from the opposing bends tend to balance each other, acting to stabilize the punch. The lateral forces acting against the die blocks, however, have no natural compensating forces. Because of this the die must be constructed in a manner that will provide adequate resistance to the side thrust imposed on the die blocks.

In some instances, the die-block positioning dowels are relied upon to resist the lateral thrust. When this is done, dowels of larger than normal diameter are installed. Minimum die-block width *W* should be at least equal to the height *H*. It will be better if $W = 1\frac{1}{2}H$, or more whenever practical. Dowel pins are subject to deflection if any appreciable lateral thrust is applied. Therefore, inaccuracies and variations may occur in production owing to "breathing" of the die blocks. In any case, the practice of depending upon dowel pins to resist lateral bending thrust must be restricted to light work. For most bending operations, some method of construction more substantial than doweling will be necessary.

Figure 3·29 Lateral forces in bending.

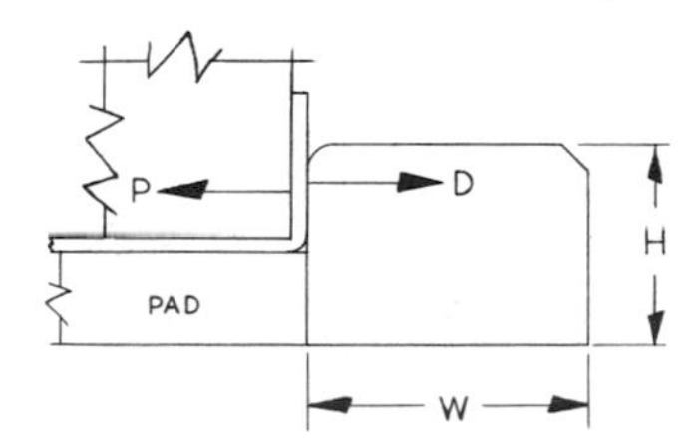

TYPICAL PAD-TYPE BENDING DIES

Some typical constructions used for bending operations are depicted in Figs. 3·30 to 3·33. In Fig. 3·30, the lower die assembly is set into the die shoe. A slot or, if necessary, a pocket is milled in the die shoe to receive the die assembly. The slot should be machined in the die shoe before completion of the width of the die blocks *W*. The die blocks should be completely finished except for surface *A*, where grinding stock should be left until ready for assembly. At assembly, the blocks are ground at surface *A* until dimension *W* is compatible with the slot width *S*. The punch should not be mounted on the punch holder until after the lower die assembly is completed. The punch is then mounted to suit the lower die assembly.

In Fig. 3·31 keys are provided in the die shoe to back up the die blocks. Key height *h* should ordinarily be 1/2 in. It should be sunk into the shoe 1/4 in. Key width *k* should be 1/2 in., min. Larger keys are required for heavy operations. The keys should be retained with screws, as shown. Distance *B* from the key to the end of the die shoe should be a minimum of $1\frac{1}{2}k$. Distance *D* between the keys should be made slightly shorter than the overall length of the die assembly: Approximately 1/32 in. shorter will do nicely. The die blocks may then be fitted between the keys in the manner described for fitting to a slotted or pocketed shoe. Another procedure is to offset the keys, as shown in the inset. Use the method which applies best to the circumstances pertaining to the particular job. With this construction also,

Figure 3·30 Die assembly set into die shoe.

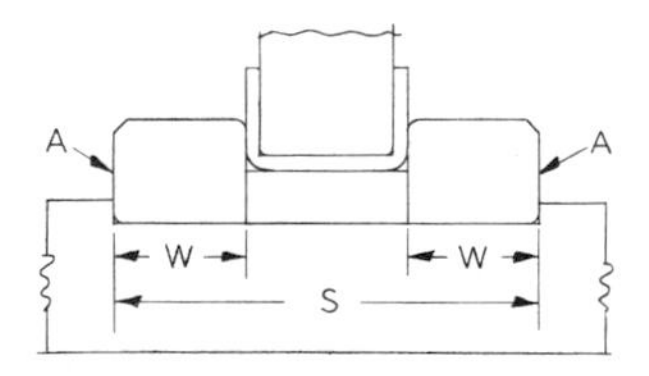

Figure 3·31 Die assembly keyed to die shoe.

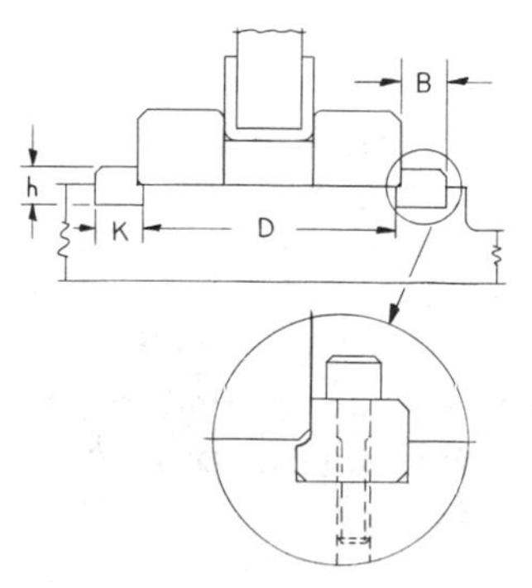

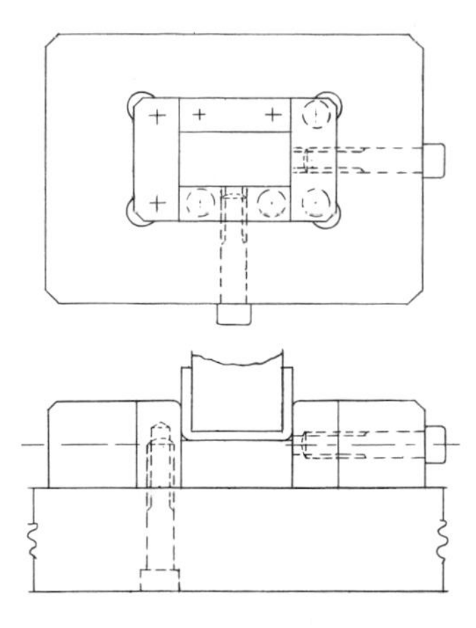

Figure 3·32 Die sections in solid yoke.

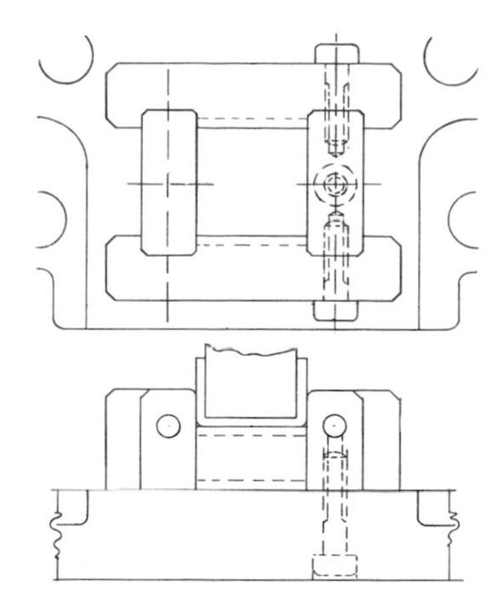

Figure 3·33 Sectional yoke.

the lower die assembly should be mounted before mounting the punch. The keys may be made of cold-drawn steel and left soft.

In Fig. 3·32, die sections are retained in a yoke block. Yokes of the type shown are often called solid yokes, because they are made from a single piece of steel. Occasionally, solid yokes are hardened, although the practice is generally restricted to severe production conditions. If a hardened yoke is required, a hardness of Rockwell C 47 to 54 is the general rule. Another procedure which is sometimes followed is to make the yoke from a pretreated alloy steel. This results in a strong, tough yoke block. Steel of this type should have a hardness range of Rockwell C 28 to 34 in order to provide toughness and yet be practical to machine. For the majority of applications, however, the yokes may be made of mild steel and left soft.

Figure 3·33 illustrates a sectional yoke construction. The die blocks are keyed into separate yoke blocks, forming a boxlike structure. Since the pad is retained between the yoke sections, they should be hardened to resist wear. Sections of this type may generally be heat-treated to Rockwell C 58 to 60. This construction, as well as the others shown, may be used for three- and four-sided bending operations. In such cases be certain that the screws which hold the die blocks to the yoke blocks are large enough to contain the lateral bending forces.

STRIPPING U BENDS

U-bending operations require two opposed stripping actions. The first is the stripping of the formed workpiece out of the die opening. As described previously, the pressure pad acts as a shedder to perform this function. The second stripping action is that of removing the formed workpiece from the punch. This function is normally performed by either of three agencies:

1. Spring-actuated plungers or the equivalent, which push the formed workpiece off the punch
2. Hook strippers, mounted as an adjunct to the lower die assembly, which rake the formed workpiece from the punch
3. Positive knock-offs, actuated by the crossbar knockout in the press slide, which knock the formed workpiece off the punch

NOTE:

The above sequence is not to be construed as an order of preference or of prevalence.

Spring-actuated plungers. Spring plungers acting as strippers are illustrated in Fig. 3·34. The right half of the punch is shown at the bottom of its stroke. The spring plungers have been forced

Figure 3·34 Stripping with spring plungers.

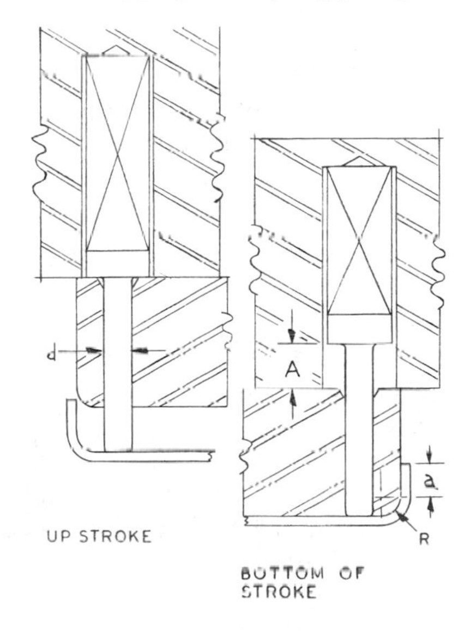

upward into the spring pockets and are compressing the springs. The left half of the punch is shown ascending during the upstroke. The plungers have traveled through distance A, stripping the formed workpiece from the punch. Plunger travel A is derived from straight dimension a of the bend leg. This relationship should be

For sharp corner	$A = a + \frac{1}{32}$ in., min.
where $R = \frac{1}{16}$ in.	$A = a + R$, min.
$R = \frac{1}{8}$ in.	$A = a + \frac{3}{4}R$, min.
$R = \frac{1}{8}$ to $\frac{3}{4}$ in.	$A = a + \frac{1}{2}R$, min.
$R = 1$ in.	$A = a + \frac{3}{8}R$, min.

Plungers should be located as close to the bend legs as sound construction practice will permit. This is necessary to avoid distorting the workpiece during stripping. Plungers should not, however, be located within the area covered by the bend radius R. This is necessary to avoid distortion of the workpiece during bending.

The most prevalent practice is to make plungers of drill rod, turned and polished, heat-treated to Rockwell C 45 to 50, with the head end slightly softer. Other practices are employed also. For example, plungers are sometimes made of cold-drawn steel, either pack-hardened or cyanide-hardened at the nose end only. For dies of a temporary nature (low production), plungers may be left soft. Plungers must operate freely. Plunger diameter d should be approximately 0.002 to 0.005 in. smaller than its installation hole. In most cases, the head diameter should be $\frac{1}{32}$ to $\frac{1}{16}$ in. smaller than the spring-pocket diameter.

Obviously, this method of stripping is restricted to operations where it is definitely apparent that both spring pressure and spring travel will be ample for the purpose. Relatively speaking, this means shallow bends in lighter-gage materials, where sufficient area is available for the installation of springs.

The action of spring plungers causes the finished workpiece to be left behind as the punch ascends. The workpiece remains lying on the pad face when the press cycle is completed. It may be removed by the operator, or some means of ejection (such as an air jet, if practical) may be provided.

Hook strippers. Hook strippers, commonly referred to as skyhooks, are an effective means of stripping. In the case illustrated in Fig. 3·35, the hooks are mounted on the die blocks. Hooks of this kind may be made of tool steel or mild steel, as desired. Hardening should be in accordance with the type of material. Localized heat-treatment procedure may be used in order to harden only the required areas of contact with the piece part.

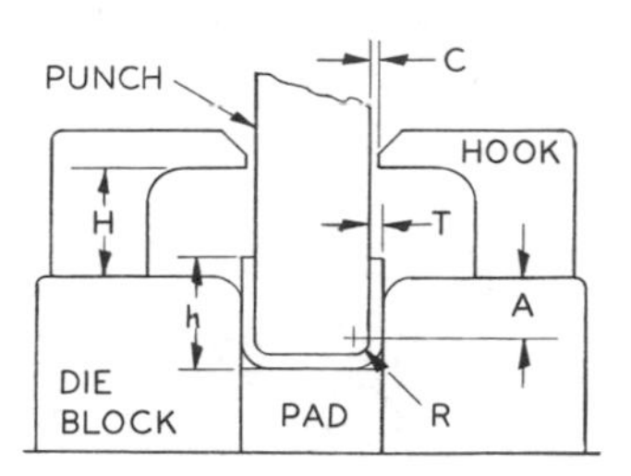

Figure 3·35 Hook stripper.

This leaves the balance of the hook soft for mounting. This procedure is not mandatory; any method which efficiently produces the desired result may be followed.

It is essential for skyhooks to be securely mounted and positioned. Clearance C between the hooks and the punch should not be more than $\frac{1}{2}T$. Clearances C of less than $\frac{1}{2}T$ are preferable. Gap height H must permit easy ejection or unloading of the formed piece part. For smaller work, the minimum H dimension should be $h + \frac{1}{4}$ in. Keep in mind that this is an arbitrary minimum figure; considerably more gap should be used for work that is large in area. Dimension A is the distance from the top of the die block to the bend-radius center. Dimensions A and H are related to the length of the press stroke. To check this relationship, add:

where $R = \frac{1}{16}$ in. or less	$H + A + \frac{1}{16}$ in.
$R = \frac{1}{8}$ in.	$H + A + \frac{3}{4}R$.
$R > \frac{1}{8}$ in.	$H + A + \frac{1}{2}R$

The sum must be less than the press stroke to assure that the piece part is disengaged from the punch. The above sums should be considered minimum required press strokes; it is practical for the press stroke to be considerably longer.

With skyhooks, the stripping force is exerted against the end of the bend leg. This is a desirable condition, since any tendency to distort the piece part is minimized. Where applicable, this method of stripping is positive and effective, especially when a great amount of stripping force is required. On the debit side, however, it must be conceded that hook strippers may present problems in relation to loading and/or unloading the workpiece in production. Because of this, it is essential for the diemaker to know and understand the relationships described.

Positive knock-offs. The simplest positive knock-off is the single-rod type shown in Fig. 3·36. Except for very low-production applications, the ends (both ends) of such rods should be hardened. Quite often, the rod is made of drill rod and the entire rod heat-treated to a hardness ranging be-

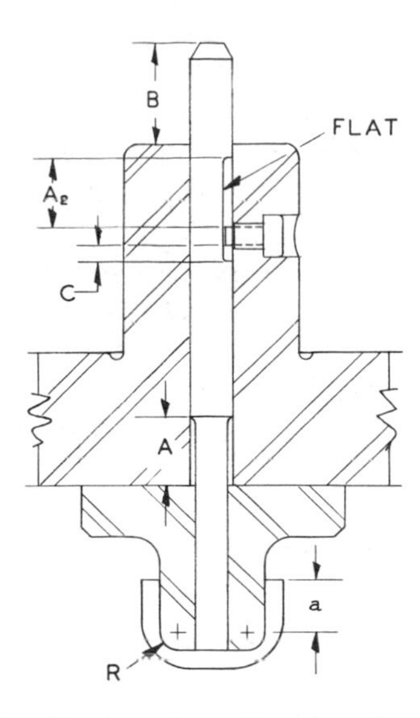

Figure 3·36 Single-rod positive knock-off.

tween Rockwell C 45 to 55. The top end of the rod should be finished with a definite chamfer to prevent peening and/or spalling.

The rod retaining screw may be a fillister head screw reworked to a dog point, if desired. If necessary because of space considerations, a button-head screw may be used instead. Check to ascertain free rod movement: The retaining screw must not cause the rod to bind. When the rod is up (as shown) at the bottom of the press stroke, be certain there is ample clearance at *C*. A minimum of 1/16 in. is suggested for *C*, and more than this is acceptable. Distance A_2 must always be greater than dimension *A*. Here too, a minimum clearance of 1/16 in. greater is suggested. Dimension *B* should be made to suit the particular press installation. If this information is not available, common practice for small and medium work is

$$B = A + 1\tfrac{1}{4} \text{ in.}$$

Rod travel *A* must assure that the piece part will be stripped clear of the punch. In a general way, this will require

where	
$R < \tfrac{1}{8}$ in.	$A = a + \tfrac{1}{8}$ in.
$R = \tfrac{1}{4}$ in.	$A = a + R$
$R = \tfrac{3}{8}$ in.	$A = a + \tfrac{3}{4}R$
$R > \tfrac{1}{2}$ in.	$A = a + \tfrac{1}{2}R$

This style of knock-off is very desirable from the standpoint of simplicity. Unfortunately, it is restricted in application. The knock-off impact is concentrated in a small area at the center of the piece part. Therefore, it has a pronounced tendency to distort the piece part (Fig. 3·37, view *A*). For this reason, such knock-offs are limited to applications where the piece part is strong (heavy) enough to resist the distorting effect. Another

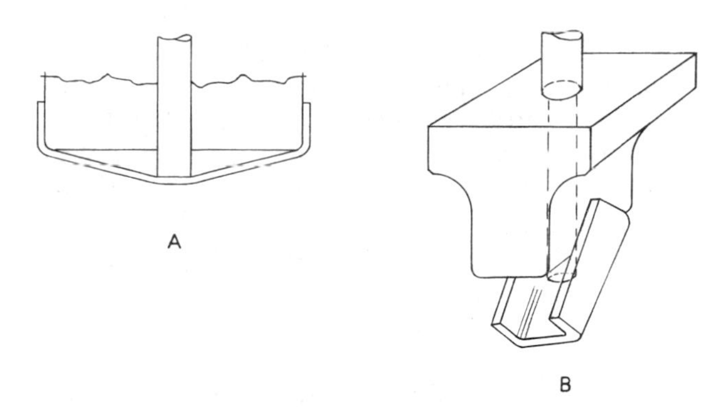

Figure 3·37 Knock-off hazards.

serious drawback is the possibility that, even without distorting, the piece part may swivel, clinging to the punch in a manner similar to that in view *B*. This could result in a "double," causing danger to the operator and/or severe die damage. It is obvious that these hazards are present for any type of stripping arrangement which does not cover a sufficient area and/or does not apply the stripping force close enough to the bends.

Knock-offs similar to that pictured in Fig. 3·38 are called "mushroom" knock-offs. These provide an enlarged periphery at the working end which distributes the stripping force more effectively than the straight rod. The force is applied closer to the bends and over a larger area, which tends to eliminate distortion and swiveling of the piece part.

In the illustration, the knock-off is retained in assembly by the lock nuts which control the travel *A*. The lock nuts are an excellent construction

Figure 3·38 Mushroom-type knock-off.

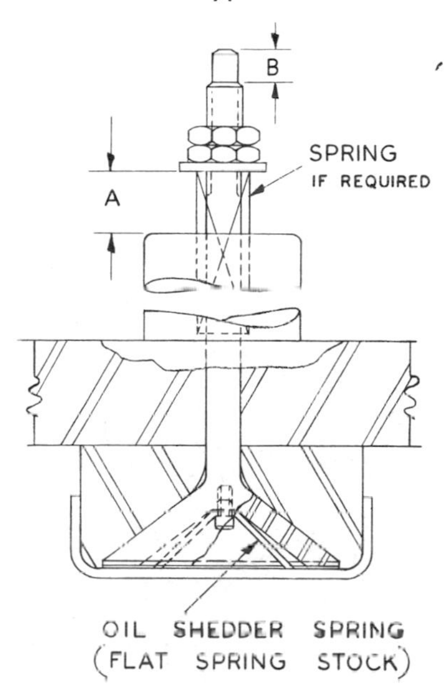

feature, providing both strength and a simple means of adjusting the travel A. Incidentally, because they are effective and safe, lock nuts are used for many and varied knock-off and knockout applications. The top end of the rod should be relieved at B to prevent thread swelling from impact. Minimum B dimension should be ¼ to ½ in., varying in proportion to rod diameter.

Due to increased area in contact with the piece part, it may be necessary to provide an oil pin or equivalent. When doubt exists about the necessity of this, it is best to play safe. In the illustration, a piece of flat spring stock is bent to a shape which exerts spring pressure against the piece part and thereby prevents the part from adhering to the face of the knock-off.

In the case illustrated, the knock-off is held in its up position by a compression spring. This spring may not be required if the piece part is strong enough and clings tightly to the punch. However, lighter-gage work especially may strip prematurely because of the weight of the knock-off unless a spring is incorporated in the manner shown.

Figure 3·39 illustrates a commonly used method of achieving effective distribution of stripping force. A spreader plate is assembled to the knockout rod. The knock-off pins are in turn assembled to the spreader plate. A suitable cavity is provided in the punch holder to accommodate the spreader plate. A bridge plate covers the cavity, providing a bearing surface for the punch. The punch is mounted to the bridge plate.

Figure 3·39 Positive knock-off with pins in spreader plate.

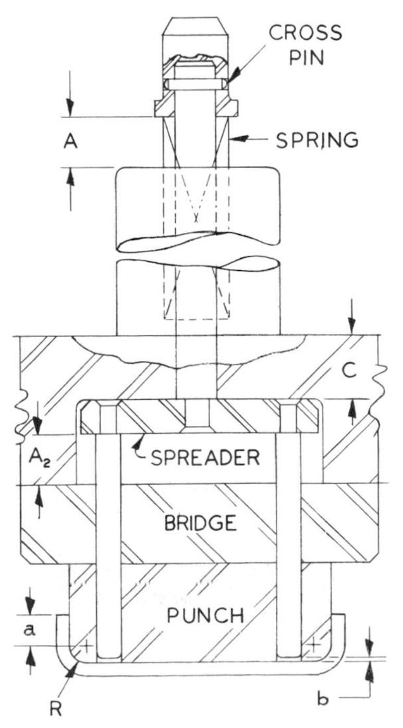

The top end of the knockout rod is fitted with a cap, which acts as a stop, limiting the travel A. The cap is secured to the rod by a cross pin. The bottom of the bore in the cap should seat against the top of the knockout rod in order to transfer the impact to the end of the rod, not to the cross pin. A compression spring acts against the cap, holding the entire knock-off assembly in its up position to prevent premature stripping. The knock-off pins must not protrude beyond the punch face when the assembly is in its up position. To assure this, the pins may retract into the punch a slight amount b.

For this type of construction the following materials and heat treatment may be employed:

Cap. Tool steel, heat-treated Rockwell C 45 to 50.

Cross Pin. Drill rod, soft. Do not heat-treat. Do not substitute a dowel pin.

Knockout Rod. Cold-drawn steel, left soft.

Spreader. If thickness is adequate, cold-drawn steel, left soft. If heat-treated plate is desired, use tool steel (ground stock), Rockwell C 45 to 50.

Knock-off Pins. Drill rod, hardened working ends only.

Bridge. Mild steel, left soft.

Die Set. Steel die sets are the rule for bending dies. However, in this case the bridge plate acts as a thrust plate, possibly permitting use of a cast-iron punch holder. Whenever possible, the shank should be cast integrally with the punch holder. For steel punch holders, the shank should be welded in place. Inserted shanks can be used if thickness C is great enough, especially if the punch holder, in addition to clamping the shank, is screw-attached to the ram. This practice, however, usually applies to large work only. For small- and medium-sized work, keep in mind that even though the shank is made an integral part of the punch holder, dimension C must not be skimped.

To assure stripping, knock-off travel A should be a minimum of

where $R < \frac{1}{16}$ in.	$A = a + b + \frac{1}{16}$ in.
$R = \frac{1}{8}$ in.	$A = a + b + R$
$R = \frac{1}{4}$ to $\frac{3}{8}$ in.	$A = a + b + \frac{3}{4}R$
$R > \frac{1}{2}$ in.	$A = a + b + \frac{1}{2}R$

(Others proportionate.)

Distance A_2 should be slightly greater than distance A. If A_2 is $\frac{1}{64}$ to $\frac{1}{32}$ in. longer than A, it will facilitate making and assembling the die.

Many variations of this knock-off arrangement

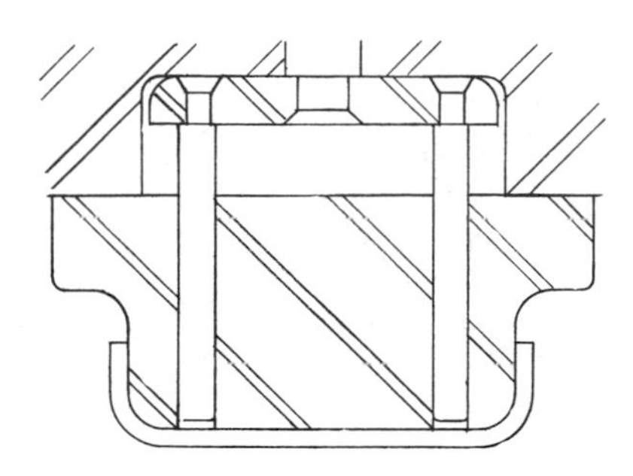

Figure **3·40** Knock-off assembly, pedestal-type punch.

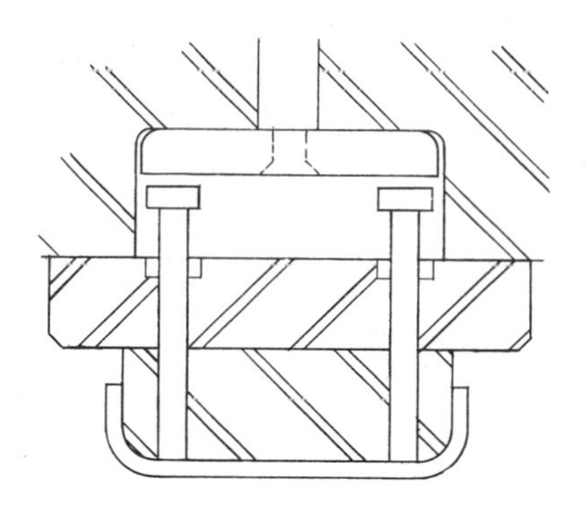

Figure **3·41** Knock-off arrangement (free pins).

are employed. For example, Figs. 3·40 and 3·41 are minor variations of Fig. 3·39.

The pedestal punch (Fig. 3·40) does not require a bridge plate. This applies, generally, to smaller work. Large work would require a massive punch, which would represent higher material and heat-treat costs and would involve more machining time.

With the free-pin construction (Fig. 3·41), be sure to heat-treat the pin plate (spreader). For best results use tool steel (ground stock) heat-treated to Rockwell C 47 to 54, making thicker plates harder than relatively thin ones. The pins should be heat-treated to Rockwell C 45 to 50 for their entire length.

Fork-type knock-offs (Fig. 3·42) apply the stripping force against the end of the bend leg. Therefore, just as with skyhooks there is little or no tendency to distort the piece part by the stripping action. As with all positive knock-offs, the piece part is stripped from the punch at the top of the press stroke. If the press is inclined, the piece part then drops out through the back of the press. This permits a faster press operation than one where the piece parts must be unloaded from the die.

Forked knock-offs must be held in the up position when they are not operating. A small clearance is necessary at *b* to avoid interference with the bending action of the piece part. In principle, heat treatment is the same as for other knock-offs: The ends which contact the piece part and the top end should be hardened. Working proportions A, A_2, etc. should be the same as those in Fig. 3·39.

TYPICAL U-BENDING DIE

An assembled U die is pictured in Fig. 3·43. The drawing is conventional, with the die shown in closed position. The bill of material for this die is shown in Fig. 3·44.

The illustrated die is equipped with skyhooks which not only perform the stripping operation but also serve as side gages: The ends of the flat blank gage against surface *A* of the skyhooks. Front (14) and back (15) gages are attached to the

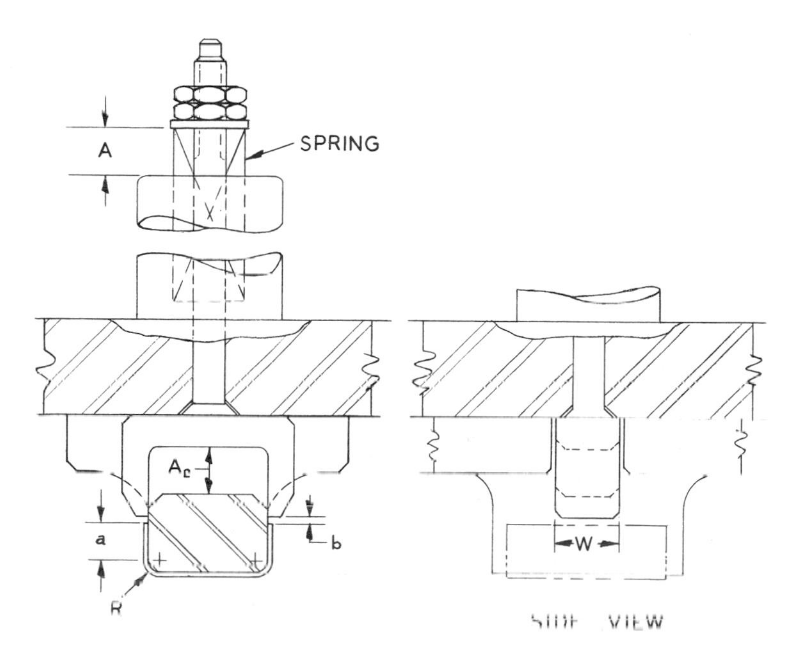

Figure **3·42** Fork-type positive knock-off.

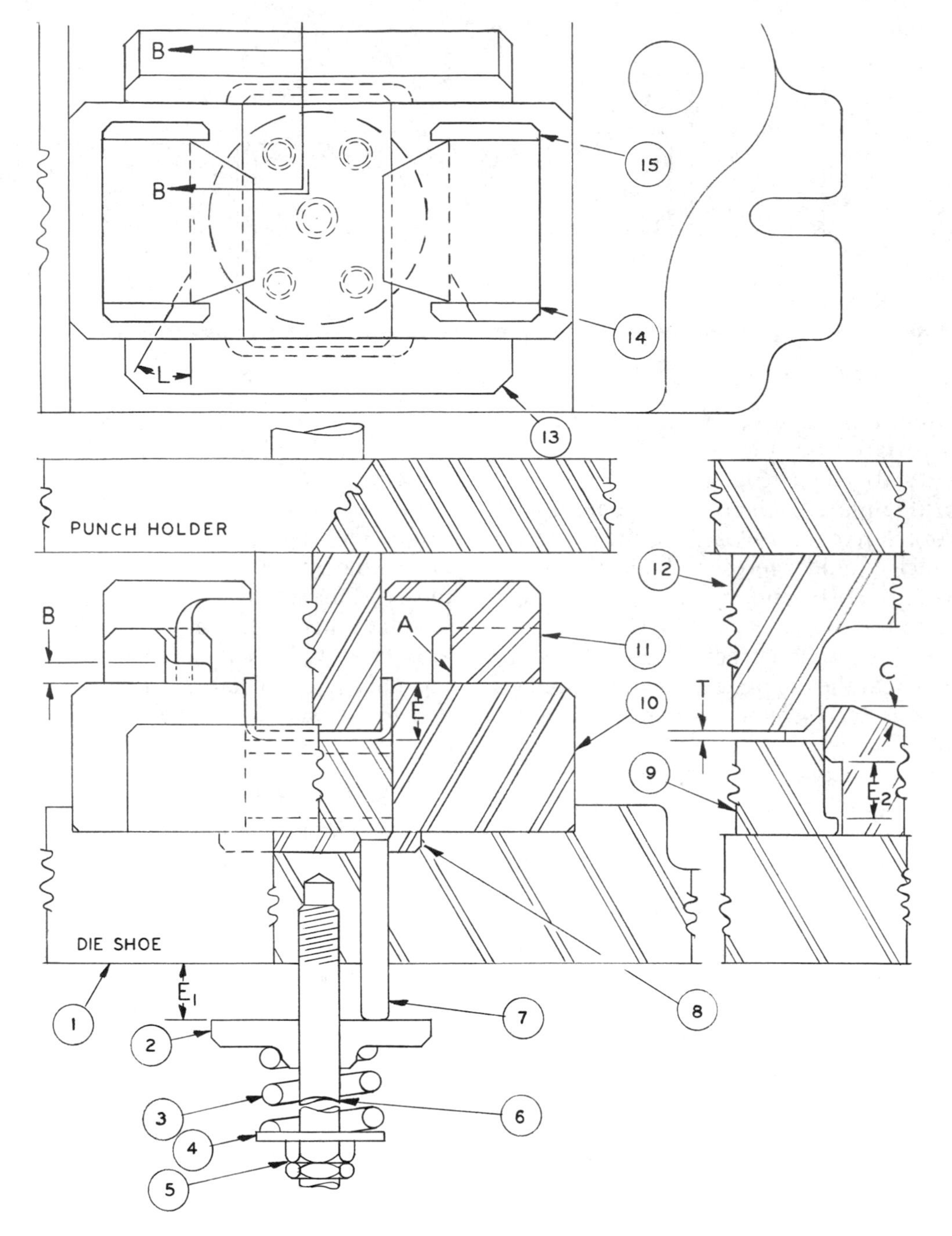

Figure 3·43 U die.

skyhooks. To facilitate loading, the front gages are offset at *B* and have a lead angle *L*. The lead angle extends into the skyhook.

After stripping, the finished workpiece lies on the top face of the pad. As the operator loads the next workpiece, the finished one is pushed out the back by the incoming workpiece. Angle *C* on the rear pad retainer block is an aid to clearing the finished workpiece from the die. Note that the pad shoulders extend into the retainer blocks, not into the bending blocks. This is preferred practice, which should be followed whenever possible.

Because the illustration depicts a complete die, it is easy to visualize the relationship of E, E_1, and E_2. It is summarized here because it is worthy of emphasis:

$$E_1 = E + 0.002 \text{ to } 0.010 \text{ in.}$$
$$E_2 = E_1 + \tfrac{1}{32} \text{ in.}$$

L BENDS ON PRESSURE PADS

As described earlier in this section, L bends are produced in V dies. They are also produced in pad-type dies. An L bend is, in essence, one side

15	2	BACK GAGE		C.D.S.
14	2	FRONT GAGE		C.D.S.
13	2	PAD RETAINER		C.D.S.
12	1	PUNCH		T.S.
11	2	SKY HOOK		T.S.
10	2	BENDING BLOCK		T.S.
9	1	PRESSURE PAD		T.S.
8	1	THRUST PLATE		T.S.
7	4	TRANSFER PIN		DR. RD.
6	1	STUD		C.D.S.
5	2	NUT	STD.	
4	1	WASHER	STD.	
3	1	SPRING	STD.	
2	1	PIN PLATE		C.D.S.
1	1	BACK POST DIE SET	STD.	STEEL
DET. NO.	NO. REQ.	DESCRIPTION		MATERIAL

Figure 3·44 Stock list for Fig. 3·43.

of a U bend. Therefore, since the other leg of the U is missing, the inherent self-equalizing qualities of the U bend are missing also. This is the source of two conditions which are of specific concern to the diemaker. The first is the one-sided lateral thrust imposed upon the punch (refer to Fig. 3·29). The second is pulling, in which the workpiece tends to pull out of the die opening in the direction of the bend.

For very light work, where the side thrust is relatively insignificant, the installation of larger punch dowels may provide adequate resistance to the lateral thrust. In these instances, be certain the punch mounting screws are large enough as well.

Most L-bending operations require a thrust member (Fig. 3·45) to support the punch. Constructionwise, this makes an L die very similar to a U die. In an L die, the side of the punch which is opposite the bend leg slides directly against the thrust member. Both the punch and the thrust member must be provided with a lead radius r to eliminate the possibility of shearing. A lead radius r of 1/32 in. should be satisfactory.

Figure 3·45 L die with pilot.

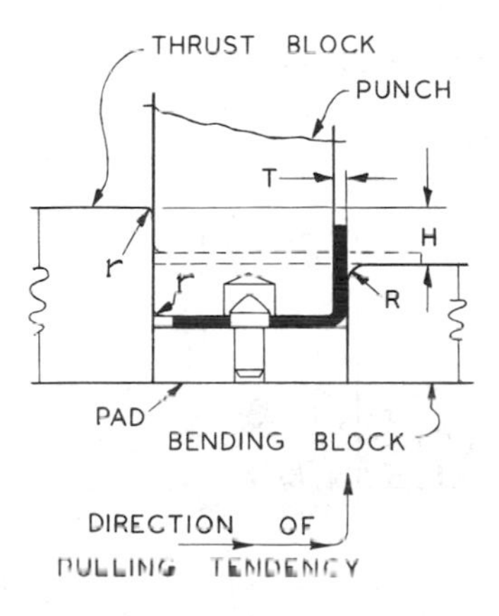

The punch and the thrust member should have their flat surfaces engaged before bending begins, as indicated by the dotted lines in the illustration. To accomplish this, there is a height differential H between the thrust member and the bending block. The thrust member is higher than the bending block, a difference of

$$H = T + r + r + 1/16 \text{ min.}$$

Thrust components of this kind should be heat-treated, generally to a hardness of Rockwell C 60 to 63. As a choice of material from which to make them, a graphitic tool steel should be considered because of the friction which is inherent in the application. The contact faces of both the thrust block and the punch should be smoothly finished to reduce friction as much as possible.

Pulling is an omnipresent hazard of L-bending operations. With pad-type bending, the workpiece tends to skid in the direction shown. It often occurs that the piece part contains an opening which may be pierced prior to the bending operation. When this is the case, take advantage of it. Install a pilot gage pin in the bending die, either in the pad, as depicted, or in the punch, if that is more desirable. (The pilot will most often be in the pad, where it can assist in gaging the workpiece.) Assuming a desirably located pilot (or pilots), this method of overcoming pulling probably requires less pad pressure than other methods. However, it should be recognized that the pulling tendency can, in some cases, be strong enough to distort the piloted opening. Since strong pad pressures tend to alleviate this condition, be sure that pad pressures are entirely adequate. Where applicable, a pilot (or pilots) combined with adequate pad pressure is a most effective means of counteracting pulling.

Keep in mind that for all L bends, the bending radius R should have the smoothest possible surface finish as an aid to alleviating the one-sided pulling tendency.

In some cases, the peripheral contour of the workpiece may be used to advantage to resist the pulling action which accompanies L bending. An example of this is pictured in Fig. 3·46, where the pad is made with two integral bosses. The shoulders of the workpiece abut the bosses and are thereby restrained from movement in the direction of pull. When the illustrated construction is employed, boss height B may be, using minimum value for T,

$$B = T - 0.002 \text{ in. for thin stock material}$$
$$= T - 0.005 \text{ in. for heavier stock material}$$

As with most diemaking practices, variations of

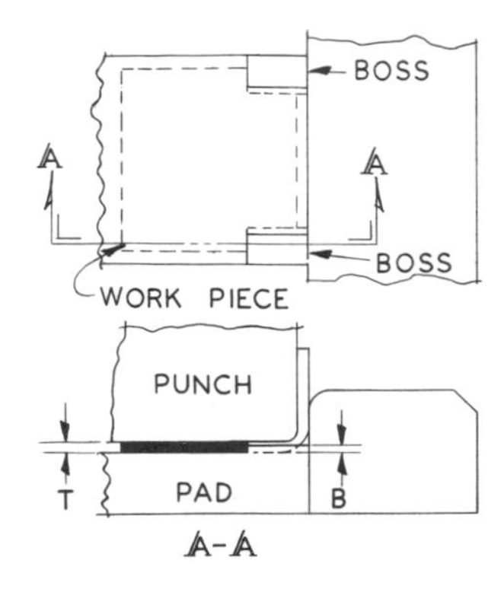

Figure **3·46** Pad bosses to resist pulling.

the customary construction are at times desirable. One such variation is to make boss height *B* higher than *T*. Clearance for the bosses is then provided in the punch. For large work the bosses may be separate plates, similar to or acting as nest plates, attached to the pad by means of screws and dowels. If the bosses are intended to function as nest gages, it is advisable to make them higher than *T*.

Unfortunately, it is not always possible to have strategically situated pilots or nesting bosses to resist pulling. Other methods must then be employed. One method sometimes applicable is indicated in Fig. 3·47, where a chisel point is installed in the punch. The chisel point digs into the workpiece, setting up resistance to the pulling action. The chisel point is positioned parallel to the bend axis. A hardened backing plate is usually required behind the chisel. In the case illustrated, the chisel is secured by a setscrew which clamps against a "whistle" notch provided in the chisel. This positions the chisel radially in addition to retaining it in the punch.

Another method of overcoming pulling is to knurl the face of the punch and/or the pad. This provides a roughened surface which tends to grip the workpiece more effectively between the punch and the pad. In lieu of knurling, tiny serrations can be ground into the face of the punch and/or the pad. This effect is similar to knurling and has the advantage of being practical to incorporate after the punch and pad are hardened. Therefore, the serrations can be used on an "add if proved necessary" basis.

Punch corner offsets (see Fig. 3·53) are generally thought of as a means for setting a bend to eliminate springback. However, it is worth remembering that offset punch corners can provide effective resistance to pulling.

OVERBENDING IN PRESSURE-PAD DIES

For L dies, an effective method of compensating for springback is represented in Fig. 3·48. The face plane of the pad is sloped (angle *A*). Punch angle *B* and die-block angle *C* are made to suit the required piece-part angle minus the developed compensating angle.

When pilots are incorporated in beveled pads, the pilot axis is perpendicular to the top (sloped) face of the pad. Therefore, once the pilot center distance *D* has been established, any change in the pad height *H* will cause a change in center distance *D*. Elevating distance *H* will shorten dimension *D*. Decreasing *H* will lengthen *D*. The amount of change at *D* will equal the amount of change at *H* times sin *A*.

The punch must be provided with a clearance hole for the pilot. The axis of the clearance hole should, normally, be parallel to the line of punch motion. If a drilled hole is used, it is best to drill the hole prior to machining the angle on the punch face.

In actuality, the construction shown combines V-bending principles with pressure-pad construction and principles. (The V-die effect, of course, varies according to angle *A*.) As a result, there is a tendency toward stabilization of the punch during bending. There is also a possibility that there may be a reduction in the tendency to pull the workpiece during bending. If desired, the included angle *B* of the punch may be made slightly

Figure **3·47** Chisel to resist pulling.

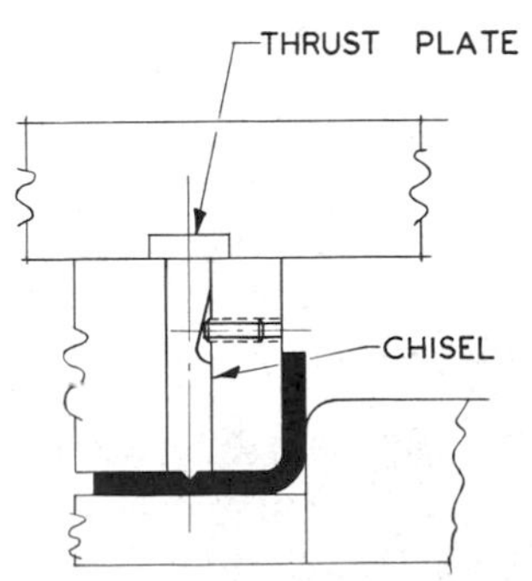

Figure **3·48** Beveled pad.

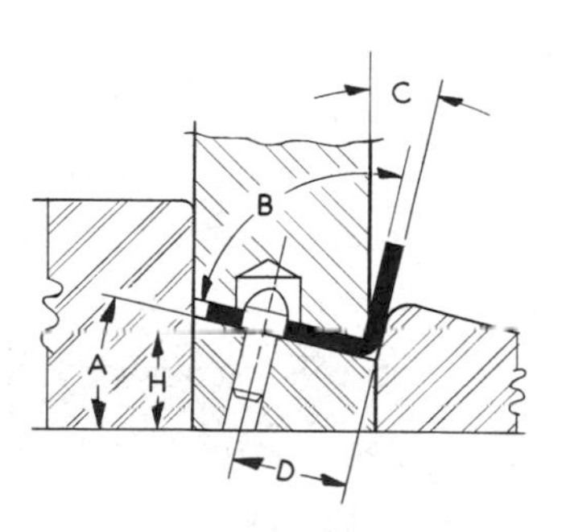

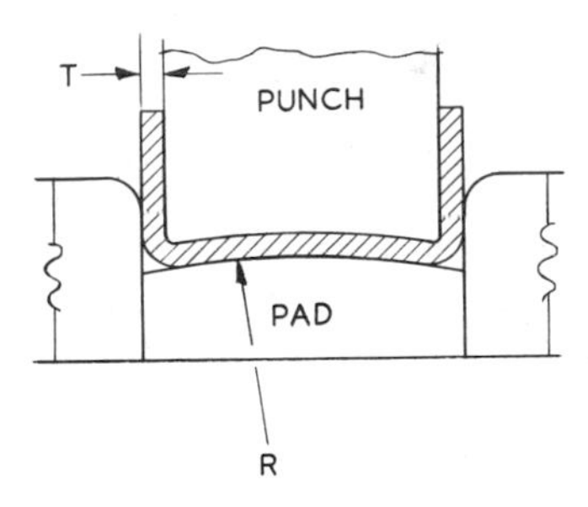

Figure **3·49** Convex pad.

less than its corresponding die-opening angle (see Fig. 3·22). This will provide corner setting which is, in principle, the same as that described for Fig. 3·22.

In U dies, overbending is accomplished in the manner indicated in Figs. 3·49 and 3·50. The two illustrations are variations of the same method, the principle involved being the utilization of springback in another area. The web (bottom) of the workpiece is formed as shown. The elastic limit of the stock material is not exceeded. Thus, when the workpiece is removed from the die, the web springs back to a flat plane. This causes the bend legs to pivot inward around the bend axes, which compensates for the springback of the bend legs.

In Fig. 3·49, the top surface of the pad is convex (radius R). Size of the radius of convexity must suit the specific springback condition, and the diemaker must determine it by development. Incidentally, radius R will always be a proportionately large radius. The mating radius of concavity in the punch is, theoretically, $R + T$. However, in practice it will usually be desirable to make the concavity in the punch face slightly less than $R + T$. The following proportions are suggested as a guide:

where $T < 1/16$ in.	Make both radii the same.
$T = 1/16$ to $1/8$ in.	$R + T - \frac{1}{2}T$
$T > 1/8$ in.	$R + T - \frac{1}{4}T$

Figure **3·50** Gable pad.

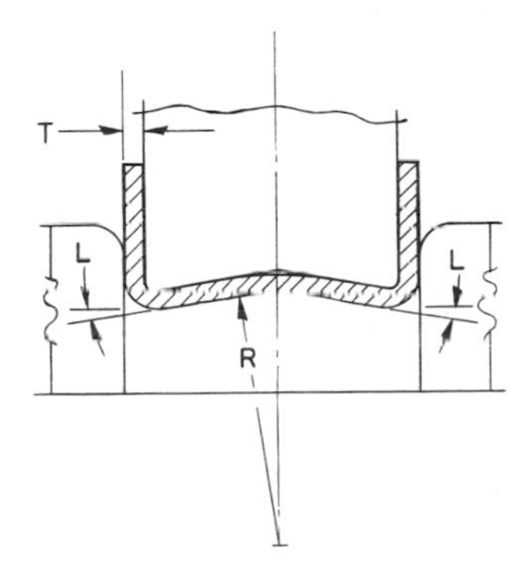

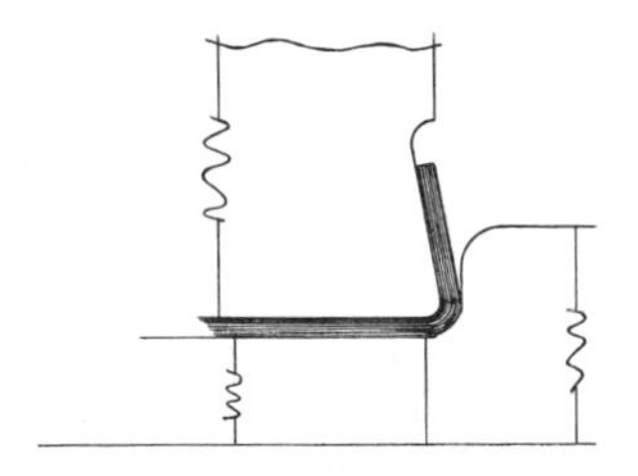

Figure **3·51** Punch sidewall relief—angular.

The proportions may have to be modified somewhat during tryout because of individual job requirements.

For large work especially, it will be simpler to proceed as indicated in Fig. 3·50. Here the top pad surface is angled slightly in the manner of a gable. The vertex of the angle is blended out with a radius R to prevent a setting condition at the center. Angles L in the punch may be equal to, or very slightly greater than, angles L on the pad. The punch does not require a matching radius for R. Here, too, the exact pad and punch contour are a result of development by the diemaker.

Where springback is not too pronounced, one of the procedures shown in Figs. 3·51 and 3·52 may be satisfactory. Utter simplicity is, of course, the keynote of these two expediencies. It should perhaps be noted that the use of relieved punch sidewalls, and the ensuing methods here described, are not restricted to either U bends or L bends.

CORNER SETTING IN PRESSURE-PAD DIES

A simple and effective method of eliminating springback is illustrated in Fig. 3·53. The offset punch corner sinks into the stock material at the bend area. The resulting coining effect kills the residual elasticity in the bend area, eliminating the tendency to springback. An offset depth d equal to 5 per cent T will be generally satisfactory for most applications. As indicated in the illustration, this method has a dimensional effect on the piece part. If the punch face were flat, piece-part height would be H. The corner offset displaces the bend leg in the direction of the offset an amount equal to the depth of offset d. The resulting piece-part height is H_2. This

Figure **3·52** Punch sidewall relief—straight undercut.

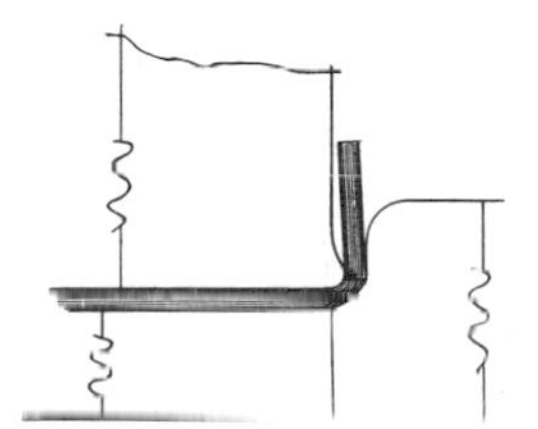

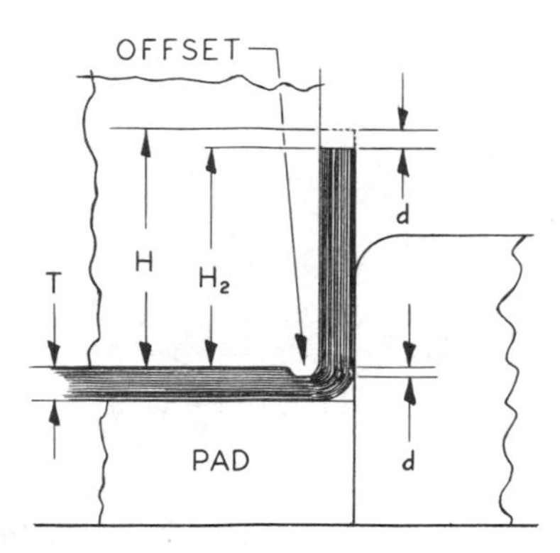

Figure 3·53 Offset punch for corner-setting.

displacement must be compensated for when determining the required flat-blank length. Compensation is made by adding an extra allowance, equal to *d*, to the normal bend allowance. This applies to V dies as well as to pressure-pad dies.

In some critical instances, this method may not be permissible because of the strength loss in the bend area of the workpiece. In many cases this is negligible, but in some it is objectionable. If the method is undesirable for a specific application, the diemaker may resort to overbending or to the die-fillet method shown in Fig. 3·54. Here, the set is achieved by squeezing the stock material against the outer surface of the bend area. This method is sometimes referred to as a developed set, because the die-fillet radius *RR* must be developed to suit the thinning of the stock material in the bend area. As a starting figure for developing radius *RR*, use

$$RR = 5/4T + R$$

Rework, if required, at tryout.

The principles of springback compensation and elimination discussed earlier in this chapter are true for all types of bending and may be referred to as required.

Figure 3·54 Die fillet for corner-setting.

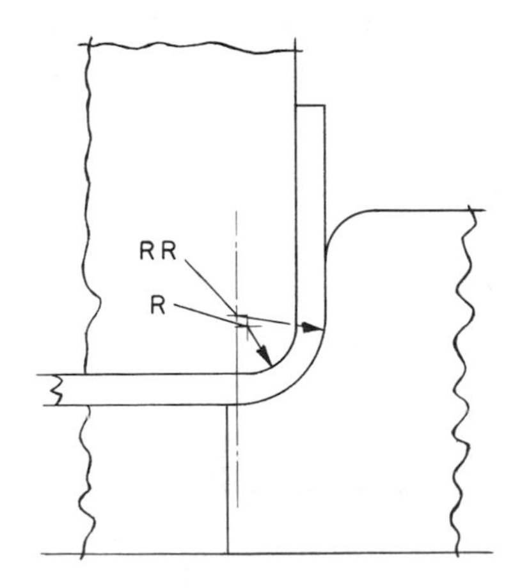

EFFECTS OF BURR SIDE

The location of the burr side of the workpiece can have an undesirable effect on bending and forming dies. It is undesirable for the burr side to be located on the outer surface of the formed piece part, because the burr drags around the bending radius and into the die opening. This condition causes excessive wear on the die members. In short order, the burrs wear grooves in the surface of the bending radius and into the walls of the die opening. If, however, the workpiece is inverted so that the burr side is located on the inner surfaces of the formed piece part, the burrs will face toward the punch. Since there is no drag between the workpiece and the punch, the burr cannot erode the punch. This is, obviously, the desired condition, to be sought whenever possible. It is the source of the following expression, which may be considered a good rule to remember:

When permissible, locate the burr side toward the inside of a form (or bend).

Z dies and the like, which produce reverse bends in a single operation, are excepted from the rule.

Important! In the above rule the emphasis is on the words *when permissible.* This is because the burr side of the piece part is often predetermined in accordance with the functional requirements of the piece part in relation to the end product. In such a case, the burr side is specified on the piece-part drawing. It is then the normal obligation of the diemaker to examine his die design drawings and verify that the die (or dies) he is about to make will produce the piece part to specifications. A burr-side specification on the piece-part drawing supersedes the possible deleterious effects of the burr on bending or forming dies.

Edge condition of the workpiece is known to be a factor which can affect the degree of bending possible for a given type of material. Smooth edges permit more severe bending than rough edges. The burr side of a blank is contiguous to the break, and therefore, if the burr side is on the outer surface of the bend (tension side), the stock material will be more susceptible to the initiation of edge fractures in the bend area. Conversely, if the burnished cut band is on the tension side of the bend, the stock material will be less susceptible to the initiation of edge fractures in the bend area. While it is true that this condition is primarily the concern of the product engineer, the diemaker should recognize it also. In some cases where bending is relatively severe, burr-side location can be the difference between fracturing and not fracturing. In doubtful cases, the burr-side effect

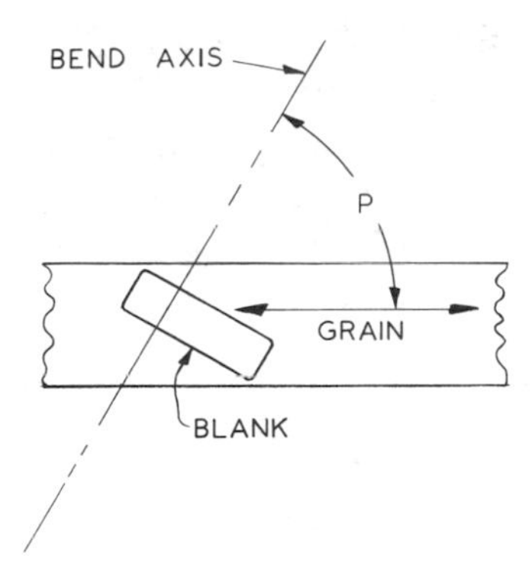

Figure 3·55 Bend axis related to grain direction.

on the bend should be verified in advance, in order to assure compatibility of the cutting and forming (bending) operations. Note that the burr-side rule cited above applies to this fracturing condition as well as to the wear effect on the die members.

GRAIN DIRECTION

Directionality of sheet or strip metal is a factor which must be considered in relation to bending and forming operations. Assuming that it is feasible to bend (or form) a given metal, the most favorable condition exists when the axis of the bend is perpendicular to the grain direction. The most severe bends practical for the type of material can be made in this direction. Conversely, the least favorable condition exists when the axis of the bend is parallel to the grain direction. The ability of the material to withstand bending strains increases as the angle P (Fig. 3·55) approaches 90°.

The minimum angle P required for bending without material failure depends upon the kind of material and the condition (usually the hardness) of the material. For cold-rolled strip steel, the practical bending potentials are indicated in Fig. 3·56.

No. 1 (Hard Temper). A very stiff, springy, cold-rolled strip intended for flatwork, where ability to withstand cold forming is not required.

Carbon 0.25 per cent or less (ladle analysis). Rockwell B 84 minimum for 0.070 in. thick and thicker, Rockwell B 90 minimum for thickness less than 0.070 in.

No. 2 (Half Hard Temper). A moderately stiff cold-rolled strip suitable for limited bending. Right-angled bends may be made at 90° to the grain direction around a radius equal to the thickness.

Carbon 0.25 per cent, maximum. Rockwell B 70 minimum, B 85 approximate maximum.

No. 3 (Quarter Hard). A medium-soft cold-rolled strip suitable for limited bending, forming, and drawing. May be bent to 180° across the grain and to 90° parallel with the grain and around a radius equal to the thickness.

Carbon 0.25 per cent, maximum. Rockwell B 60 minimum, B 75 approximate maximum.

No. 4 (Skin-rolled Temper). A soft, ductile, cold-rolled strip suitable for fairly deep drawing operations where surface disturbances such as stretcher strains are objectionable. Strip of this temper is capable of being bent flat upon itself in any direction. Skin-rolled, planish-rolled, and pinch-passed are equivalent terms with respect to temper.

Carbon 0.15 per cent maximum. Rockwell B 65 maximum, approximate.

No. 5 (Dead Soft Temper). A soft, ductile, cold-rolled strip produced without definite control of stretcher strains and fluting. It is suitable for difficult draw applications where such surface disturbances may be tolerated. It is suitable for bending flat upon itself in any direction.

Carbon 0.15 per cent, maximum. Rockwell B 55 maximum, approximate.

Red Brass and Cartridge Brass. Up to half hard temper and 0.050 in. thick; sharp-cornered 90° bends may be made in any direction without fracturing.

High Brass and Yellow Brass. Up to half hard temper and 0.090 in. thick; sharp-cornered 90°

Figure 3·56 Potential bends for which the various temper numbers of cold-rolled carbon steel strip are suited.

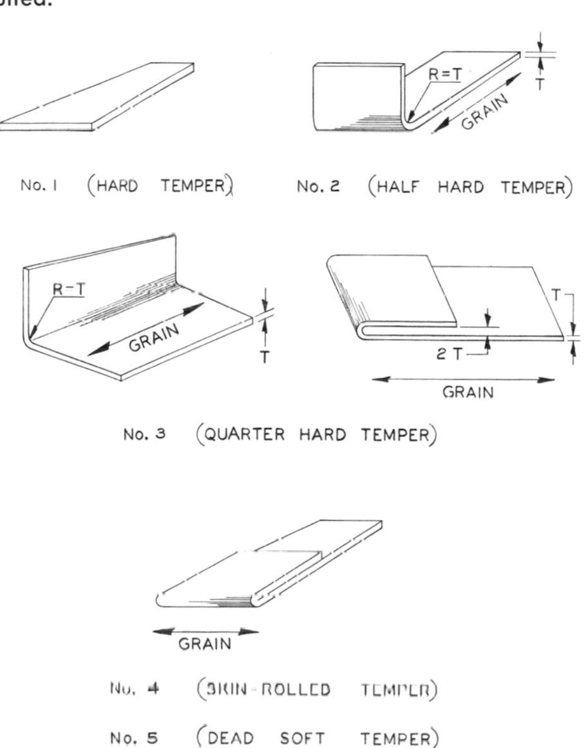

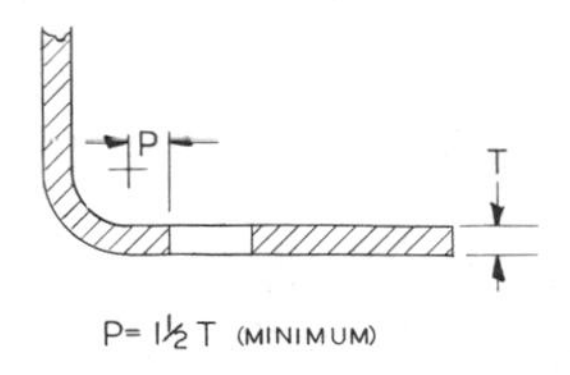

Figure 3·57 Proximity of bend line to edge of pierced hole.

bends may be made in any direction without fracturing.

Beryllium Copper. Up to half hard temper and 0.040 in. thick; sharp-cornered 90° bends may be made in any direction without fracturing.

Phosphor Bronze (5 Per Cent Tin). Up to half hard temper and 0.070 in. thick; sharp-cornered 90° bends may be made in any direction without fracture.

If there is any doubt about the feasibility of a bending operation, the feasibility should be proved by experiment before the dies which produce the flat blanks (or equivalent) are made.

BENDING IN PROXIMITY TO PIERCED HOLES

Holes pierced before bending will be distorted if they are too close to the bend area. As a rule, distortion will be minimized if distance *P* (Fig. 3·57) is held to a minimum of 1½*T*.

CHAPTER 4

SCREW HOLES AND DOWEL HOLES

The subject of screw threads is a broad and varied one. The information available is extensive in range and thorough in detail. If a toolmaker is concerned with the production of threaded parts, he can refer to mechanical handbooks such as the *Tool Engineers Handbook* (McGraw-Hill Book Co.), etc. He has recourse also to specialized books on the subject, including books on thread measuring and checking. If necessary, he can secure information from manufacturers and suppliers of threading equipment.

Not all this information, naturally, is required by the diemaker who is making stamping dies on an individual-die basis. His problems regarding tapped holes used to assemble die components are but a small and specialized facet of the subject. As a result, it is somewhat difficult for the diemaker to determine, from available reference information, a practical approach to adequate provision of mounting holes in die members.

The information in this section is not intended to apply to any other phase of thread tapping. It is rather a discourse on the salient features of a specialized subject, for the convenience of diemakers. It is presented as a safe, practical approach to problems of providing mounting holes in the components that make up a quality die. The information is generalized at the quality-die level. When it is practical to do so, departures from the practices recommended can and should be made, especially in the case of dies for low production and light duty.

In this section, the terms "thread depth" and "depth of thread" refer to the thread in profile. They indicate the difference between the radii of the major and minor diameters of the thread. This dimension is at right angles to the axis of the screw.

The term "tapped depth," or "depth of tapping," indicates the distance into the die component from the base surface that a usable thread has been produced by the tap. This distance is parallel to the axis of the screw.

TAPPED HOLES IN DIE COMPONENTS

Where feasible, the preferred method of providing tapped holes for mounting screws is as shown in Fig. 4 · 1, view *A*. The tap drill is drilled through the component and the hole is tapped from the base or mounting surface to a suitable depth.

Advantages of drilling through. For die components that require heat treatment, the practice of drilling through with the tap drill is usually desirable because it allows more even heating and also better circulation of the quenching medium. The depth of tapping does not have to be so carefully controlled as in the case of blind holes, which expedites the tapping operation. Chips produced by the tap are free to drop through the hole. Therefore, the hazards of thread damage and tap breakage due to impacted chips are considerably lessened.

Drilling through for tapped holes often expedites mounting procedure because it facilitates accurate transfer of the hole location to the adjoining die member. To transfer the screw holes, the die component can be clamped in its proper location and a tap-size drill inserted through the tapped holes. The component acts as a drill guide during spot-drilling of the mating die member.

Depth of spot drilling is a factor which influences accuracy of hole transfer. A drilled spot produced by the point of the drill alone is not

Figure 4 · 1 Representation drawing of typical tapped holes in cross section.

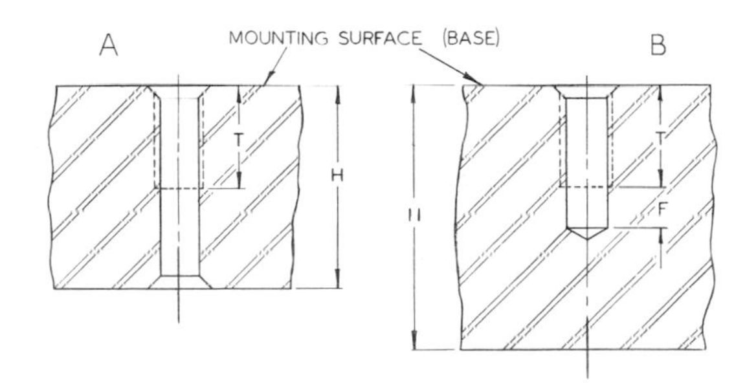

adequate. The full diameter of the drill should enter the adjoining member a distance of 1/16 in. or more.

Tapping depth. The usable thread must be tapped to a depth *T* (Fig. 4·1) which will allow the mounting screws to enter deep enough to secure the component adequately throughout the life of the die. Normally, this distance should be equal to, or greater than, 1½ times the diameter of the screw. Many dies are subject to extreme and varying work loads, including high-impact conditions. For heavier work especially, it is safer to have the screws enter the component a distance of twice the screw diameter. Wherever possible, the usable thread should be tapped ¼ in. deeper than the screw entry. For the smaller sizes, over-tapping should be a distance equal to the major diameter. Overtapping avoids time-consuming and exacting control of counterbore depths in the mating parts, as well as of the tapping depth in the component. Also, normal variations in the nominal screw length are readily absorbed if the tapping depth is sufficient.

Die life in relation to tapping depth must be kept in mind. The diemaker should be certain that the tapped hole is adequate for the life of the die.

Blind-tapped holes for thick sections. The thickness or height *H* of the component is a factor which, at times, determines the method used to provide tapped holes. In cases where the height of the component is great enough to make through-drilling impractical, blind holes are used, as shown in Fig. 4·1, view *B*.

The usable tapped-length depth *T* should be the same as for a through-drilled hole. The total depth of the tap size drilled hole should include end clearance *E*. The minimum end clearance required to avoid striking the bottom of the hole when using a standard plug tap is six times the pitch distance. Keep in mind that this figure is a minimum; generally, the allowance should be seven times the pitch distance, or more, if feasible (see table, Fig. 4·2).

The following is a practical shop method often used to determine drilling depth for blind-tapped holes: Scale-measure the chamfer length on the tap to be used; to this figure add a safety factor of approximately two threads; then add the desired usable tapped depth.

In Fig. 4·1, view *B*, the end clearance *E* is equal to the tap chamfer plus safety factor. *T* is the desired usable tapped depth. $T + E =$ drilled depth.

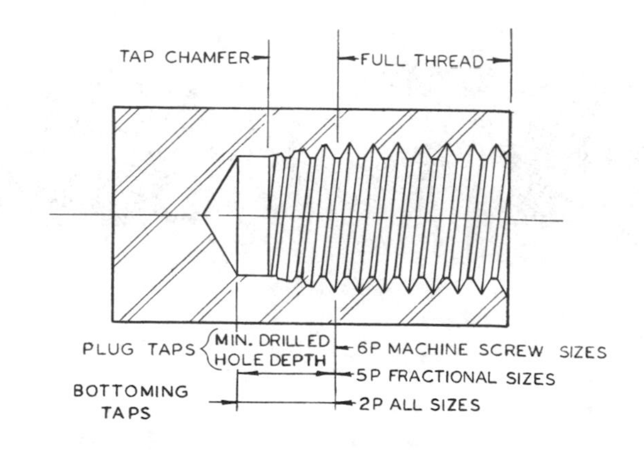

THREADS PER IN.	MINIMUM DRILLED HOLE DEPTH IN.			
	1P	2P	6P	7P
80	0.0125	0.025	0.075	0.088
72	0.0139	0.028	0.083	0.097
64	0.0156	0.031	0.094	0.109
56	0.0179	0.036	0.107	0.125
50	0.020	0.040	0.120	0.140
48	0.0208	0.042	0.125	0.146
44	0.0227	0.045	0.136	0.159
40	0.025	0.050	0.150	0.175
36	0.0278	0.055	0.167	0.194
34	0.0294	0.059	0.176	0.206
32	0.0312	0.062	0.187	0.219
30	0.0333	0.067	0.200	0.233
28	0.0357	0.071	0.214	0.245
27	0.037	0.074	0.222	0.259
24	0.0417	0.083	0.250	0.292
22	0.0455	0.091	0.273	0.318
20	0.050	0.100	0.300	0.350
19	0.0526	0.105	0.316	0.368
18	0.0556	0.111	0.333	0.389
16	0.0625	0.125	0.375	0.438
14	0.0714	0.143	0.429	0.500
13	0.0769	0.154	0.461	0.538
12	0.0834	0.167	0.500	0.583
11	0.0909	0.182	0.545	0.636
10	0.100	0.200	0.600	0.700
9	0.1111	0.222	0.667	0.778
8	0.125	0.250	0.750	0.875
7	0.1429	0.286	0.875	1.000
6	0.1667	0.333	1.000	1.167

Figure 4·2 Table for tap-drill depth.

Adequate end clearance, by eliminating the need for overly precise control of the tapped depth, simplifies the tapping operation. The end clearance also provides a chip pocket. When blind holes are being tapped, the chips can accumulate in the hole. If the accumulation becomes too great, the chip mass damages the threads being cut and also causes the tap to break. To avoid this, the tap must be backed out of the hole and both the tap and the hole cleared of chips, as required. The chip pocket reduces the time and effort necessary for clearing the tap and the hole of chips. It is also a safety factor against thread damage and tap breakage. Caution! Bear in mind that in any blind hole, as the tap progresses into the hole, the chip space becomes less and less.

Blind-tapped holes for smooth working faces. The working face of a die component is the surface opposite its base. There are many occasions where the working face, or a considerable portion of it, must be smooth and unbroken. If mounting screws are located within this type of working area, blind-tapped holes must be used to accommodate them. If possible, the same considerations which apply to blind holes in thick sections should be used. However, this cannot always be done, since the height of the component may be restricted because of its relationship to other factors of the die. If the height is limited and the working face of the component must be solid, the depth of the tapped holes may have to be restricted. Under these circumstances, the end clearance *E* is reduced, and a bottoming tap is used.

Bottom tapping normally requires two tapping operations. First, the hole is threaded as deep as possible with a plug tap. Second, the thread is finished out with the bottoming tap. The standard chamfer on a bottom tap ranges from 1 to 1½ threads. Consequently, full threads can be tapped closer to the bottom of the drilled hole. The minimum end clearance *E* required for a bottoming tap is twice the pitch distance (see table, Fig. 4·2). The usable tapped depth *T* may have to be reduced and overtapping held to a minimum.

When hole depths are limited, there is, of course, less chip space. Because of this, the diemaker must proceed more cautiously, to forestall the risks of tap breakage and thread damage.

Transferring blind screw holes. Occasionally, screw holes are located in die components and their adjoining die members by means of a jig borer or equivalent machine. This practice is usually restricted to high-precision dies, where interchangeability or replacement of sections is a factor. In the great majority of cases, blind-tapped holes are transferred to the adjoining member by the use of transfer screws.

TAP-DRILL SIZES

A tap drill is any drill used to produce a hole which is to be tapped. The choice of drill diameters used for tap-drilling is determined by compromise. For die construction purposes, the conditions which influence the choice of tap-drill diameters are: thread strength, tapping effort, and standard drill sizes. The possibility of tap breakage is closely associated with proportional tapping effort, especially for the smaller sizes.

Optimum tap-drill sizes vary according to the material being tapped. It is practical to use full threads in most nonmetallic materials or compositions; it is not practical to use full threads in most metals, unless the thickness of the material is less than half the tap diameter. For small holes in relatively tough materials, tap-drill diameters are frequently increased to leave only 50 per cent thread in tapped holes. For holding power, a high percentage of thread is desirable. However, a smaller percentage of thread depth reduces tapping effort and risk. These opposing circumstances require that a compromise be made between maximum holding strength and minimum tapping effort. A good compromise is important to the diemaker because of the tapping depths often used in die construction.

Percentage of thread in tapped holes is controlled by using larger or smaller tap-drill diameters as required. In order that standard drills may be used for tapped holes, the percentage of thread depth must vary a slight amount. Tap-drill charts or tables based on using standard drills show drill sizes which are the nearest approximation to the specified thread percentage.

A full thread is only about 5 per cent stronger than a 75 per cent thread, while it requires three times as much power to cut. In fact, tests have shown that any increase in the percentage of thread over 60 per cent does not appreciably increase the thread strength. However, it must be borne in mind that a shallow thread is more apt to strip the mating bolt or screw.

For common tapping practice in most metals, experience has proved the most generally desirable thread depth to be 75 per cent. These threads are close to maximum strength. They do not tend to strip the mating thread, and they are practical to tap. This percentage is so widely accepted that mechanical handbook tables for tap-drill sizes are predicated on it.

Simple rule for tap-hole size. Tap-hole diameters for American threads may be obtained by subtracting the pitch distance from the major diameter. The difference is the tap-hole size for approximately a 75 per cent thread. For drilling, use the nearest standard drill size. (To find the pitch distance, divide 1.000 in. by the number of threads per inch.)

This method of determining tap-hole diameters may be expressed as a formula:

$$C = A - B$$

where A = major diam
B = pitch distance
C = tap hole diam

In general, 75 per cent thread is used for tapped holes required in die construction. However, some of the tool steels used for die components are relatively tough. In these materials, when the tapping depth is greater than twice the tap diameter, the thread depth should be reduced to approximately 60 per cent where possible. This is especially true for the smaller sizes up to and including ¼-in. taps. The diemaker can estimate the toughness of the material when he is drilling it. If the drilling action indicates that the material is relatively difficult to cut, he may then open up the hole to reduce the thread percentage.

Drills may normally be expected to cut slightly oversize: The resulting drilled hole will be somewhat larger than the drill which produces it. Drilling tests have shown that the amount of oversize varies progressively, ranging on the diameter from approximately 0.0015 in. for a no. 56 drill to 0.005 probable for a ½-in.-diam drill. These figures assume that drills are sharpened properly and that correct drilling practice is followed. A skilled diemaker is aware of this condition and keeps it in mind, especially when he is using larger tap drills in order to reduce tapping effort in tougher materials.

Most existing tables showing tap-drill sizes do not take into consideration the amount oversize that a drill may normally be expected to cut. In Fig. 4·3 this allowance has been made, as indicated. The thread percentages shown are derived from the probable hole size, assuming that proper drilling practice is followed.

In production work, special tap drills are often used in order to produce threads which are exactly a required percentage of thread depth. However, such exacting control of thread percentage is not necessary for die construction purposes. Standard-size drills are entirely adequate for tapped holes in die components.

Thread percentages must vary slightly from the optimum in order to use standard drills for tap drilling. Because of this, the percentages shown in Fig. 4·3 are not necessarily optimum theoretical percentages. In each case, the table indicates a drill size for approximately 75 per cent thread, to be used for average tapping conditions. It also indicates larger tap drills for reduced percentages, to be used for tough materials. The tap-drill sizes recommended have been proved practical for die construction purposes.

COUNTERSINKING TAPPED HOLES

When a tap is started in a hole, it has a tendency to lift or force the material up from the starting surface. If this occurs, the first thread may be thicker than it should be, which may cause interference with the screw. In addition to the possibility of a thick starting thread, the material that is raised or lifted by the tap causes a hump or swelling on the surface of the component. The hump can, of course, be filed or machined off. However, removing the hump can cause the first thread to be distorted a slight amount from cutting pressure, or there may be an objectionable burr left on the starting thread. To avoid these undesirable effects, the tap size hole should be relieved at the starting surface before tapping.

Countersinking, shown in Fig. 4·4, is an effective means of relieving holes for tapping. The included angle L should be either 82° or 90°. This allows the use of standard countersinks and at the same time compensates nicely for any thread distortion that may take place when starting the tap. The diameter of countersink at the surface should be 1/16 to 3/32 in. greater than the major diameter of the screw M. If the surface is to be ground or machined after countersinking, the C diameter should be proportionally larger to allow for the stock removal.

TAPPED HOLE CHECKOUT

After tapping, it is a good idea to check each hole with a proper screw. Be certain that the screw turns freely and enters the component to the required depth. Important! This check is imperative in the case of components which are to be heat-treated.

CLEANING TAPPED HOLES AFTER HEAT-TREATING

After the component has been heat-treated, check again that a proper screw enters the hole easily and to the required depth. Usually, checking the hole in this manner also clears the threads of any slight heat-treat residue remaining in them. If there is objectionable binding due to residual heat-treat deposits, it will be necessary to clean the threads more thoroughly. A standard screw may be used for cleaning out stubborn residual deposits: Simply grind a longitudinal notch or flute (similar to a tap flute) in the screw. Don't be fussy about the notch; offhand grinding will be good enough. Then, turn the screw in and out of the hole a few times. It is a good idea to keep the prepared screw on hand for future use.

TAP SIZE	STD. DRILL DR. SIZE	STD. DRILL DEC. EQUIV.	PROBABLE HOLE SIZE	% THD.
0-80	56	.0465	.0480	74
	3/64	.0469	.0484	71
1-64	54	.0550	.0565	81
	53	.0595	.0610	59
1-72	53	.0595	.0610	67
	1/16	.0625	.0640	50
2-56	51	.0670	.0687	74
	50	.0700	.0717	62
2-64	50	.0700	.0717	70
	49	.0730	.0747	56
3-48	48	.0760	.0779	78
	46	.0810	.0829	60
3-56	46	.0810	.0829	69
	45	.0820	.0839	65
4-40	44	.0860	.0880	74
	43	.0890	.0910	65
4-48	43	.0890	.0910	77
	3/32	.0938	.0958	60
5-40	40	.0980	.1003	76
	38	.1015	.1038	65
5-44	38	.1015	.1038	72
	37	.1040	.1063	63
6-32	36	.1065	.1088	72
	34	.1110	.1136	60
6-40	34	.1110	.1136	75
	32	.1160	.1186	60
8-32	29	.1360	.1389	62
	28	.1405	.1434	51
8-36	29	.1360	.1389	70
	9/64	.1406	.1435	57
10-24	26	.1470	.1502	74
	23	.1540	.1572	61
10-32	22	.1570	.1602	73
	20	.1610	.1642	64
12-24	11/64	.1719	.1754	75
	15	.1800	.1835	60
12-28	15	.1800	.1835	70
	13	.1850	.1885	59

TAP SIZE	STD. DRILL DR. SIZE	STD. DRILL DEC. EQUIV.	PROBABLE HOLE SIZE	% THD
1/4-20	8	.1990	.2028	73
	4	.2090	.2128	57
1/4-28	3	.2130	.2168	72
	7/32	.2188	.2226	59
5/16-18	F	.2570	.2608	72
	17/64	.2656	.2697	59
5/16-24	I	.2720	.2761	67
	J	.2770	.2811	58
3/8-16	5/16	.3125	.3169	72
	P	.3230	.3274	59
3/8-24	Q	.3320	.3364	71
	R	.3390	.3434	58
7/16-14	U	.3680	.3726	70
	V	.3770	.3816	60
7/16-20	W	.3860	.3906	72
	25/64	.3906	.3952	65
1/2-13	27/64	.4219	.4266	73
	7/16	.4375	.4422	58
1/2-20	29/64	.4531	.4578	65
9/16-12	15/32	.4688	.4736	82
	31/64	.4844	.4892	68
9/16-18	1/2	.5000	.5048	80
	33/64	.5156	.5204	58
5/8-11	17/32	.5313	.5362	75
	35/64	.5469	.5518	62
5/8-18	9/16	.5625	.5674	80
	37/64	.5781	.5831	58
3/4-10	41/64	.6406	.6456	80
	21/32	.6563	.6613	68
3/4-16	11/16	.6875	.6925	71
7/8-9	49/64	.7656	.7708	72
	25/32	.7812	.7864	61
7/8-14	51/64	.7969	.8021	79
	13/16	.8125	.8177	62
1-8	7/8	.8750	.8809	73
	57/64	.8906	.8965	64
1-12	59/64	.9219	.9279	67
	15/16	.9375	.9435	52

Figure 4·3 Tap-drill diameters.

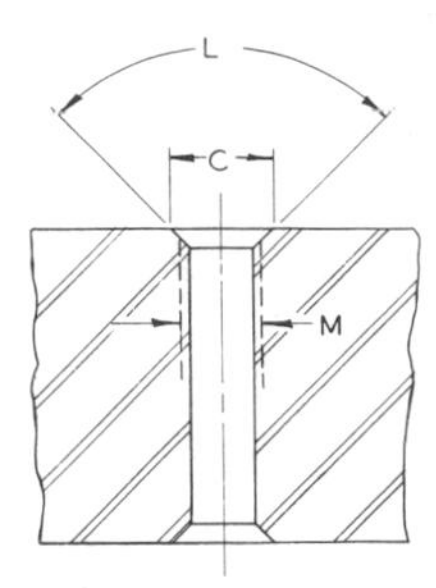

Figure 4·4 Typical countersunk relief for tapped hole.

If the binding is due to shrinkage of the hole, use lapping compound on the screw. In the case of severe hole shrinkage, a thread lap and lapping compound may have to be used to bring the threaded hole up to size. A thread lap may be made by die-threading a soft cold-rolled steel rod slightly undersize so that it freely enters the shrunken threaded hole. Such a lap, used with lapping compound, will quickly enlarge a threaded hole. After the hole has been mechanically cleaned or lapped, wash it out with a suitable medium (such as kerosene). Caution! Never use a tap to clean out a threaded hole after heat-treating the component.

FUNCTION OF MOUNTING SCREWS AND DOWELS

Any construction method that securely holds a die component in assembly and simultaneously assures accurate positioning of the component throughout the life of the die may be said to be satisfactory. If, in addition, the method provides ease of necessary maintenance and is not needlessly expensive to make originally, it can then be considered good diemaking practice.

It is common procedure to use a combination of screws and dowels to provide for proper mounting (assembling) of the various components that make up a die. The function of the dowels is to provide and maintain accurate positioning of the component. The function of the screws is to retain the component securely in the doweled position.

Disadvantages of doweled construction. The most obvious disadvantage is that dowels do not lend themselves to use in small die members; a component must be large enough to contain the dowels without being weakened by them. Another, and not so obvious, disadvantage is that doweled construction is not sufficiently rigid to resist lateral deflection if any great amount of unbalanced side thrust is imposed on the component.

Advantages of doweled construction. Simplicity of die construction is the greatest advantage of using dowels for locating die members. From the diemaker's viewpoint, doweling is a comparatively easy and effective method for accurate positioning of die components. The comparative ease of doweling also makes doweled construction economically desirable. Another distinct advantage is availability. Dowels are commercially available in a wide range of standard sizes at low cost. Standard commercial dowels are finished to 0.0002 in. larger than the nominal diameter with a tolerance of plus or minus 0.0001 in.

Oversize dowels. In the event that a dowel hole is inadvertently made slightly too large, oversize dowel pins are commercially available at 0.001 and 0.002 in. over the nominal diameter. The tolerance on commercial oversize dowels is the same as the tolerance for standard dowels. Oversize dowels are required for die maintenance and repair work, where dowel holes often become enlarged because of removal and replacement of the dowels. When oversize dowels are used, it is a good idea to identify the hole and dowel clearly to avoid needless and time-consuming checking of dowels and dowel holes.

Quality specifications for dowel holes. If a diemaker is employed by a manufacturer, he will often find that his company has set up standards to which various types and classes of dies are to be made. These standards will usually specify the quality and accuracy requirements for dowel holes. If he works for a jobber, he should check with his foreman to ascertain customer requirements for doweling.

Quality requirements for dowel holes. The degree of quality required for dowel holes generally varies directly in accordance with the overall quality and accuracy requirements of the die as a whole. Do not construe this statement to mean that dowel holes can be treated carelessly. However, a low-production die that has relatively high clearances and less exacting accuracy requirements does not require the same dowel-hole treatment that a high-quality die requires.

The diemaker should analyze the purpose and function of each die and use good judgment about the nicety of his doweling procedure for the job. His decision will be influenced by such factors as the fits and clearances required for the die, accuracy specifications of the piece part, production requirements (die life), and maintenance considerations.

For heat-treated die components, dowel-hole quality can generally be divided into three major categories: (*a*) holes which are ground and lapped (or ground and honed), (*b*) lapped (or honed) holes, and (*c*) polished holes.

Ground and Lapped Dowel Holes. Grinding a dowel hole will give accuracy of location, if required. It will also produce a hole that is, for all practical purposes, perfectly round. In addition, grinding will assure perpendicularity of the hole to the base surface. If a ground hole is honed or lapped slightly for final finishing, it can be considered the ultimate in dowel holes. This procedure is usually restricted to dies of highest quality and/or to dies where interchangeability or replacement of components is an important factor in proper function of the die. Occasionally, such holes are ground to size, omitting the lapping or honing operation. However, the fit on the dowel will be more permanent if the hole is honed or lapped slightly.

Lapped Dowel Holes. For the majority of work, the most practical method of finishing dowel holes in heat-treated die components is lapping. Skill and judgment are required on the part of the diemaker. The finished hole should have a straight, smooth, cylindrical surface, held accurately to the desired dowel fit. The depth *DD* of this nicely fitted portion of the hole (Fig. 4·5) should extend from the mounting or base surface a minimum distance equal to 1½ times the dowel diameter *D*. A distance *DD* equal to twice the diameter *D* can be considered optimum. The balance of the hole can be made slightly looser in fit in order to save time when lapping and to provide a slight lead for starting the dowel in the hole. The fitted depth *DD* must be suitably increased if it is intended to grind off the base surface of the component for die maintenance.

If an excessive amount of hardening scale is present in the hole, it will sometimes be advantageous to clean out the hole with emery cloth before lapping. A handy tool for cleaning or polishing out holes can be made as shown in Fig. 4·6. It is simply a rod with a slot sawed in one end. A suitable piece of emery cloth is inserted in the slot with the grit exposed and is wound

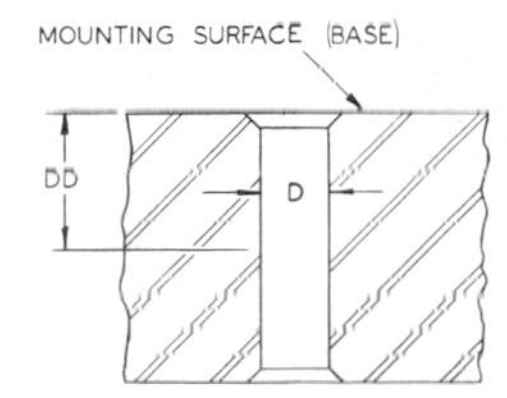

Figure 4·5 Relationship for lapping dowel holes.

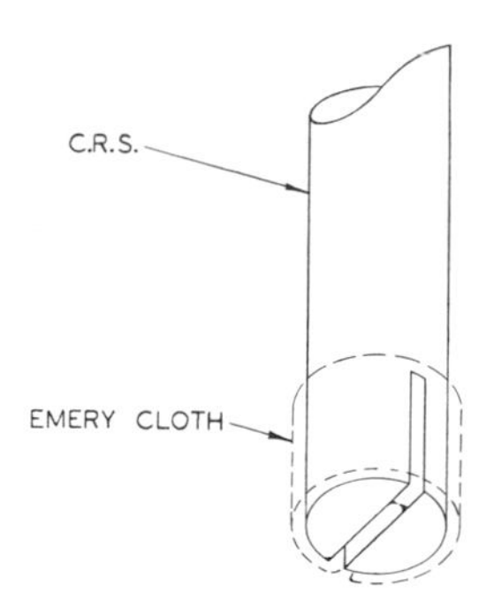

Figure 4·6 Polishing tool for dowel holes.

clockwise. The opposite end is chucked in the drill press chuck. The cleaning or polishing is then carried out in a manner similar to lapping.

Polished Dowel Holes. In this method the hole is reamed in the soft component to leave a minimum of polishing stock in the hole. Keep in mind that different types of tool steel react differently to heat treating: In some steels, the reamed hole may shrink more than in others. After heat treatment, the hole is polished with emery cloth in the same manner as that previously described under "Lapped Dowel Holes." Polished dowel holes are used primarily for reasons of economy, in situations where the doweling requirements allow the use of relatively low-quality dowel holes.

Relieved Dowel Holes. When the height dimension *H* (Fig. 4·7) is greater than four times the dowel diameter, it is good practice to relieve dowel holes as shown. The fitted length, or depth *DD,* should equal at least twice the dowel diameter *D*. The relief diameter may be 1/64 to 1/32 in. larger than the dowel diameter. Relieving holes in this manner is good practice applicable to all classes or types of dowel holes. In cases where the grinding life is on the base surface of the component, the fitted depth *DD* must be made correspondingly deeper.

The relief should be drilled before the hole is reamed. Caution! Before drilling the relief, check and be certain that it is being done from the proper side.

Figure 4·7 Dowel-hole relief drilled to minimize lapping and fitting time.

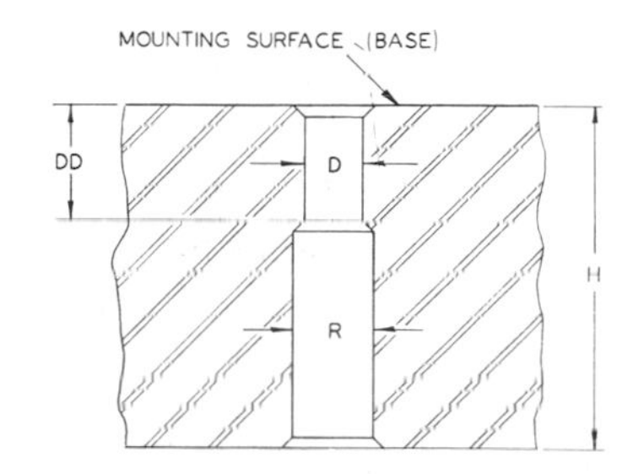

CHAPTER 5

DIE LIFE

The production potential of a die is considered in terms of the number of piece parts which the die can produce from the time it is new until it is worn out. It is obvious that the production potential of a die must vary according to the kind of stock material from which the piece part is made. For example, assume that a given die is capable of producing approximately one million piece parts from AISI No. 1020 cold-rolled steel strip. It is conceivable that the same die could produce approximately twice as many piece parts if the stock material were a free-working brass instead of steel.

It is equally obvious that the production potential will vary according to the kind of material from which the die components are made. For example, assume a piece part to be a transformer lamination stamped from silicon steel. If the cutting components of the die were made from:

1. Conventional oil-hardening tool steel, then the production potential would be so limited that the die would not be practical.
2. Oil-hardening high-carbon, high-chrome tool steel, then a normal production potential could be approximately five million hits.
3. Sintered carbide, then the normal potential could be fifty million hits.

Generally speaking, the potential production estimates given are conservative. Actual production may often be considerably greater. However, it must be kept in mind that individual situations may be encountered where actual production would be less than that indicated above.

PUNCH LIFE AND DIE-BLOCK LIFE

The term die life is synonymous with production potential. The terms punch life and die-block life are expressions which refer to the production potential of specific die components. Punch life and die block life are also used in a dimensional sense to denote the usable portion of a punch or a die block.

When a die is running and producing, cutting edges wear and become dull. They must be renewed by sharpening from time to time as required. Most commonly, cutting components are sharpened by grinding away the cutting face. Therefore, heights H and H_1 (Fig. 5 · 1) are reduced a corresponding amount each time the components are sharpened. Vertical dimensions are reduced accordingly. At each sharpening, for example, dimensions G and G_1 are reduced by the same amount as H and H_1. The usable portion of the punch is equal to G; G_1 represents the usable portion of the die block.

Eventually after repeated sharpenings, total

Figure 5 · 1 Schematic illustrating die life.

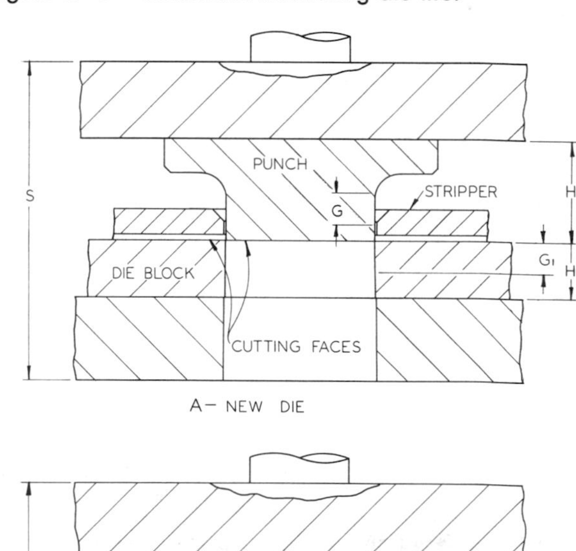

amounts equal to G and G_1 are removed from the punch and die block, respectively. The die then appears as shown in view *B*. Punch height is now equal to $H - G$, and die-block thickness is $H_1 - G_1$. The shut height S has been reduced by an amount equal to $G + G_1$. Normally, at this stage the die is considered used up and, if further production is required, is replaced with a new die. Occasionally it is practical to rework the punch, as indicated by the dotted outline at *E*. This will extend the life of the punch only. If the cutting clearance has become too great, the die block also will have to be reworked in some manner to decrease the cutting clearance. Such reworking of the die block is not usually practical except in special cases.

If, as a result of sharpening, more than the necessary amount of material is removed from a punch or die block, the ultimate life of the component is needlessly curtailed. The amount of material removed per sharpening should be limited to the minimum necessary for satisfactory renewal of the cutting edges.

Relative life: punches and die blocks. If records are kept for a sufficiently large group of different cutting dies, it becomes evident that punch life varies in relation to die block life. In some dies, the punches will wear at approximately the same rate as the mating die-block cut edges. In other dies, the punches will wear perhaps twice as fast as the mating die blocks. It will be observed also that the attrition is usually greater for piercing punches than for blanking punches.

When a blank or a slug is blanked or pierced through a die opening, it enters the die and goes on through. Thus it moves in one direction only. A punch, however, is forced through the stock material during the cutting action and stripped from it on the return press stroke. Thus the punch is subject to about the same wear conditions as the die insofar as cutting action is concerned, plus the additional wear induced by stripping friction. The punch therefore wears faster than the die cut edge by an amount directly in proportion to the stripping friction. If a punch strips freely, it tends to wear at approximately the same rate as the die cut edge. If stripping friction is high, the punch wears correspondingly faster.

It has been observed that, under identical service conditions, blanking punches often show less wear than piercing punches. This is because blanking punches as a rule are freer-stripping than piercing punches, owing to the difference in the proportional area of stock material which must be stripped. Since the stock material which surrounds an entered blanking punch is usually scrap, it is of course held to a minimum. Its relatively small area makes it relatively weak, so that it causes a minimum of stripping friction.

Sidewall finish. Any factor which affects the amount of friction between a punch or die opening and the stock material also affects the efficiency and life of the component. Thus the quality of the sidewall finish has a direct influence on the degree of friction. A smooth finish is essential to satisfactory punch and die function. In fact, even the lay of the finish can be an important factor in punch and die efficiency. (The lay of the finish is the direction of the visible pattern of surface finish irregularities.) The efficiency and life of the component will be maximum for the quality of the finish if the lay is parallel to the direction of punch travel.

EXCESSIVE WEAR

Abnormal wear may be caused by any of the following conditions:

1. Cutting clearance: Insufficient or excessive.
2. Entry: Punches enter too far into die openings.
3. Crowding: Punches are too close to other punches.
4. Punch height: Vertical punch length is too great in relation to cross-sectional area of the punch.
5. Hardness: Is component hard enough?
6. Finish: Are sidewalls of punch or die opening smooth enough? Is lay of finish parallel to direction of punch travel?
7. Material compatibility: Is component made from a material suitable for working the required type of stock material?
8. Mounting: Components must be securely mounted. Mounting surfaces must be clean as well as flat and perpendicular to the punching axis. Punches must be properly aligned with their mating die openings.
9. Stripping: Is stripping action even?
10. Excessive runs: Do not attempt to produce too many pieces between sharpenings.
11. Press conditions: Dies cannot be expected to produce unless presses are in good condition.
12. Careless setup: Check particularly for security and parallelism of bolsters. Also, die must be adequately secured in press.

The list can be extended and refined almost indefinitely. However, in most instances premature wear and/or failure can be associated with one or more of the conditions given.

CHARACTERISTIC CUTTING WEAR

The typical manner in which cutting components wear is exhibited in Fig. 5·2. Effects of excessive wear on both the work and the components are depicted also. Views *A*, *B*, and *C* present the normal running situation. The work produced has a smooth cut band of even height *H*. The break is clean, displaying a consistent crystalline pattern.

Height *H* is influenced by a number of factors, such as physical properties of the stock material, punching speed, lubrication, and others. It is especially influenced by the cutting clearance. Smaller cutting clearances produce higher cut bands *H*. However, tighter cutting clearances tend also to accelerate the wear on the cutting edges of the components. Optimum cutting clearances normally produce a condition where $H = \frac{1}{4}T$ to $\frac{1}{2}T$. This is generally the best compromise, considering such factors as:

1. Quality of piece part, including degree of burr
2. Slug pulling
3. Punch and die life

Newly sharpened cut edges are dead sharp (inset, view *B*). The dead-sharp condition begins to deteriorate as soon as the die starts to run. After a relatively short time, the initial edge breakdown produces a working radius *R* (view *C*) on the cut edges. This is the optimum running condition. Size of working radii is a variable, influenced by:

1. Type of material from which component is made
2. Hardness of component
3. Physical properties of stock material
4. Thickness of stock material *T*

A "normal" working-edge radius *R* for average conditions will be quite minute, generally 0.0002 to 0.005 in. Heavy-gage stock material and exceptional circumstances may produce working radii larger than indicated above. However, these would not be average applications.

As the initial working radius develops, the cutting edge attains its maximum resistance to wear. It is during this stage that the practical production potential of the component is realized. As production continues, the size of the working radius increases. Eventually, the corner profile begins to assume a parabolic shape (view *E*). When this stage is reached, component wear accelerates rapidly. If production is continued, the deleterious effects shown in view *D* begin to appear. The height of the cut band *H* becomes erratic. The normally smooth surface of the cut band may assume a rough, striated appearance. Burr effects become more evident, often showing an irregular ridge around the periphery of the opening. Friction between the stock material and the components is radically increased, which tends to instigate and support galling. The ultimate result is work spoilage and complete failure (breaking) of the components.

Preventive maintenance is indispensable to maximum component life. Components should be sharpened before the parabolic wear stage becomes pronounced, a condition which can generally be anticipated by watching the burr on the work. If the components are sharpened before the burr becomes excessive, sharpening is possible without excessive grinding. If, on the other hand, sharpening is put off until burr conditions are severe, an inordinate amount of grinding will be required to renew the cutting edges. If excessive grinding is necessary for sharpening, ultimate component life is drastically shortened.

Whenever possible, sharpening should be timed to require stock removal of only 0.005 to $\frac{1}{64}$ in. This applies to the majority of die applications. In a minority of cases it may not be possible to proceed as described. Heavy, scaly, and/or otherwise abrasive stock material may cause component wear requiring the removal of larger amounts when sharpening. In fact, extreme instances may be encountered where dies are maintained by replacing components instead of

Figure 5·2 Typical cutting punch and die wear with resulting effects on the stock material.

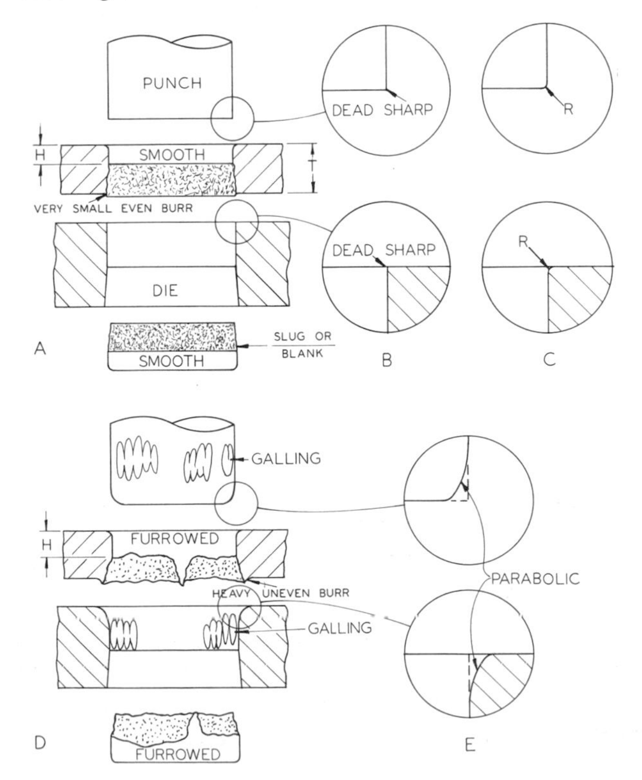

sharpening. An example of this may be the piercing of relatively small holes through boiler plate, etc.

CHIPPING

Persistent chipping along the cutting edges will seriously detract from the potential life of the component. In some cases where the die design does not permit adjustment of individual components, the life potential of the entire die will be correspondingly curtailed. Chipping may occur for a number of different reasons. Among these are:

1. Poor housekeeping. Foreign material enters the die, usually with the stock strip or workpiece. This is cited here simply because it occurs all too often. The solution is obvious.
2. Grinding cracks. Shallow invisible or nearly invisible cracks are induced by localized grinding heat. The occurrence of grinding cracks can generally be attributed to inadequate grinding equipment and procedures, especially carelessness when grinding. If spalling originates in the face surface of the component, resharpening to the depth of the chipped-out portion will often eliminate the affected area. It is assumed, of course, that resharpening will be carefully performed with a suitable grinding wheel.
3. Insufficient cutting clearance.
4. Hardness (brittleness) of the component.

In regard to the third and fourth conditions, it is not always practical to increase the cutting clearance or reduce the hardness of the component. Chipping can often be prevented by eliminating the dead-sharp corner condition—see Fig 5 • 2, view *B*—after sharpening. This can be accomplished by stoning or lapping the cutting edge to produce the working radius shown in view *C*. Caution! Use maximum care with this procedure; do not overdo it. Do not assume that this is a standard practice applicable to all dies but remember it for what it is: a method for coping with more or less extreme circumstances. Carbide components, however, are somewhat of an exception. They normally require that dead-sharp cutting edges be touched off before the die is placed in operation. It should also be remembered that eliminating dead-sharp corners may often be an effective means of counteracting slug pulling.

COMPENSATING FOR DIFFERENTIAL WEAR

Components which serve different functions have different rates of attrition. For example, in Fig. 5 • 3 the parting punch 2 requires periodic

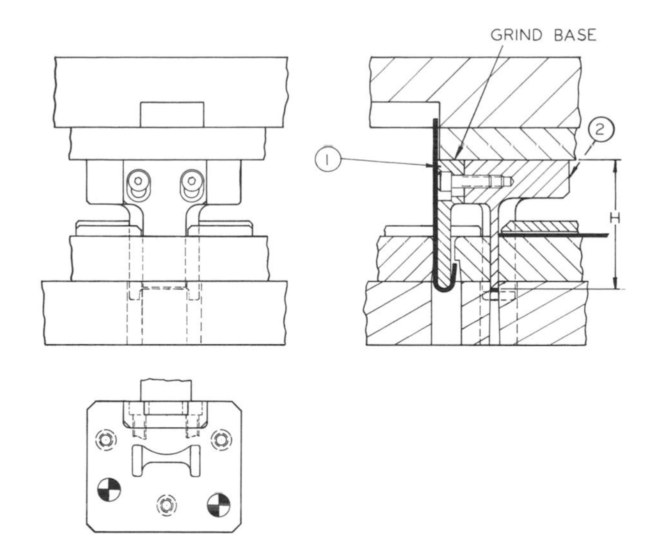

Figure 5 • 3 Punches mounted to permit compensation grinding for maintenance purposes.

sharpening. However, it is quite possible that the forming punch 1 may not require re-dressing at all during the entire life of the die. Sharpening the parting punch reduces its height *H*, which necessitates making a compensatory adjustment in the height of the forming punch. Contour-grinding the nose end of the forming punch would entail needless and time-consuming maintenance work. Therefore, this die was designed to allow shortening of the forming punch simply by grinding its flat base surface, as indicated. The forming punch is bolted to the parting punch and is provided with elongated clearance holes for the mounting screws. The elongated holes permit the necessary vertical adjustment between the two punches.

In Fig. 5 • 4, the base surface of the forming die block 5 can be ground to compensate for reduc-

Figure 5 • 4 Provisions for compensation grinding.

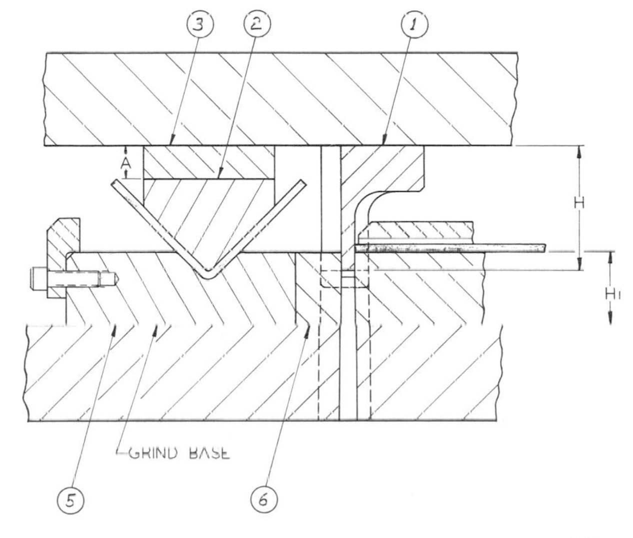

tions in the height H_1 of the parting die block 6 due to sharpening. However, because of its profile, some extra setup work would be required to grind the base of the forming punch 2. To eliminate the need for grinding the forming punch, it is mounted on a compensator block 3. When height H of the cutting punch 1 is lowered by sharpening, it is a simple matter to surface-grind the same amount from the thickness A of the compensator.

Provision for compensation grinding is primarily a maintenance consideration. However, the benefits are not limited to maintenance only; in most cases such provisions will also facilitate the original building of the die. The importance of compensating provisions is in direct ratio to the complexity of the die. Such provisions are, practically speaking, indispensable to both the making and the maintenance of intricate progressive dies.

CHAPTER 6

PUNCHES

CLASSIFYING PUNCHES

As previously defined, a punch is a male member of a complete die. It supplements or complements the female die in order to produce a desired effect upon the material that is being worked. It is both convenient and practical to classify punches in three categories, each composed of two major groups which include the numerous punch types and variations.

Punch categories. Categorical divisions are predicated on the basic functional requirements of a punch.

1. Cutting Punches. In this category are the punches which blank, pierce, notch, trim, shave, or otherwise effect cutting of the stock material.

2. Noncutting Punches. This category includes punches which bend, swage, draw, extrude, or otherwise act to form or deform the stock material.

3. Hybrid Punches. Both cutting and noncutting functions are combined in the same punch. Some typical hybrid punches are shear-form punches, semipierce punches, stab-pierce punches, and pinch-trim punches.

Punch groups. The two major punch groups are distinguished by the manner in which the punches are mounted.

Segregated Punches. These punches are self-mounted in the sense that they are positioned and retained by means of self-contained screws and dowels. Mechanical integration with other components such as punch plates is not necessary. As far as mounting is concerned, these punches are self-sufficient as to both positioning and retention. This tends to simplify mounting procedures and to minimize the amount of work required for mounting a given punch.

Integrated Punches. Punches in this group depend upon some other component, such as a punch plate, to locate and position them. In most cases, they depend upon this other component to retain them also. When assembled, these punches are, in effect, an integral unit with their locating and retaining components.

Punches belonging to either of these groups may require additional support or guidance. In this event, they may be supported or guided by a stripper or some other appropriate die member.

Punch types. It is not practical to attempt to be overly definitive in regard to punch types. In a general way, punches can be typed according to their characteristic structural configuration. However, for the sake of expediency, almost any distinguishing feature or combination of features may be commonly used to typify a given punch. In most cases, punches are specified by giving a condensed description of the punch in question.

Ordinarily, punches are referred to in accordance with their function. For example: blanking punch, forming punch, shear-form punch, etc. This is actually the categorical description. If greater specificity is desired, the punch-type description may be included. For example: plain blanking punch, pedestal-type form punch, clamp-type shear-form punch, etc.

Generalized punch type group relationship. Broadly speaking, the first group (segregated punches) is composed essentially of two basic punch types: plain punches and pedestal punches. Other punch types will generally be associated with the second group (integrated punches).

Punch know-how. The most practical way to learn about punches is to study specific punches and punch applications. Information derived from such studies can, by associative reasoning, be applied to other punch requirements. Even though new applications may be radically different, the basic principles previously studied may be expanded and modified to satisfy new requirements.

The punch constructions which are discussed

in the following pages were selected because they involve simple, basic, and rather typical punches.

PLAIN PUNCHES

As described earlier (see in Chap. 1 "Basic Blanking or Piercing Operation"), the mechanics of blanking and piercing are identical. The purpose of the operation performed determines whether the punch is a blanking punch or a piercing punch. Of course, it is mandatory that the diemaker keep in mind that pierced openings are sized by the punch and that die openings impart their sizes to piece parts produced by blanking (refer in Chap. 1 to "Relationship of Piece-Part Sizes to Punch and Die Sizes").

Punches made as shown in Fig. 6·1 are commonly called "plain" punches. As indicated, the cross-sectional configuration of these punches is rectangular. The cutting contour is, of course, determined by the work requirements and may be quite intricate, as indicated to some extent in the illustration. The sidewalls of the punch follow the cutting contour, originating at the cutting edge and extending straight through to the base surface. Punches made this way are called "straight-through" punches. A plain punch is actually a self-mounting (segregated group) straight-through punch.

Advantages of plain punches. Economy in die construction is a major advantage associated with plain punches. As a result, they are often used in low-production dies. However, the fact that this type of punch is so often economically desirable for low-run, and even for temporary, dies is not an indication of inferior construction. If suited to the particular application, plain punches can be appropriate components in a die of the very best quality.

There are several factors which contribute to the economy and general desirability of plain punches. One of these is economy of material. For any given contour, the minimum amount of material is required to make a plain punch. This in itself is usually a minor consideration, but it is directly associated with another more significant saving: a saving of machining time. Since a minimum of material is required to make a plain punch, it follows that there is also a minimum of excess stock to be removed in the process of making the punch. Therefore, the machining time required for the removal of superfluous material is also minimized.

Ease of setup for machining is a generally desirable characteristic of plain punches. The absence of projecting flanges or shoulders can, in many instances, simplify setting up the punch for required machining and grinding operations.

Simplicity of mounting procedure is an important attribute of plain punches. This is essentially due to the fact that they are self-mounting; that is, a plain punch does not need to be fitted to other components, such as punch plates or yokes, for mounting purposes. It stands on its own. It may be mounted on the face surface of a punch plate, on a thrust plate, or directly on the die set. It is located and retained by means of its own screws and dowels. In cases where two or more such punches are required in the same complete die, each punch can be mounted individually in relation to its mating die component.

Another major advantage of plain punches is the fact that they may be machined and ground straight through. This can effect a considerable saving in machining and grinding time. The more complex the cut-edge contour, the greater the comparative saving. When a punch is machined or ground straight through, the directional effect, or lay, of the surface finish agrees with the direction of punch travel. If the sidewalls are machined in the direction of punch travel, it is easier to polish them in the same direction. The need for polishing the sidewalls of a punch can often be eliminated by grinding to a smooth finish. This is especially true when the lay of grinding finish is parallel to the direction of punch travel.

Figure 6·1 Plain blanking punch.

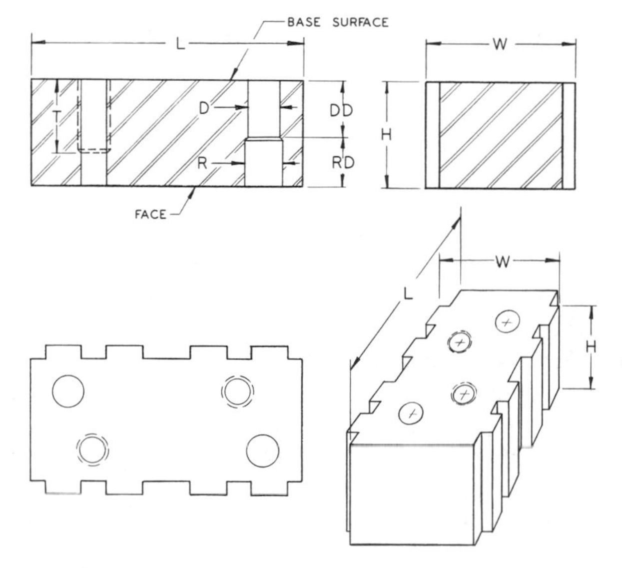

Proportions of plain punches. Plain punches are mounted by means of screws and dowels which are located within the area of the punch proper. The face area of the punch must, therefore, be large enough to accommodate the necessary screws and dowels.

Overall height *H* of plain punches can be less than for other types of punches without sacrificing potential punch life. This can be a desirable feature in situations where vertical space is restricted.

It should be remembered that plain punches not only *can* be made lower in height, but in many instances *must* be lower than other punch types in order to be proportionally adequate. This is especially noteworthy with smaller punches. Dimensional proportions of a plain punch must be such that it is stable. Generally, to be considered stable, length of the punch *L* and width *W* should be equal to or greater than height *H*. Exceptions to these proportions can be made when conditions permit. Where the work performed is not too heavy or unbalanced, and the punch is proportionally long, the width can be somewhat less than the height. However, when dimensional proportions are dubious, any decision to use a plain punch should not be lightly made. If the punch proportions are not favorable, consideration should be given to the following:

1. Assembling the punch into a yoke or a punch plate to provide support.
2. Supporting or guiding the punch in the stripper or other suitable die members.
3. Using a pedestal-type punch instead of the plain punch. A pedestal punch may be economically more practical than a straight-through punch which requires stripper guidance or support from a punch plate.

The base area of a plain punch is inherently the same as the area of the face. The mounting screws and dowels must be contained within this area, which imposes a size requirement for plain punches. A plain punch must be large enough to safely contain a sufficient number of suitably located mounting screws and dowels. Moreover, the screws and dowels must be adequate in size. Since the mounting holes tend to weaken the punch, the base area must be large enough to contain them without endangering the strength and durability of the punch.

In most cases, punch height *H* is such that it is advantageous to relieve the dowel holes as indicated at *R*. Relief depth *RD* should permit a fitted depth *DD* equal to, or greater than, the dowel diameter *D*. Caution! Relieve as shown—from the face surface of the punch, *not* from the base surface. Thread depth *T* should normally be at least ¼ in. deeper than the desired screw entry depth (see Chap. 4, *Screw Holes and Dowel Holes*). Be sure to tap the screw holes from the *base* surface, as shown. In the case of forming punches, etc., compensatory die-life grinding may be performed on the base surface. If so, thread depth *T* and fitted dowel depth *DD* must be correspondingly deeper.

On occasion, if the work is heavy or severe, especially if the cutting force is unbalanced as well, it may be necessary to provide a hardened thrust plate to back up the punch. Area of the thrust plate must, of course, be adequately larger than the base area of the punch.

With larger punches, a shorter *H* dimension can effect considerable savings in material and heat-treating costs. In addition, handling of the punch during construction will be facilitated. To secure a shorter *H* dimension, it may be necessary to use a laminated construction, as represented in Fig. 6 • 2. Here, the required overall punch height *H2* is attained by mounting the punch on a spacer plate of suitable thickness *S*. For really large punches, not only should height *H* be restricted to a practical minimum, but punches should be appropriately sectioned, as suggested in the plan view.

Mounting plain punches. Because of the highly individual nature of much die work, job-to-job mounting procedures can and do vary quite radically. However, certain instances, similar to the ensuing illustrations, are common enough to be considered rather conventional. These should be studied and assimilated, then modified and expanded upon to suit individual applications.

Figure 6 • 3. A spotting punch, commonly called a "spotter," is being used to transfer the mounting screw centers to the punch holder. For this kind of mounting procedure, the spotter diameter

Figure **6 • 2** Large plain punch, laminated and sectioned.

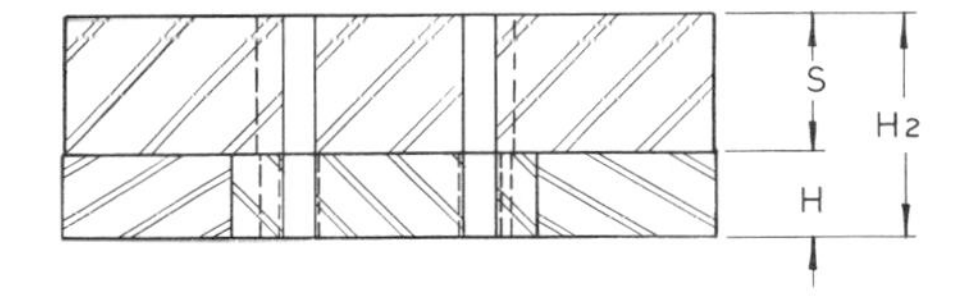

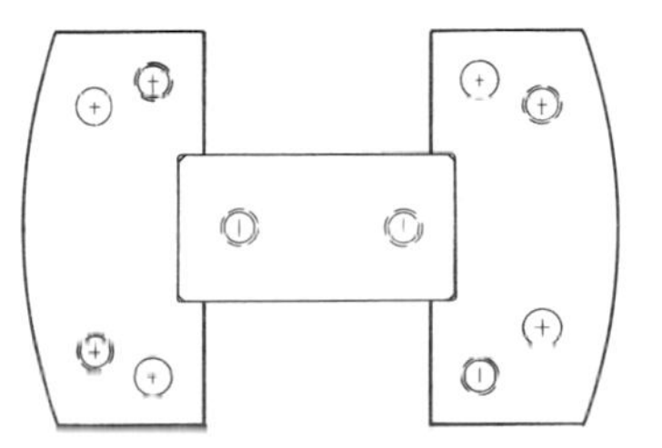

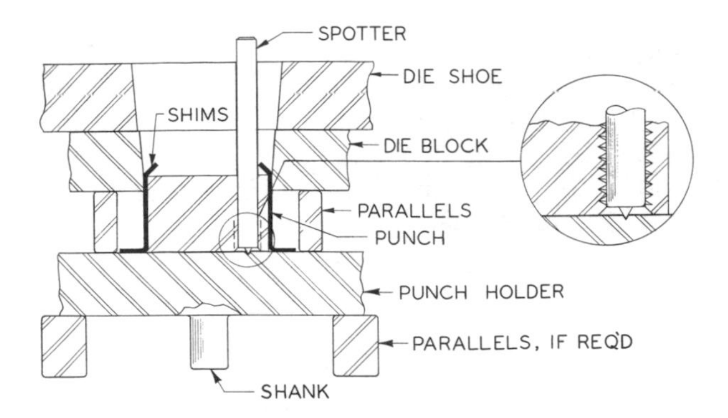

Figure 6·3 Spotting screw hole locations.

is a slip fit to the tap-drill size. In this case the work sequence requires that the die block be mounted first. Then the die set is inverted. If the punch holder has a shank, place it on parallels, as shown. Bring the assembly together, centering the punch in the die opening with shims, if necessary. Shims will not be required if clearance between punch and die is less than, say, 0.002 in. per side. In order to expedite punch entry it may be helpful to secure the shims with a rubber band or masking tape, etc. Another trick is to bend the shims in some manner to stabilize them, as indicated in the illustration.

A set of parallels of suitable height (lower than the punch) is used to space the assembly, assuring optimum punch entry and permitting the assembly to be clamped together, if desired. Insert the spotter through the screw holes, as shown. Using a small hammer, lightly tap the spotter three or four times, applying a random rotary motion to the spotter between taps. Do not attempt to secure a deep center impression for drilling. Do not attempt to spot dowel holes. Next, separate the assembly, exposing the center marks. Enlarge the center impressions with a center punch and drill the screw holes, counterboring, etc., as required.

Figure 6·4. Die block is mounted first. A suitable parallel block (or blocks) is placed in the die opening to support the punch. Standard transfer screws (enlarged view) are screwed into the punch base. If required, shims are used to center the punch within the die opening. The punch holder is then brought into contact with the transfer screws and tapped upon with enough force to secure an impression from each transfer screw.

If the die opening has a corresponding clearance opening, then place the die set on a flat plate and use a correspondingly higher parallel block. If this is not feasible, mount the die block temporarily with low parallels inserted between the die block and die shoe to provide support for the

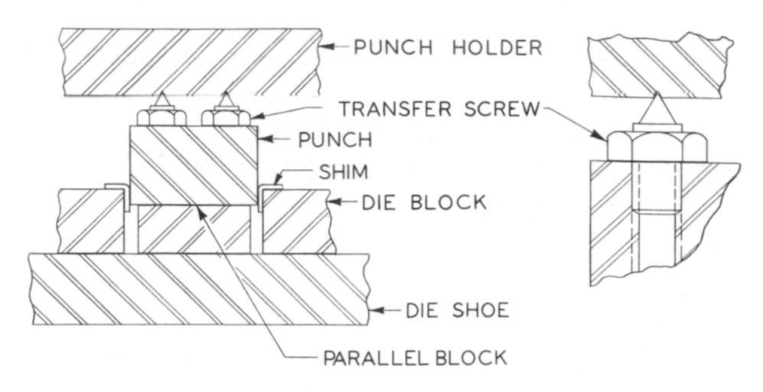

Figure 6·4 Setup for transferring screw hole locations.

parallel block. However, the method shown here is best accomplished if there is no clearance opening in the die shoe. Therefore, when such a clearance hole is required, it may be advantageous to spot the mounting screws before making the clearance opening, provided, of course, that there isn't some better reason for making the clearance hole first.

Figure 6·5. Here, the punch is shown sheared a slight distance into the die opening in order to transfer the die-opening contour to the punch. The punch may then be removed and finished to the shear size less the required punch and die clearance. The punch may be finished to size before hardening, or it may be ground to the shear mark after hardening, depending upon individual job requirements. In any case, once the initial shear has been taken, it is possible and convenient to transfer the mounting screw locations, as depicted in the illustration.

Doweling plain punches. After the screw holes have been transferred, drilled, counterbored, and/or otherwise made ready, the punch may be doweled in place. To dowel, position the punch in the die opening, using shims if required. Tighten the screws not quite to locking tightness, but tight enough for security. Then, using a soft drift and a light hammer, tap the punch laterally as required, until it is exactly positioned. Tighten the screws all the way. This tightening may disturb the punch position slightly; check and correct, if

Figure 6·5 Transferring screw hole locations (sheared punch).

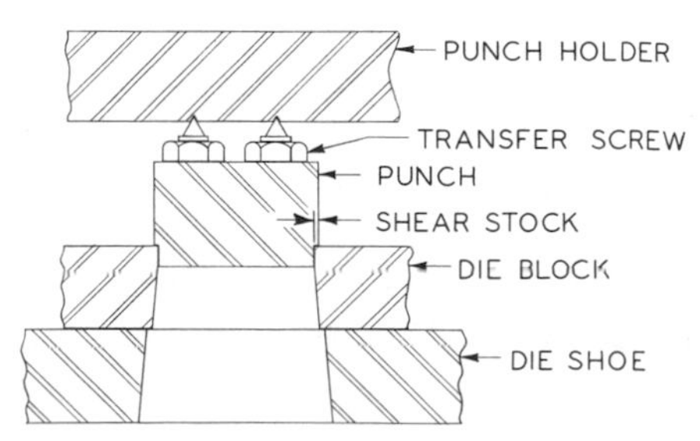

necessary. Then drill and ream for the dowels, using the punch as a jig.

Where it can be employed, visual checking is the best method for final alignment of cutting punches to doweling position. To check visually, enter the punch a minimum distance into the die opening. Arrange for light to shine into the juncture. Then sight through the die opening from the bottom. This method is applicable to blank-through and similar operations where the die opening has a clearance opening below it.

For dies without clearance openings, the punch can be accurately positioned by making trial cuts in some appropriate material. The condition of the trial blank can be examined and analyzed to ascertain the existing punch and die edge relationship. Use some material that will tend to exaggerate the punch and die relationship. For example, use tissue paper for low clearance conditions. Use a heavier, but soft, paper for medium clearances, etc.

Bending and other forming punches can, at times, be brought to dowel position by means of shims. Another method sometimes appropriate is to form a piece of soft wire (solder wire, for example) between the punch and die. The flattened wire section can then be measured to determine the existing relative punch and die positions. When bending or other forming conditions are well balanced or supported, it is sometimes feasible to form a workpiece with the punch screws snugged. After forming, and with the die kept closed, the screws are tightened to lock. Then another piece is formed as a check, and if satisfactory, the punch is doweled.

When the punch holder has a shank, extra forethought is required of the diemaker. From the very beginning, he must anticipate the shank location and avoid the possibility of screws and/or dowels which interfere with the circumference of the shank. If at all possible, screws and dowels should be located safely outside the shank. If not, then they should be located adequately within the shank area. Avoid locations which result in screws or dowels that are partly in the shank. In rare and extreme circumstances, this may be considered unavoidable. If so, check and be certain that the locations will not interfere with mounting the assembled die in the press.

This brief discussion of punch mounting is not by any means all-inclusive. However, if thoroughly studied and assimilated, it can be directly of value for conventional mounting situations. In practice, the methods discussed can be modified and developed for other situations. Actually, each different die is an individual case, some extremely so. The skill and ingenuity of the diemaker is relied upon to select, derive, and create, if necessary, the most logical and efficient procedures for each given set of circumstances.

PEDESTAL PUNCHES

As far as nomenclature is concerned, these punches have been referred to by a number of different terms—for example, flanged punches, shoulder punches, base-type punches, etc. However, a very appropriate term, and probably the least confusing, is pedestal punches.

These punches are made as shown in Fig. 6•6. In at least one cross section, the punch will exhibit the characteristic pedestal configuration. The area of the base surface is larger than the area of the cutting face by an amount equal to the area of the flanged portion. The flanged portion is an integral part of the punch. Thus, a pedestal punch is a solid unit which has a base area larger than its cutting-face area.

Stability of pedestal punches. Stability is a necessary attribute for any type of punch. If properly designed and proportioned, pedestal punches are inherently stable. Stability is inherent because of the relatively broad base surface characteristic of the pedestallike configuration. The large base area makes an ideal foundation surface for the punch.

Load distribution. The load-distributing qualities of pedestal punches are excellent. In all types of punches the cutting force is concentrated along the cut-edge profile. The force is then transferred through the mass of the punch to the mounting surface. Since the base area of a pedestal-type punch is larger, the cutting force load is dispersed through a greater area of mounting surface. The force exerted against each square inch of mounting surface is correspondingly reduced. This ability to disperse cutting loads is especially advantageous for heavier work.

Figure 6•6 Pedestal-type punch.

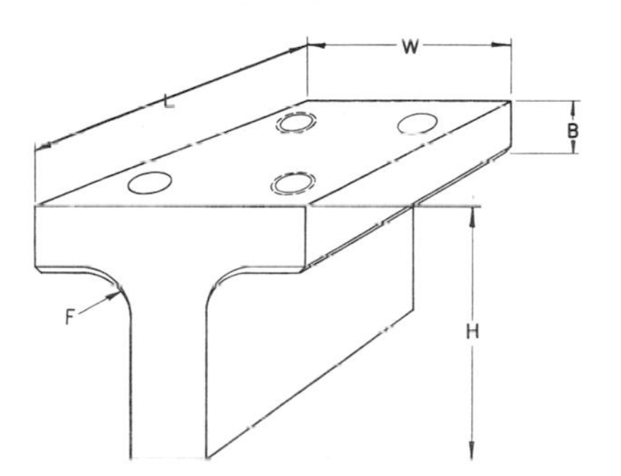

Mounting advantages. Like plain punches, pedestal punches are self-mounting, and each punch is mounted individually. In addition, the diemaker will find that the flanged portion of pedestal punches can be a very real help to him when he is mounting his punches. On smaller dies, the flanges provide an ideal ledge for clamping the punch in position while transferring the screw holes by drilling.

In many cases, however, relatively small punches must be mounted on large die sets. If the punches are too far in from the edges of the die set, it becomes impractical to hold the punches with clamps while transferring the screw hole locations. In such cases, a convenient method of transfer is that shown in Fig. 6 • 7, where the punch is entered in the die opening and supported by parallels which are inserted between the flanges and the die block. The punch holder is then brought down on the points of the transfer screws. The points leave their impression in the surface of the punch holder, to be picked up for drilling.

When the punch is entered in the die opening as described above, it is advisable to make certain that the punch is adequately positioned within the opening. When punch and die clearance is more than 0.002 in. per side, shims should be placed between the punch and die walls, as described earlier in this chapter under "Mounting plain punches." The shims should be strategically disposed around the contour in such a manner that the punch is properly positioned in relation to the die opening.

Doweling procedure should be generally the same as that described for plain punches. However, with pedestal punches on smaller dies, it is sometimes practical to secure final doweling position immediately, using clamps to hold the punch in place. Both the screws and the dowels can then be drilled, reamed, etc., while the punch is secured by the clamps. Keep in mind that, while this procedure may be faster, it is also riskier. If it is used, be careful.

Figure **6 • 7** Transferring screw holes from pedestal-type punch.

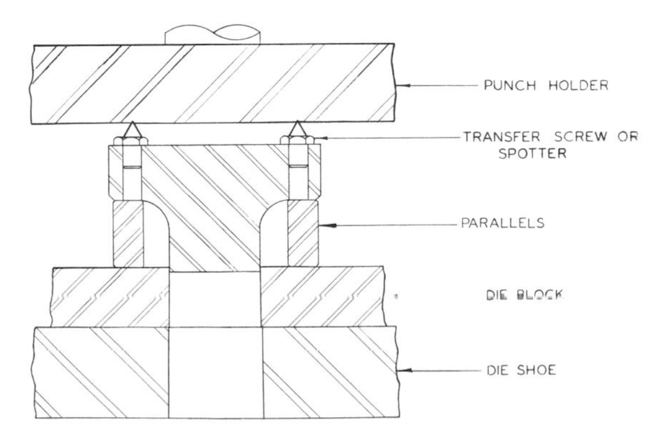

The larger base area of pedestal punches tends to simplify the problem of providing a sufficient number of screws and dowels which are both adequate in size and strategically located. The minimum base size of a pedestal punch must be such that the punch is stable. If necessary, the minimum base area can as a rule be considerably enlarged for the sole purpose of providing a desirable combination of mounting screws and dowels. This is of particular advantage when punches must be mounted in the area covered by the punch-holder shank.

Proportions of pedestal punches. The overall dimensional requirements are the same as for plain punches. Refer to Fig. 6 • 6: the length L and the width W should be greater than, or at least equal to, the height H.

The flange thickness B and the fillet F must be large enough in proportion to the other dimensions of any given punch. There is no set rule for determining flange proportions, because they are influenced by a number of variable circumstances such as overall dimensions of the punch, size relationship between the cutting portion of a punch and the flanged portion, severity of the punching operation, stripper thickness, punch life considerations, and die space available. Although there is no rule for determining flange sizes, keep in mind that the flanges must be strong enough and that heavier work will require stronger flanges. Skimpy flange thicknesses are not ordinarily desirable and, in many cases, will be definitely inadequate.

Properly made fillets are important to the structural qualities of pedestal punches. It seems that one of the most common errors in the construction of pedestal punches is fillets that are too small and/or too rough. The purpose of a fillet is to blend the flanges into the cutting portion of a punch. If a fillet is too small, there is an abrupt transition instead of a blend. An abrupt change in profile will cause a stress concentration area at the junction of the flange and the rest of the punch. A concentration of machining stresses there will be followed by a further concentration due to heat treating. When the punch is in service, the cutting and stripping forces will also tend to be more concentrated at the flange junction instead of being evenly dispersed through the entire mass of the punch. An inadequate fillet is therefore a potential source of punch failure.

An adequate fillet will, practically speaking, eliminate undesirable machining and heat-treating

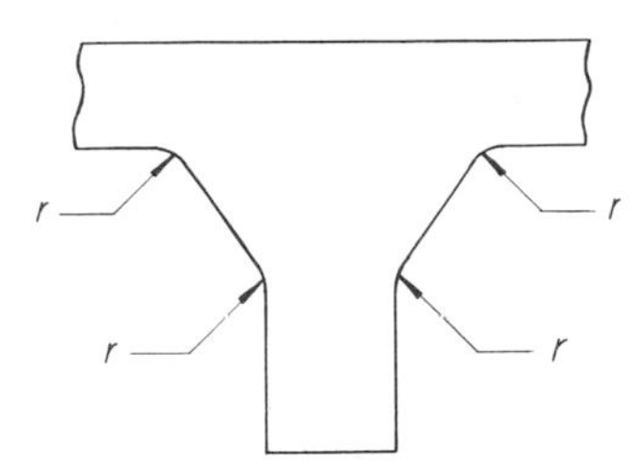

Figure 6·8 Angular fillet.

effects. In addition, when the punch is in service, the forces acting on it are more evenly distributed throughout the entire volume of the punch.

The surface of a fillet should be smooth. Any gouges, ridges, steps, etc., should be blended into a smooth-flowing surface. This does not mean that the fillet should be regarded as a fussy, time-consuming job. It simply means that fillets should be reasonably smooth and free of sharp irregularities, which are always a source of detrimental heat-treat strains and can, especially under severe service conditions, develop into fractures.

Usually, fillets are thought of as concave in shape, and generally the concave shape is structurally strong and desirable. However, in some cases, such as with relatively narrow punches, an angular fillet may be stronger and more desirable. Also, there will be instances where an angular fillet is more convenient to make. The ease with which a fillet can be machined will often decide which type is used. Caution! If an angular fillet is used, be certain that the flank of the fillet blends smoothly into the punch sidewalls and flanges. Do not allow a sharp intersection. A close inspection of the shape of an angular fillet will reveal that it consists of an angular flank blended into the other surface of the punch by smaller concave fillets, in the manner shown in Fig. 6·8.

Strength of pedestal punches. Strength is a basic quality of the pedestal configuration. If this type of punch is properly proportioned and finished, it will be inherently strong. The size and contour of the cutting profile are determined by piece-part requirements. Therefore, the cutting portion of the punch is restricted to this size and shape. The base area, however, can usually be made any practical size and shape. Thus, when the cutting portion of the punch is of necessity weak, it can be blended into the flanged base. If this is properly done, the weak section derives strength from the strong section, and the composite strength of the punch will always be good. The cutting profile of the punch shown in Fig. 6·9 is unfavorable as far as punch strength is concerned. However, blending the weak section into the base

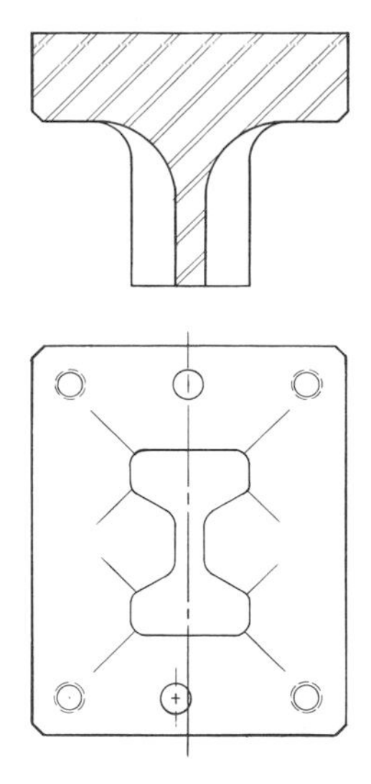

Figure 6·9 Weak area of punch blended into base for strength.

improves the condition considerably, producing a punch that is practical to heat-treat and that has very good strength characteristics.

The difference in mass between the cutting portion and the base of a punch can be so large that special consideration must be given to blending out the disparity. This is a frequent occurrence with small work and with long, narrow punches. For these applications, it is good practice to pyramid or double-step the punch, as shown in Fig. 6·10. These configurations improve the cross-sectional proportions of the punch, making it practical for heat treating as well as giving it strength and rigidity.

Offset pedestal. Pedestal punches are often made with the base offset, similar to that in Fig. 6·11. There are two reasons for offsetting the base of a punch:

1. Space Consideration. It is, on occasion, necessary to eliminate the flanges from one or more sides of a punch in order to provide space for other components in the punch assembly.

2. Machining and Grinding Accessibility. At times, the required cutting contour may be such that it becomes desirable to eliminate part of the flange area in order to permit straight-through

Figure 6·10 Pyramiding of pedestal punches.

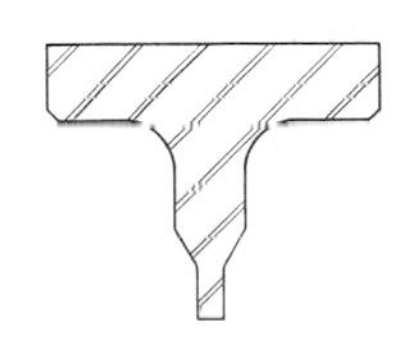

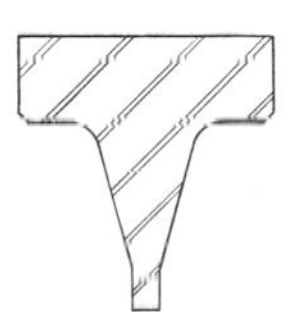

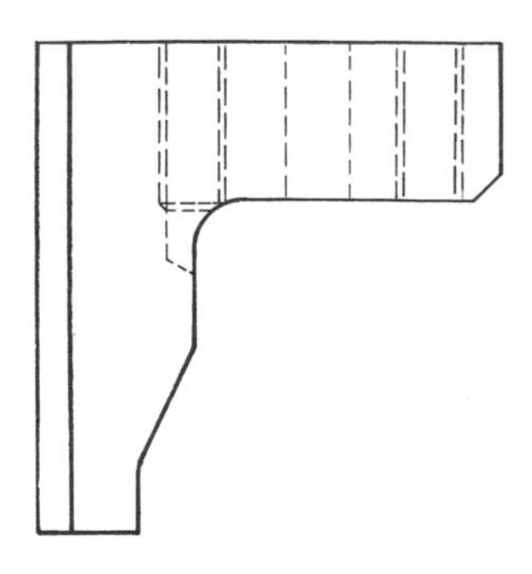

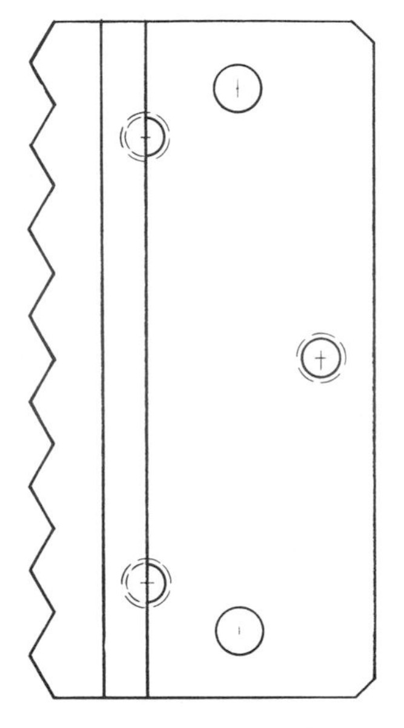

Figure 6·11 Offset pedestal punch.

grinding and/or machining. The punch illustrated is an example of this.

Offset bases and heavy work. When the base is set off to one side, the load distribution is unbalanced and the cutting force is not dispersed equally throughout the base area. Instead, a large percentage of the force is concentrated at the mounting surface directly above the cutting portion of the punch. As a result, heavy work may cause this side of the punch base to set into the mating punch-holder surface. If this occurs, the punch will be tilted out of perpendicular. The resulting misalignment can cause shearing and possible damage to the punch and its mating die opening. Such setting in and resultant tilting caused by heavy, unbalanced punching forces can be prevented by providing hardened thrust plates to back up punches of this type.

Warping tendency of offset punches. Because of their one-sided configuration, these punches are quite susceptible to hardening warpage. An example of this kind of warping is shown, exaggerated for clarity, in Fig. 6·12.

The typical cross-sectional shape of an offset punch is similar to the letter L. It seems that a large percentage of L-shaped pieces will tend to warp open, as illustrated, and that a somewhat smaller percentage will tend to close in, or warp in the opposite direction. The amount of warpage is also a variable condition. If an offset punch has a relatively slender cross-sectional shape, it will have a pronounced tendency to warp. Conversely, a punch whose cross section is comparatively massive may warp only a slight amount. Some L-shaped components, especially die sections or punches which have L-shaped cutting contours, may require special corrective treatment by the heat-treater if they have warped excessively. This is, of course, also true for U shapes and other unfavorably irregular contours. However, this discussion is concerned exclusively with the warpage which occurs between the base surface and the sidewalls of a punch.

In the large majority of instances, it is practical to correct this kind of warpage by grinding the base surface to restore its perpendicularity with the punch sidewalls. Figure 6·12 illustrates, in principle, a setup for such surface grinding. One of the sidewalls is clamped against a precision angle plate. This assures that the base surface will be ground at right angles to this sidewall. The punch must also be squared up in the other direction when it is clamped to the angle plate, in order that the entire plane of the base surface will be ground perpendicular to the cutting axis. Usually this can be accomplished by placing the punch on a parallel as shown, checking, of course, to be sure that the punch rests evenly on the parallel. However, it is a good idea to check the

Figure 6·12 Setup for restoring perpendicularity (warpage exaggerated).

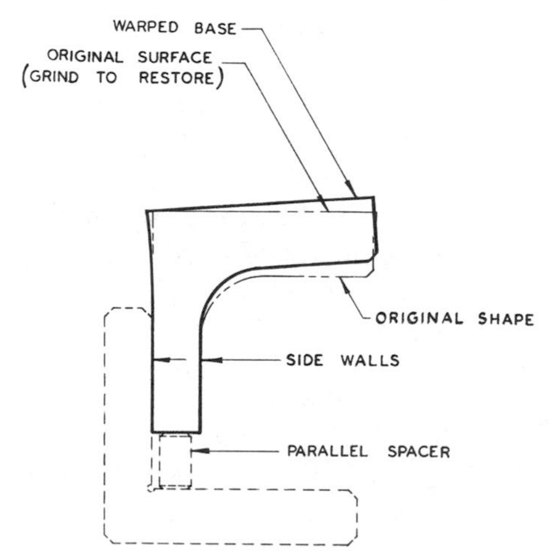

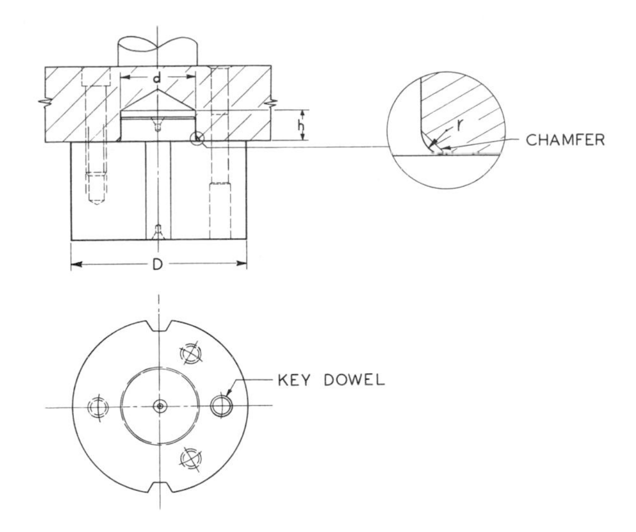

Figure **6·13** Punch with integral positioning boss.

setup, using a diemaker's square. For smaller punches a toolmaker's vise may be used instead of an angle plate, etc. In fact, any variation of the illustrated setup is acceptable, provided only that the desired result is attained in a safe and efficient manner.

Perpendicularity of the dowel holes is affected by warpage. This is usually insignificant, at least for average work. However, as a possible alleviation for this condition, the dowel holes should be lapped after the base surface has been restored to perpendicular. Lapping can then be done with the punch resting on its base. Since the base is now perpendicular, the lap will *tend* to improve perpendicularity of the dowel holes. Keep in mind, though, that this procedure is not considered adequate where absolutely perpendicular dowel holes are required. For such cases it is necessary to grind the dowel holes. Here, too, the holes should be finished after the base surface has been restored, since this will in most cases facilitate setting up the punch for perpendicular grinding.

BOSSED PUNCH

Punches are sometimes made with an integral positioning boss similar to that in Fig. 6·13. Boss diameter *d* should be made a standard nominal dimension. Do not make *d* too small; generally, it should be the largest diameter practical for the given set of circumstances. Height *h* should be restricted. As a rule, a good proportion is $h = d/2$, approximately. Heights *h* of less than ¼ in. are not generally practical. Avoid undercuts at the junction of the boss with the base surface. Instead, make a small fillet *r* and countersink the mating surface to clear the fillet.

Ideally, boss diameter *d* should be a wring fit to its mounting hole. A small lead should be provided on the boss. A lead diameter equal to $d - 0.0005$ to 0.001 in. is generally satisfactory. A lead length of more than 3⁄16 in. should not be necessary, and less than 1⁄16 in. is not practical.

If the work contour is irregular, the punch must be keyed to prevent rotation. In most cases, radial orientation may be secured with a dowel, as shown.

If the positioning boss were not present, this punch would be a plain punch. In an associative manner, it can be described as a plain punch complicated by the incorporation of an integral positioning boss. In principle, these punches are mechanically sound. If properly made, they are quite strong and stable. However, for many applications they are not as simple to make or to mount as some other punch types. This accounts for their rather restricted application to situations for which they are peculiarly suitable.

FLANGED PUNCHES

Figure 6·14 depicts a flanged punch. It is essentially a pedestal punch which incorporates an integral positioning boss. The extended base area may be necessary to provide space for attachment screws, or it may be required for stability. In most cases, both considerations probably influence the choice of this configuration. Except for the flange, this punch is the same as the preceding punch (Fig. 6·13).

Another flanged-punch version is pictured in Fig. 6·15. Here, the boss is proportionately much larger. It no longer acts solely as a means of positioning. It has become a shank which is used to retain the punch in addition to positioning it.

Figure **6·14** Flanged punch.

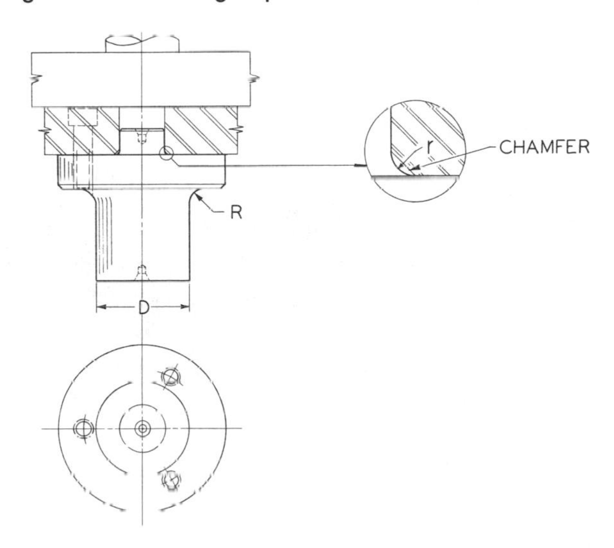

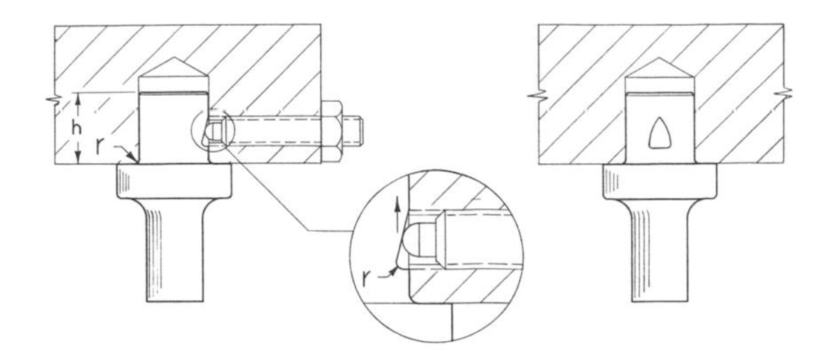

Figure **6·15** Flanged punch with whistle notch.

Height *h* is of necessity higher, to provide space for the whistle notch. Whistle notches are made as indicated (inset). Oval-point or modified cone-point setscrews are generally used in conjunction with the notch. The screw pressure acting against the notch slope tends to force the punch in the direction shown, thereby seating the punch base. Side pressure exerted by the screw may cause a slight lateral displacement of the punch, but actual displacement is very slight and can be disregarded for ordinary work. However, the possibility should be kept in mind, as it can be objectionable where closely fitted punches are required.

HEELED PUNCHES

A typical heeled-punch condition is illustrated in Fig. 6·16. Specifically, the illustrated punch is a notching punch. However, principles relating to the heel function will be much the same for other punches as well.

Here, the heel is fitted in a manner commonly used for progressive die applications. The nature of the notching operation is such that cutting force at the front of the punch is unopposed and thus tends to displace the punch away from the front edge. Also, since strip stock is fed through the die, either the punch may be required to cut partial notches at the lead or tail ends of the strip, or partial notches may occur inadvertently as a result of misfeeding. Partial notching will tend to displace the punch in a direction parallel to the feed direction.

Figure **6·16** Fitted punch heel.

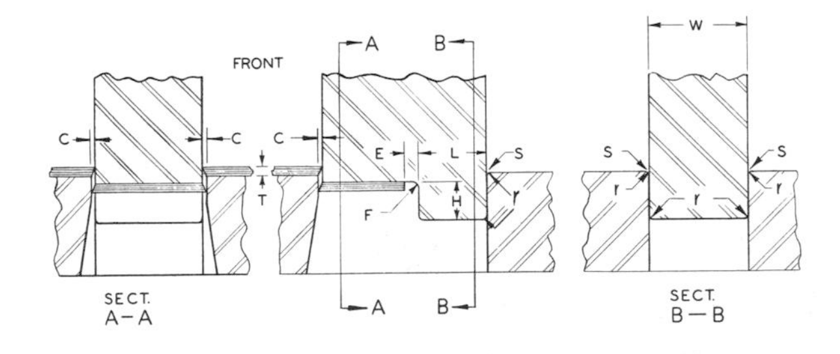

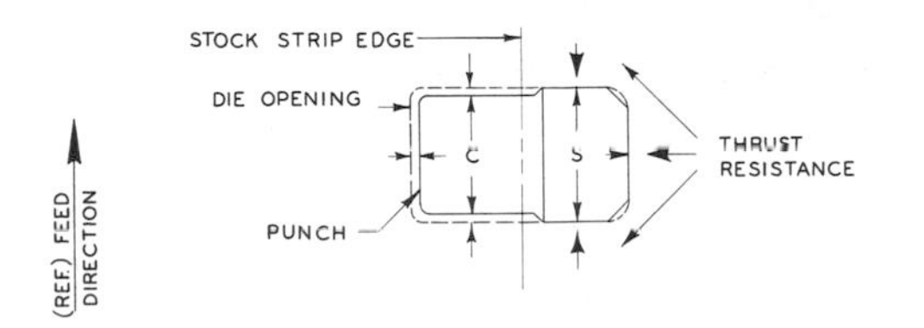

The purpose of a heel is to support the punch by resisting lateral displacement. This type of heel is an integral boss extending beyond the working face of the punch. The heeled portion is made a sliding fit in the die opening on three sides, as indicated at *S* in the illustration. Therefore, the heel affords lateral thrust resistance along any vector included within the three directions shown.

Heel proportions. Width of the heel *W* is made a sliding fit to the die opening. Its actual size depends upon the required die-opening width. Length *L* is made long enough to provide ample bearing surface along the sides, which will in most cases provide an adequate H/L ratio for heel strength. A fillet radius *F* should be incorporated at the junction of the front edge of the heel and the working face of the punch. For small, light work a fillet of $\frac{1}{32}$ in. radius will do. This should be increased in proportion to heel size and severity of work performed. Clearance *E* should be larger than *F* by an amount which will at least safely accommodate the widest anticipated stock strip.

Lead radii *r* should be provided on the heel edges and on the matching die-opening edges. File and polish or stone these radii, whichever is appropriate. Proceed with care to avoid overrunning the cutting edges. Do not make lead radii too large, as this will decrease the heel-bearing area. Lead radii *r* of $\frac{1}{32}$ in. should be sufficient for practically all occasions. Height *H* must be high enough to secure ample bearing before the work begins. The following minimum heights are suggested:

For stock thickness *T* less than $\frac{1}{16}$ in.

$$H = 2r + T + \tfrac{1}{16} \text{ in.}$$

For *T* equal to $\frac{1}{16}$ to $\frac{1}{8}$ in.

$$H = 2r + T + \tfrac{3}{16} \text{ in.}$$

For *T* equal to $\frac{1}{8}$ to $\frac{1}{4}$ in.

$$H = 2r + T + \tfrac{5}{16} \text{ in.}$$

Cutting clearance *C* must, of course, be made as required for the type and thickness of the stock material. Cutting clearance and die taper extend only around the required cutting periphery (plus a small safety margin). Die-opening walls are made straight (no taper clearance) at the heeled periphery. Quality of surface finish on

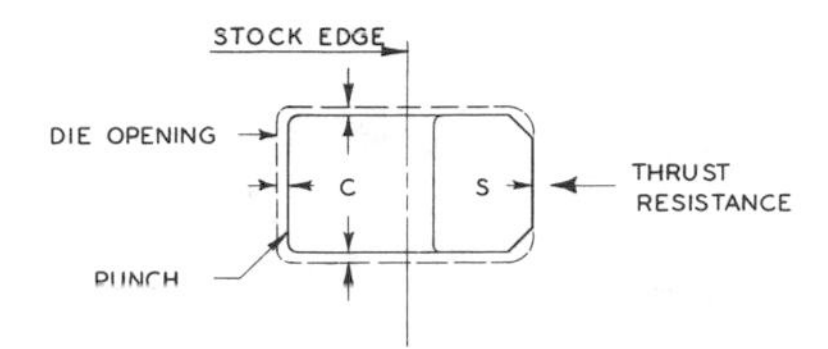

Figure 6·17 Semifitted punch heel.

both punch and die sidewalls, at the heeled periphery as well as the working periphery, should not be neglected. The heeled area of the punch is always in sliding contact with the die walls and may be subject to high intermittent lateral thrust forces, as well. Therefore, the contacting surfaces must be well finished to reduce friction with its attendant potential galling and seizing proclivities.

For applications where partial notching is not anticipated (for example, with unit stock or secondary operations), punch construction can be simplified in the manner indicated in Fig. 6·17. The heeled punch surface is in sliding contact with the die at the back surface only. This provides unidirectional thrust resistance to lateral displacement resulting from work forces at the front of the punch.

Die openings fitted to punch heels. It is assumed in Fig. 6·16 that it is easier to make the punch contour to suit the die opening. While true in many instances, this is obviously not true for all. There are at least as many cases where it is more logical to make the die-opening contour to suit the punch. This is especially true for sectioned die-block construction. Two rather typical notch die openings are pictured in Fig. 6·18.

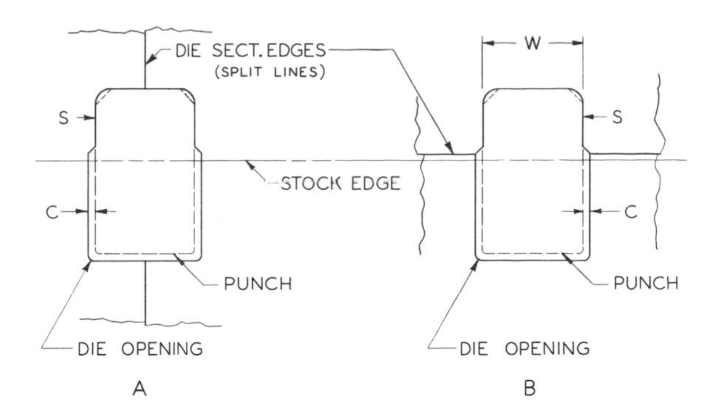

Figure 6·18 Die openings fitted to punch heels.

View A: Vertical Split Line. Die opening is ground offset an amount equal to the required cutting clearance *C*, to produce sliding fit to the heel *S*.

View B: Horizontal Split Line. Work section is made to required die size, including cutting clearance *C*. Heel section dimension *W* is made the required amount smaller (in this case, an amount equal to 2*C*), thus providing a sliding fit *S* for the punch heel.

Precisely speaking, a heeled punch is not a type of punch. The kind of heel described above may be applied to many different types of punches belonging to any of the groups and categories mentioned earlier. For example, heels may be necessary on cutting punches, forming punches, and/or hybrid punches. They may be incorporated in segregated punches or integrated punches, etc.

PUNCHES MOUNTED IN PUNCH PLATES

A punch plate is so called because its function is to retain and/or position a punch or punches. Punches assembled into punch plates are, for many applications, the most practical method of punch mounting. This is especially true for smaller punches and for punches whose proportions are such that they will benefit from the support provided by a punch plate. Punches assembled in this manner are classified as members of the integrated group, as defined in the previous chapter.

A number of different methods are commonly used to assemble punches into their respective punch plates. The choice of methods varies in accordance with individual requirements dictated by each job.

HEADLESS PUNCHES

One method of assembling a punch and punch-plate combination is pictured in Fig. 7·1. Considered alone (without the punch plate), this is essentially a plain punch except for the fact that it does not contain dowels. It does, however, contain mounting screws, which are used to fasten the punch within the confines of the opening provided in the punch plate. With this type of construction, the punch plate is not used to hold the punch in assembly but only to provide an effective means for accurately positioning and supporting the punch. To ensure accuracy and stability, the punch must fit snugly in the punch-plate opening. The punch is then confined laterally by the punch plate and thus prevented from shifting or being displaced sideways.

Figure 7·1 Headless punch.

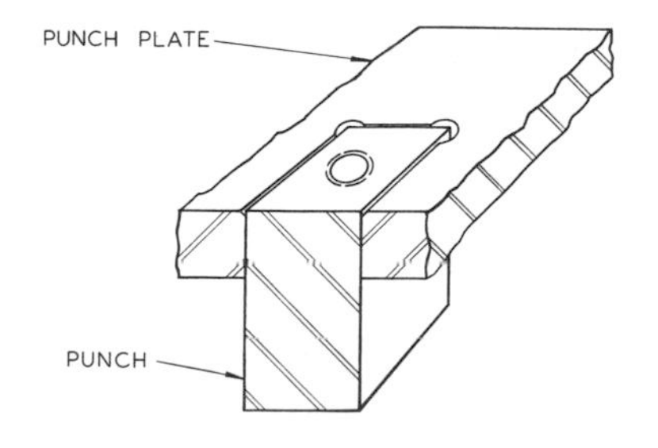

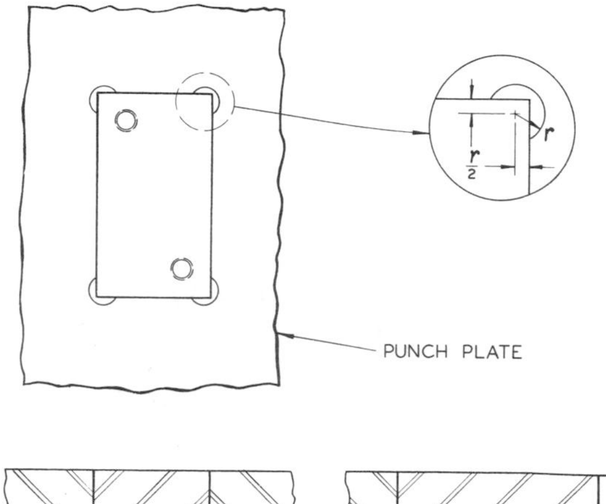

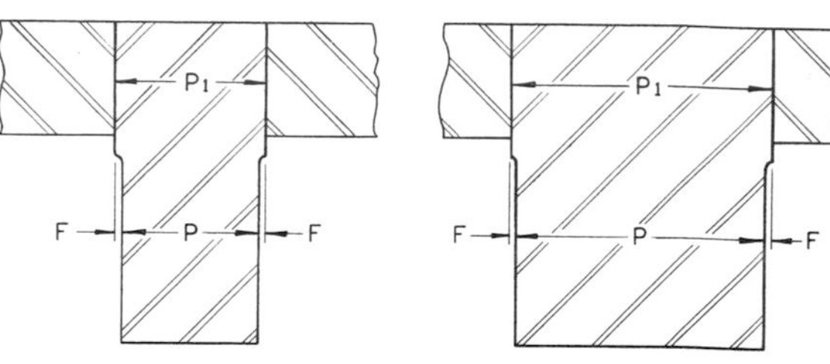

Figure 7·2 Punch assembly fitting using principle of lead provided on punch.

The punch in the illustration is typical of straight-through punches: Its shape and size are predetermined by the shape and size requirements of the work it is intended to produce. Therefore, the punch is not made to suit the punch-plate opening but the punch-plate opening to suit the punch. One method of fitting this type of punch and punch-plate assembly is pictured in Fig. 7·2.

Fitting for assembly. Fitting of the punch-plate opening can often be facilitated with the procedure indicated in Fig. 7·2. Here, a lead effect is incorporated in the punch. The punch is made as shown, with $P_1 = P + 2F$, where P is the required punch dimension and F is the fit allowance per side. The opening in the punch plate is made a close push fit for dimension P. Dimension P_1 then provides an interference fit when the punch is pressed home. Heavy press fits are not required and should not be used. As a guide to fitting, the following fit allowances are generally satisfactory for average conditions (it is assumed here that the punch-plate opening is exactly equal to P):

where $P = 0$ to 1 in.	$F = 0.0002$ to 0.0005 in.
$P = 1$ to 2 in.	$F = 0.0004$ to 0.0010 in.
$P = 2$ to 4 in.	$F = 0.0008$ to 0.0012 in.
$P = 4$ to 6 in.	$F = 0.0010$ to 0.0020 in.
$P = 6$ to 10 in.	$F = 0.0015$ to 0.0030 in.
$P = 10$ to 16 in.	$F = 0.0020$ to 0.0040 in.

These fit allowances are intended for soft punch plates. However, they may be applied to punch plates which are heat-treated to Rockwell C 48 or less. For punch plates harder than Rockwell C 48, the fit allowances should be only 10 to 20 per cent of the allowances indicated above, with the smaller percentage applied to the larger size. The foregoing procedure and allowances may be applied to round punches also. For round punches, P is the required work diameter. Diameter $P_1 = P + 2F$, for pressing into the punch plate.

NOTE:

To preclude the possibility of scoring and attendant troubles, a lubricant must be employed when assembling interference fits. Lubricant should be applied to both the punch and the punch opening.

Clearances facilitate punch assembly. To avoid needless and time-consuming fitting at corners, drill clearance holes as shown in the plan view. The holes are, of course, drilled before sawing the opening. The most generally practical center location for corner clearance holes is indicated in the enlarged inset view of Fig. 7·2.

When providing corner clearance holes for narrow punches, avoid the condition shown in Fig. 7·3, view *A*. In these circumstances, corner clearance holes should be provided in the manner depicted in view *B*.

Needless fitting should be avoided. Wherever practical, punch-plate openings should be cleared to eliminate exacting fitting to the punch. This is exemplified at *C* in Fig. 7·4, views *A* and *B*. It is assumed, of course, that clearing will not be overdone to the point where punch stability will be endangered because of inadequate bearing surface between the punch and punch plate.

Assembling pressed-in punches. If feasible, pressed-in punches should be provided with a lead (see Fig. 7·2). The lead will simplify as-

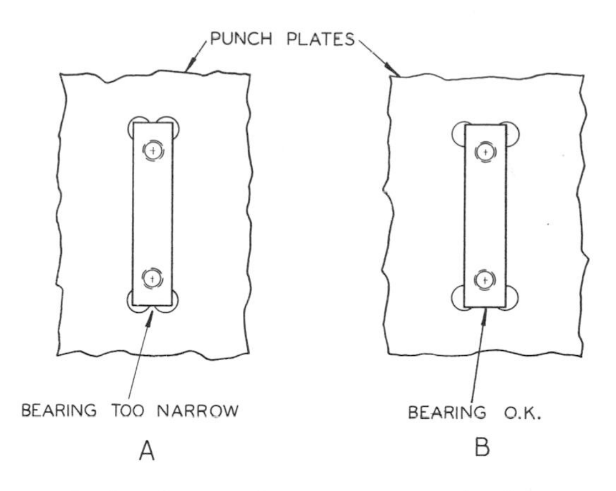

Figure 7·3 Corner clearance in punch plates.

sembly of the punch into the punch plate, and a minimum of care will be required to assure that the punch is perpendicular to the plane of the punch plate. However, for various reasons (most often because of the punch contour), it is not always practical to provide lead on the punch. Such an instance is pictured in Fig. 7·5.

This illustration also includes two small but important points which apply to pressing punches into punch plates:

1. Provide a chamfer around the edge of the punch-plate opening on the punch entry side. In most cases, this will be on the back (mounting) surface of the punch plate, as shown. Chamfer size will generally range from 1/64 to 1/16 in. in proportion to punch-plate thickness. Chamfer should be consistent on all edges in order to facilitate a perpendicular start for pressing the punch into the punch plate.
2. Do not attempt to press a sharp punch into a punch plate. Use a hand stone to dull the cutting edges slightly in order to preclude the possibility of shaving or gouging the punch-

Figure 7·4 Clearances in punch-plate openings to expedite fitting.

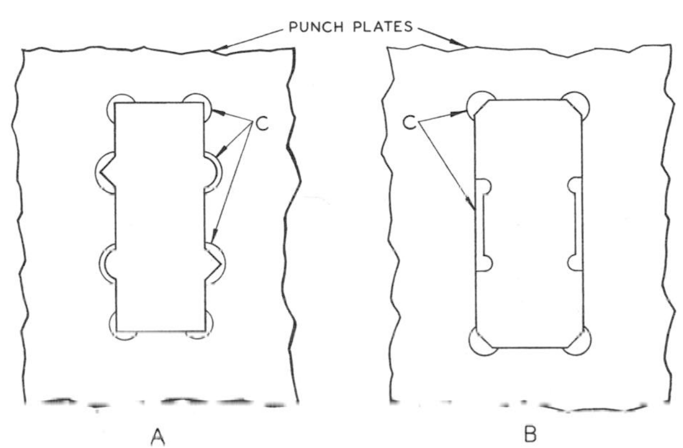

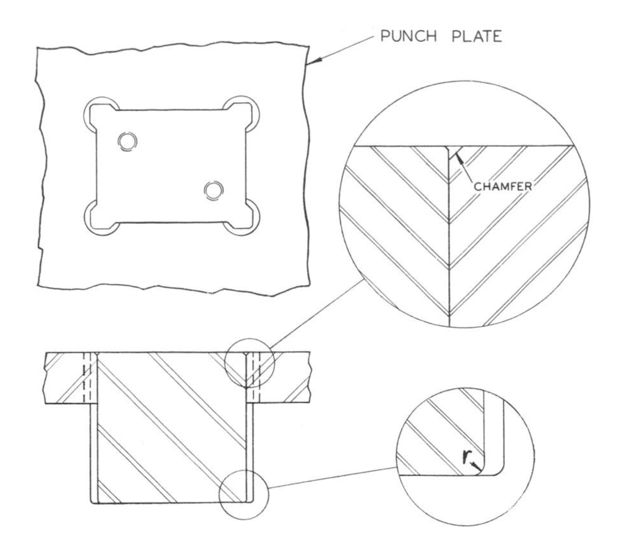

Figure 7·5 Fitting punch plate to punch (no lead provided on punch).

plate opening. A very slight radius r of 0.002 to 0.005 in. will do quite nicely for most situations. The punch is then sharpened at the most convenient time after assembly.

These procedures are not restricted solely to assembly purposes. They apply also to situations where the punch is used as a broach to shear the opening into the punch plate. The chamfer is an aid to starting the broaching sequence. Slightly dull cutting edges r will tend to prevent the punch from broaching the opening larger than would be desirable for the interference fit. Remember that this practice applies to assembling and/or broaching any punch which is pressed into a punch plate or into any other die component which may be serving as a punch plate.

STEP-HEAD PUNCHES

In addition to being positioned by punch plates, punches of this general type are retained by punch plates. They are assembled as shown (Fig. 7·6), pressed into a suitable punch-plate opening. Again, heavy forced fits are not necessary and in most cases will be objectionable. Size differential

Figure 7·6 Step-head punch.

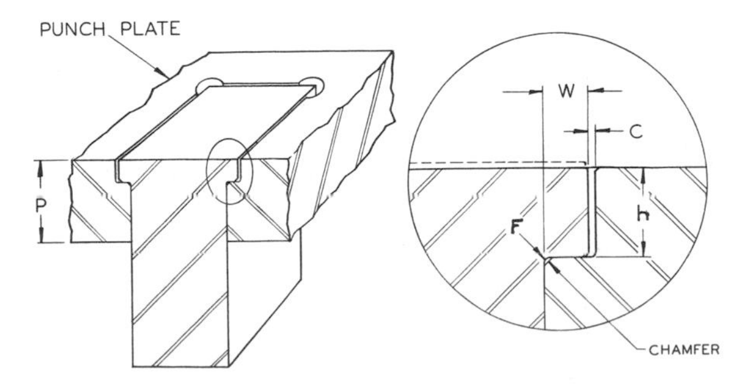

between punch-plate opening and punch should normally provide an interference approximating that recommended for light to medium drive fits. When determining fit allowance, keep in mind that the allowances given earlier for headless punches are generally suitable for any punch which is pressed into a punch plate.

A fillet F should be provided at the junction of the head and sidewalls, so do not use a sharp corner there. The fillet may be made very small if necessary, but do not eliminate it entirely, since it is important to punch strength. Use reasonable care in making the chamfer: It must be large enough to avoid interference with the fillet but no larger than necessary for this purpose.

The chamfer also facilitates pressing the punch into the punch-plate opening. Remember to dull the cut edges slightly before beginning this operation. To eliminate unnecessary finishing and fitting, clearance C should be provided between the punch head and the offset part of the punch-plate opening. Be sure of the clearance, because interference in this area can force the punch out of perpendicular. Head height requirements h depend upon the severity of the work, especially the severity of the stripping effort. For these punches the punch-plate thickness P will have to be greater, generally by an amount equal to h, than for headless punches.

In many instances it is a good idea to allow grinding stock on the h dimension, as indicated by the dotted line in the inset. This surface is then ground flush with the mounting surface of the punch plate in assembly. Caution! When following this practice, press the punch home carefully to avoid the possibility of breaking it and/or distorting the punch plate. Much punch breakage both in assembling and in service can be eliminated by tempering the punch head (headed portion only) down to Rockwell C 48 or lower.

Step width W should not be larger than h and preferably should be smaller. Normally, a well-proportioned head for this type of punch will have a step width W of $\frac{1}{2}h$ to $\frac{3}{4}h$. Actual head dimensions for average work will usually require heights h of $\frac{1}{8}$ to $\frac{3}{16}$ in. and widths W of $\frac{1}{16}$ to $\frac{1}{8}$ in. in proportion. As a rule, it is pointless to make the heads smaller than this, even though the work is light. On occasion for very heavy work, stronger punch heads and thicker punch plates will be required.

BEVEL-HEAD PUNCHES

If the punching contour is considered practical for such a procedure, punches can be machined and ground to the bevel-head configuration in a manner similar to that used for step-head punches.

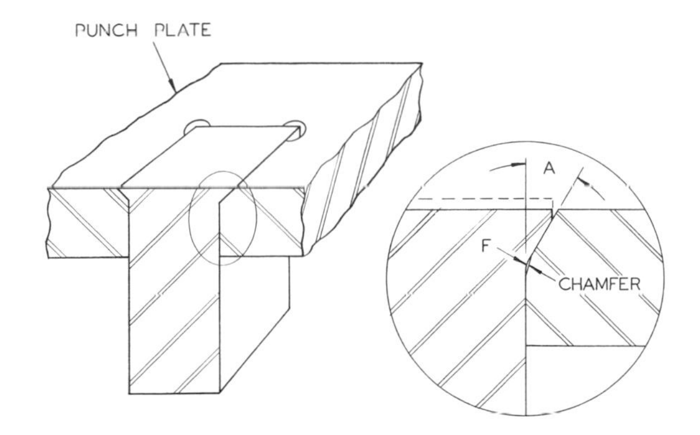

Figure 7·7 Bevel-head punch.

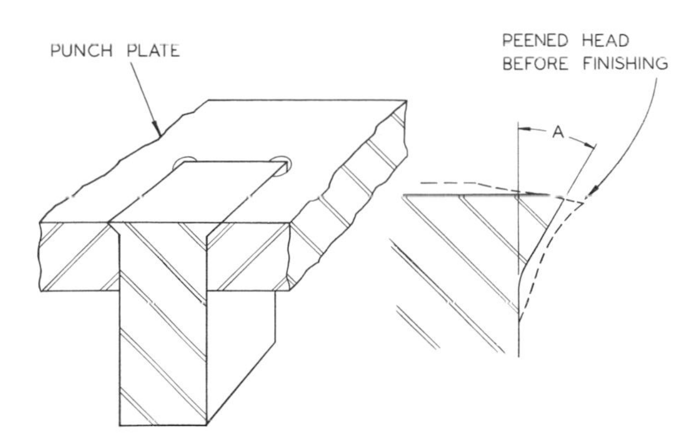

Figure 7·8 Peened bevel-head punch.

In trade parlance this is called making the head from the solid. Procedures will then be much the same for both types, the only difference being in the shape of the head (see Fig. 7·7).

Bevel-head punches require more exactitude in fitting the head and the punch plate. A nice matched fit is necessary along the plane of the bevel angle *A*. Generally, bevel angles may be made any angle from 30 to 45°, depending upon what is most convenient for the diemaker. For exceptional conditions, however, keep in mind that an angle of 30° can give stronger head proportions than a greater angle. As with any angular plane junction, a fillet *F* is desirable. It may be necessary to make the fillet quite small, but remember that a fillet of only 0.005 in. will be considerably better than a sharp corner. A chamfer is employed to clear the fillet. Here, too, the chamfer must be minimized in order to provide maximum support for the punch head. Grinding stock should be provided on the back (mounting surface) of the punch as indicated by the dotted lines, in order to permit grinding this surface flush in assembly.

Peened-head punches. As pictured in Fig. 7·8, a peened-head punch is a bevel-head punch. It is specifically called a peened-head punch to describe the method of producing the head on the punch. As opposed to other headed punches, a peened-head punch offers the advantage of permitting straight-through machining of its contour, since the head is peened on after the punch contour has been finished. This is the sole advantage of peened-head punches.

PROCEDURE:

1. Finish the punch to its required contour and size.
2. Displace, by peening, the material necessary to form the head. A bevel angle *A* of 30° is predominantly the optimum angle for punches made this way. Caution! Before starting to peen, double-check to be certain that you are about to peen the proper end of the punch. After peening, the head will have the general appearance indicated by the dotted lines in the enlarged view.
3. Finish the head to the desired bevel angle, blending smoothly into the punch sidewalls. Check for any bulges or irregularities which may affect perpendicularity or alignment.
4. Heat-treat. Head portion should not be harder than Rockwell C 48.
5. Grind back surface square with punching axis. Leave grinding stock for grinding flush with punch-plate assembly. If the punch is a cutting punch, dull the cut edges as described earlier.
6. Apply lubricant and press punch home in punch plate.
7. Check for perpendicularity. If punch is slightly out of perpendicular, apply localized pressure at the appropriate spot on the back surface of the punch (see Fig. 7·9).
8. Grind back of punch flush with punch-plate mounting surface.

MULTIPLE PRESSED-IN PUNCHES

Whenever possible, avoid having more than one pressed-in punch of irregular contour mounted in a single punch plate. Round punches and/or punches with round shanks are excepted

Figure 7·9 Squaring up.

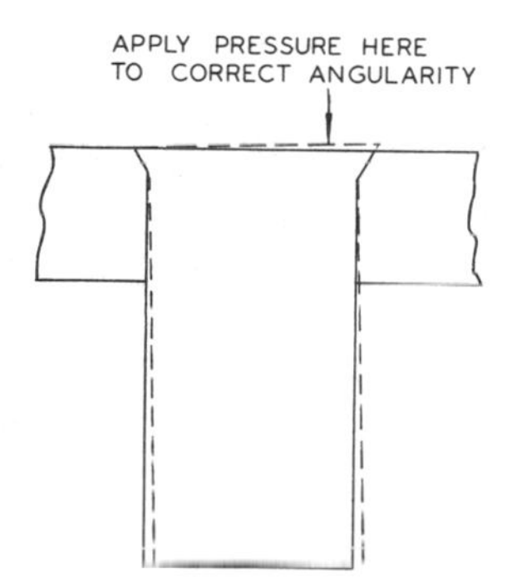

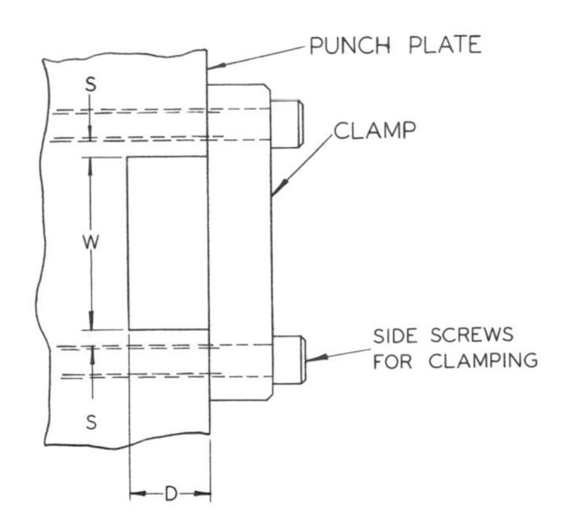

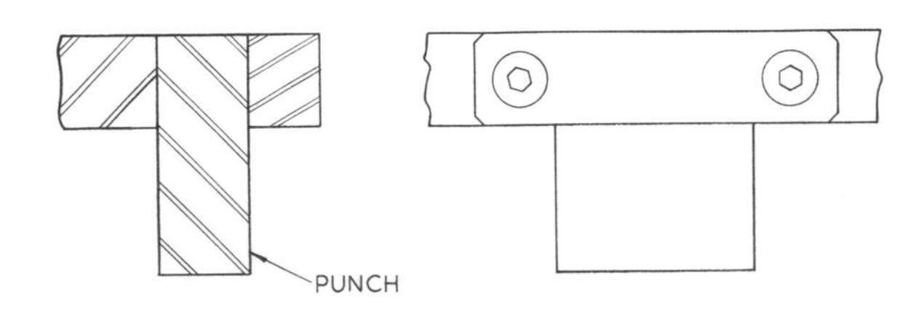

Figure **7 · 10** Clamped punch (plain).

from this recommendation, since it is a simple matter to mount groups of round punches in the same punch plate.

CLAMPED PUNCHES

A simple clamped-punch assembly is illustrated in Fig. 7 · 10. A suitable open slot is machined in the punch plate to receive the punch. The punch is held in the opening by pressure from the side screws acting on the clamp. For most assemblies of this kind, slot width *W* should provide a tap fit or a light drive fit for the punch. Slot depth *D* should normally be 0.001 to 0.002 in. less than the punch dimension to assure clamping pressure. To be effective, the clamping screws must be relatively close to the punch. For the average application, spacing *S* between the punch and the clamp screw should be 3/32 to 5/32 in. One method of determining the clamp-screw locations is to make dimension *S* approximately 1/3 the nominal diameter of the clamp screw.

Larger punches will require one or more additional clamping screws. The additional screws are installed through the punch as pictured in Fig. 7 · 11. Note that the clamp-screw heads are shown counterbored into the clamp. This is done only when necessary (usually because of mounting space considerations).

Clamp-type punch assemblies are sometimes employed in order to make it possible to remove the punch while the assembled die is mounted in the press. In such cases, there must be either sufficient space to permit complete withdrawal of clamping screws which extend through the punch, or some other provision which permits punch removal, such as the slot in the illustrated punch. The slot should not be provided unless it is needed. Note: If the punch is designed to be removable, as described, it should not fit too tightly in the punch plate. A snug wring fit should be used for such applications.

For situations where it is not deemed advisable to depend only on side clamping screws for punch security, some other means of retaining the punch may be used in addition to the clamp. Some possibilities are depicted in Fig. 7 · 12.

View A. Tapped holes (one or more, as required) are provided in the punch to receive the mounting screws. The mounting area behind the punch must permit proper installation of these screws.

View B. A step head is provided on the punch.

View C. The clamp is made so as to key the punch to the punch plate. To preclude needless fitting for the key, slight clearances should be provided where shown at *C*. Clearances approximating 0.005 to 0.010 in. are normally appropriate for such applications.

FLOATING PUNCHES

Floating punches are not positioned exactly by the punch plate. They are guided in the stripper, which gives them their exact positioning. In most cases they require the use of a spring stripper, since such punches should not be completely withdrawn from the stripper during the punching cycle.

Figure **7 · 11** Clamp screw through punch.

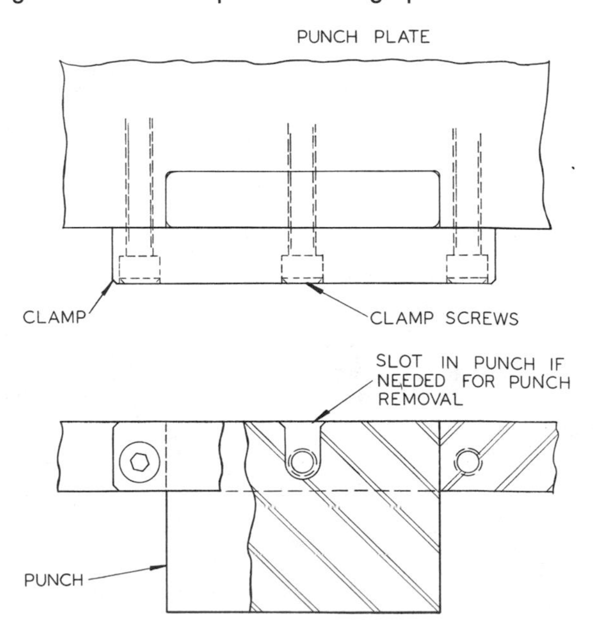

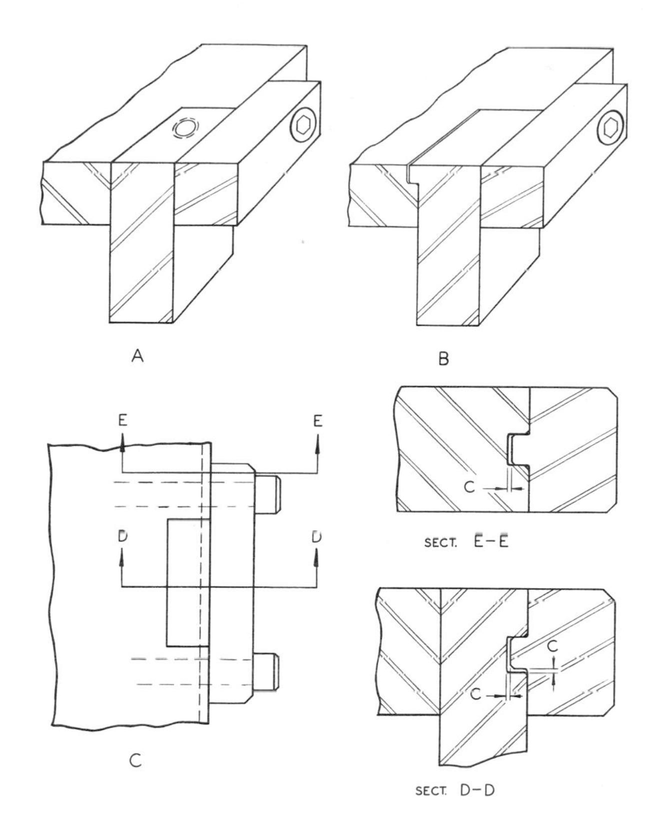

Figure 7·12 Yoked punches.

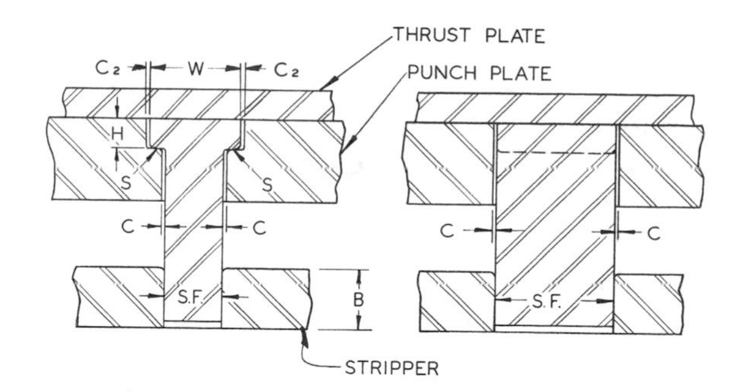

Figure 7·13 Floating punch for blanking or piercing.

One version of a floating punch is illustrated in Fig. 7·13. The punch-plate opening is slightly larger than the punch, permitting the punch to float laterally in any direction. Vertical float, however, is not desirable; head height *H* should be a nice sliding fit allowing lateral movement only. Keep in mind, of course, that there should be no binding. The back of the punch should be smoothly finished. A hardened thrust plate, smoothly finished also, should be provided behind the punch. It is a good idea to lap these surfaces lightly to the degree necessary to provide an adequate low-friction bearing surface. Surfaces *S* must also be nicely finished to facilitate lateral float.

Head width *W* must be wide enough to supply adequate bearing surface at *S*. Clearance *C* will most commonly range between 0.002 and 0.010 in., depending on the specific application. In some cases more clearance may be permissible, but it should not normally be necessary. One procedure applicable to many situations of an average nature is to make *C* approximately equal to the cutting clearance. Extra clearance should be provided in the punch plate for the punch head. To prevent interference, dimension C_2 must be larger than *C*. Adequate stripper guidance and support is vital for punches of this type. Stripper thickness *B* must provide ample bearing for the punch.

PYRAMID-TYPE PUNCH PLATE

Although punch plates are commonly thought of as flat plates, this is not always the case. By way of an example, see Fig. 7·14. In this instance, the punch plate has been contoured to achieve a pyramid effect in its third dimension. Specifically, this is a clamp-type punch plate which provides extra-deep support for the punch. The drawing is self-explanatory. The unit is subject to the conditions and procedures described earlier for other clamp-type punch and punch-plate assemblies.

MATERIALS FOR PUNCH PLATES

Material requirements for punch plates vary to suit the specific occasion. Some common factors affecting the choice of material are:

1. Production requirements. These are usually tied in with die cost and quality.

Figure 7·14 Pyramid-type punch plate.

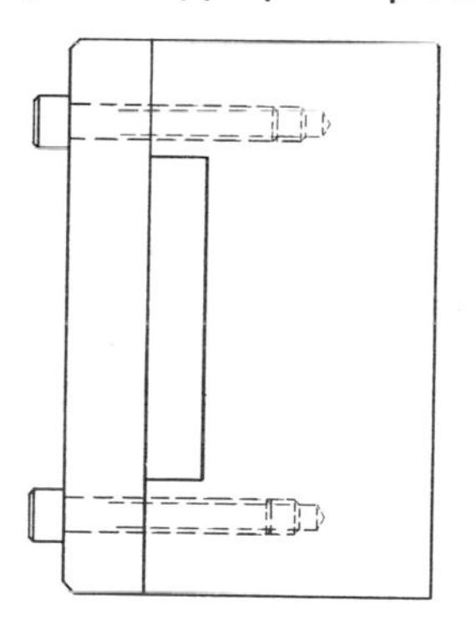

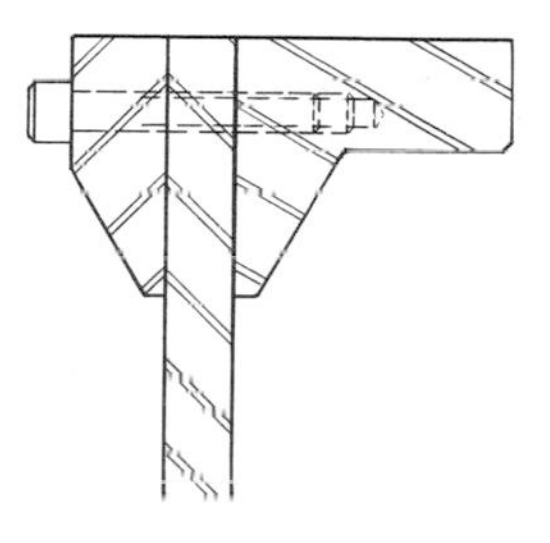

2. Complexity. The more complex the punch plate, the more stringent the material requirement.
3. Severity of the application. Severe work requirements will necessitate more rugged punch plates made of better material.
4. Precision requirements. These strongly influence quality of all die components, including punch plates.
5. Maintenance considerations. Fragile punches, punches subject to unusual wear, interchangeable punches, etc., can influence the choice of material for a punch plate.

Cold-drawn (Cold-rolled) Steel. Cold-drawn steel is often adequate for simple punch plates, especially those used to retain round perforators. This material is considered desirable from the cost standpoint and is, therefore, used rather extensively for punch plates on relatively low-cost dies. Cold-rolled is notoriously subject to warpage from machining. If a punch plate requires the machining of key slots or is in some other way more complex, it will be better not to use cold-rolled steel. Cold-rolled can, of course, be stress-relieved before machining, but this tends to cancel the original cost advantage.

Machine Steel. Where warpage is apt to be a factor, machine steel (hot-rolled mild steel) is preferable to cold-rolled steel as a material for making punch plates. The use of machine steel commercially available as ground stock should be considered.

Tool Steel. Tool steel is often employed for punch plates where greater strength or higher quality is desired. If the plate is to be left soft, either water-hardening or oil-hardening tool steel will be an appropriate choice of material. The prime consideration for soft tool-steel punch plates is machinability: Select a steel that is known to be free-machining. For punch plates which are to be hardened, the material should be chosen for its heat-treating stability, and consequently a high-quality steel, either oil-hardening or air-hardening, is a logical choice. Degree of hardness suitable for most punch-plate applications should be within the range Rockwell C 47 to 56.

PUNCH-PLATE THICKNESS REQUIREMENTS

There is no fixed rule for determining optimum punch-plate thickness, which varies according to the specific application. The following factors influence thickness requirements:

1. Punch height: Punch-plate thickness should be in direct proportion to overall punch height. Higher punches will require thicker punch plates.
2. Punch strength, which is derived from general punch configuration. This is a relative situation often influenced by factors other than the punch to punch-plate relationship. Generally, punches with lesser strength proportions will require relatively more punch-plate support.
3. Severity of work performed, which also is a relative consideration associated with punch strength. Normally, heavier work requires heavier punch plates.
4. Unbalanced work: Where unequal lateral forces are induced or applied, the punch will require more support from the punch plate.
5. Punches guided or supported by other components require less support from the punch plate.
6. Step-head punches commonly require heavier punch plates than equivalent headless punches.
7. Clamp-type punch-plate assemblies will normally be thicker than equivalent "solid" punch plates.

Probably the majority of punch plates for average conditions will have a plate thickness P which is approximately equal to $H/3$. This proportion is not a fixed rule; it is simply the result of examining many existing dies. It is mentioned here as a guide to punch-plate thickness proportions which may often apply in the absence of other criteria (H = overall punch height).

PERFORATOR-TYPE PUNCHES

A perforator-type punch can be described as a cutting punch one inch or less in diameter, if round. If not round, then its contour may be circumscribed by a circle whose diameter is one inch or less. For the sake of convenience, these punches are commonly called perforators whether or not their function is strictly one of perforating. As a rule, perforators are mounted in punch plates. Perforators whose working contours are other than round are often made with round shanks, and these must be secured against rotation. They are normally keyed to the punch plate in order to assure proper orientation.

Types of perforators. The differences and variations in perforator design may not be endless, but they certainly are numerous. However, a study of the features of some basic types should enable the diemaker to cope with any perforator construction by association and extrapolation.

Step-head perforators. For general application, the step-head perforator shown in Fig. 7 • 15 is probably the most widely used perforator type. The typical configuration is as follows:

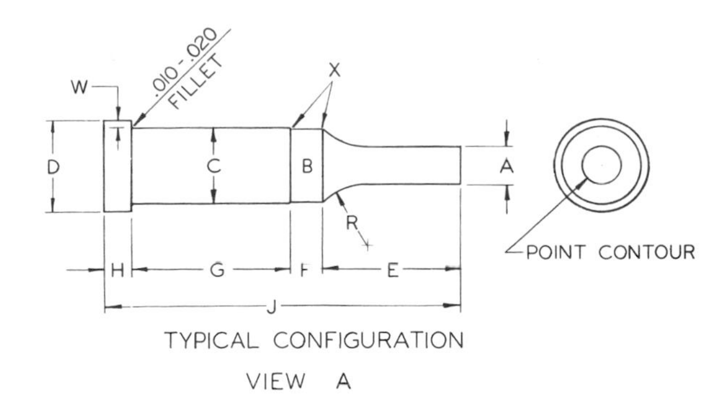

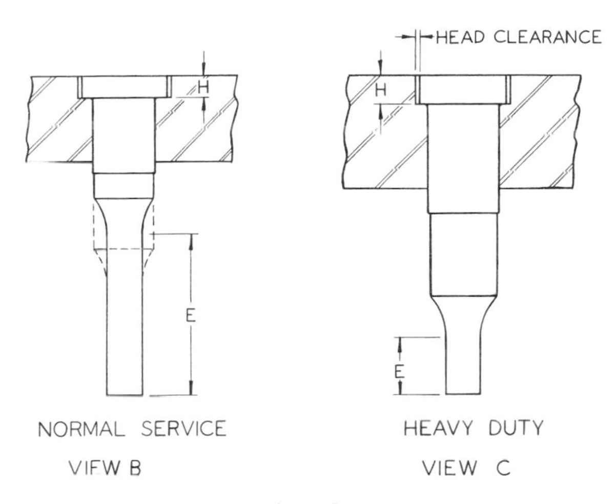

Figure 7·15 Step-head perforators, shank type.

1. Point Diameter. Must be as required for the application. For piercing applications, the point diameter is normally made 0.001 in. larger than the size of the opening it is intended to pierce. Point contour is also dictated by the application. The point in the illustration is round, but points may be oval, triangular, rectangular, hexagonal, etc., as required.

2. Lead Diameter. This diameter *B* should be 0.001 in. smaller than the shank diameter *C*. The lead is a definite asset for assembling the punch into the punch plate. Eliminate sharp corners at *X* to prevent possible shearing of the punch-plate opening during assembly. It should be self-evident that a lead is not needed to facilitate installing floating punches or slip-fit punches.

3. Shank Diameter. Should be a diameter that will permit the use of a standard reamer for sizing the mating hole in the punch plate. *C* diameter is commonly made 0.0002 to 0.0005 in. larger than the nominal reamer size. Also, shank diameter *C* should be in reasonable proportion to point size *A* and overall punch length *J*. For most applications a tap fit to a light drive fit at *C* is appropriate. Heavier fits should be used for heavier applications, but even these should not be overdone.

4. Head Diameter. Should be ⅛ in. larger than *C*, with a tolerance of minus 0.010 in. This will give a step width *W* of 0.057 to 1/16 in. This is the optimum proportionate head diameter for most applications. Larger step widths should not normally be necessary for punches with shank diameters less than 1 in. Head diameters smaller than those indicated above are best avoided.

5. Point Length. Should be in direct proportion to point size *A*. Larger point sizes permit longer point lengths *E*. With smaller sizes, the most common error is to make dimension *E* too long in an attempt to provide maximum grinding life.

6. Lead Length. To be effective, length *F* must be a minimum of 1/16 in. Permissible lead length is limited by the individual situation. In a general way, lead length should be the longest practical for the given set of conditions (views *B* and *C*).

7. Fitted Length. Should permit shank diameter *C* to extend through the punch plate, at least. Longer *G* dimensions serve no purpose and are undesirable when the punch is being pressed into the punch-plate opening. If the shank length ($G + F$) extends considerably beyond the punch plate, the lead length *F* should be made longer (see view *C*).

8. Head Thickness. Should be a minimum of $2W$ (⅛ in.). This should be increased to $3W$ for larger perforators and/or heavier work. A fillet radius of 0.010 to 0.020 in. should be provided at the junction of shank and head. Do not undercut at this junction.

9. Overall Punch Length. This is directly proportional to point size *A* and shank diameter *C*. Punches with larger *A* and *C* dimensions may be made longer at *J*. Actual punch length *J* can be influenced by a number of factors. For example, a thicker punch plate may permit a longer punch length. A punch which is guided or supported in the stripper may be made longer than the equivalent unguided punch.

10. Radius R. This is tangent to *A*, intersecting diameter *B* at *F*. The surface generated by this radius should be smoothly finished, blending to a nice tangency at *A*. For smaller punches, a radius *R* equal to the shank diameter *C* is satisfactory. For larger punches and/or heavy-duty punches, *R* should be ½ to ⅝ in.—assuming that a reasonably normal size relationship exists between *A* and *C*.

Hardness requirements will vary according to the material from which the perforator is made, the material being punched, and the relative strength proportions of the perforator. Generally, punches made of oil-hardening tool steel may be hardened Rockwell C 59 to 61. High-carbon, high-chrome punches may be Rockwell C 60 to 63. High-speed steel punches may be Rockwell C 60 to 64. Important! Punch heads must be softer;

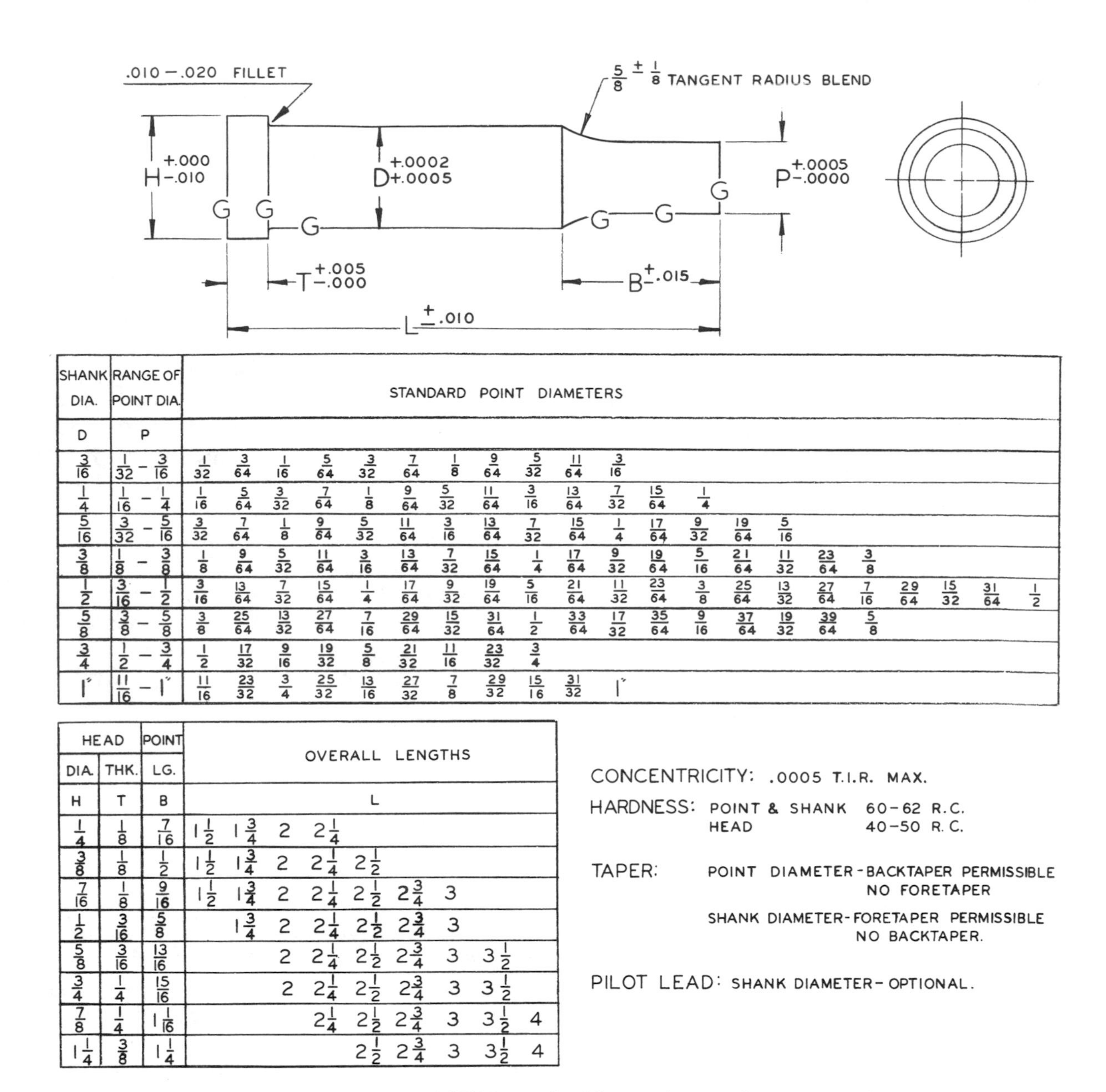

SHANK DIA. D	RANGE OF POINT DIA. P	STANDARD POINT DIAMETERS
3/16	1/32 – 3/16	1/32, 3/64, 1/16, 5/64, 3/32, 7/64, 1/8, 9/64, 5/32, 11/64, 3/16
1/4	1/16 – 1/4	1/16, 5/64, 3/32, 7/64, 1/8, 9/64, 5/32, 11/64, 3/16, 13/64, 7/32, 15/64, 1/4
5/16	3/32 – 5/16	3/32, 7/64, 1/8, 9/64, 5/32, 11/64, 3/16, 13/64, 7/32, 15/64, 1/4, 17/64, 9/32, 19/64, 5/16
3/8	1/8 – 3/8	1/8, 9/64, 5/32, 11/64, 3/16, 13/64, 7/32, 15/64, 1/4, 17/64, 9/32, 19/64, 5/16, 21/64, 11/32, 23/64, 3/8
1/2	3/16 – 1/2	3/16, 13/64, 7/32, 15/64, 1/4, 17/64, 9/32, 19/64, 5/16, 21/64, 11/32, 23/64, 3/8, 25/64, 13/32, 27/64, 7/16, 29/64, 15/32, 31/64, 1/2
5/8	3/8 – 5/8	3/8, 25/64, 13/32, 27/64, 7/16, 29/64, 15/32, 31/64, 1/2, 33/64, 17/32, 35/64, 9/16, 37/64, 19/32, 39/64, 5/8
3/4	1/2 – 3/4	1/2, 17/32, 9/16, 19/32, 5/8, 21/32, 11/16, 23/32, 3/4
1"	11/16 – 1"	11/16, 23/32, 3/4, 25/32, 13/16, 27/32, 7/8, 29/32, 15/16, 31/32, 1"

HEAD DIA. H	HEAD THK. T	POINT LG. B	OVERALL LENGTHS L								
1/4	1/8	7/16	1 1/2	1 3/4	2	2 1/4					
3/8	1/8	1/2	1 1/2	1 3/4	2	2 1/4	2 1/2				
7/16	1/8	9/16	1 1/2	1 3/4	2	2 1/4	2 1/2	2 3/4	3		
1/2	3/16	5/8		1 3/4	2	2 1/4	2 1/2	2 3/4	3		
5/8	3/16	13/16			2	2 1/4	2 1/2	2 3/4	3	3 1/2	
3/4	1/4	15/16			2	2 1/4	2 1/2	2 3/4	3	3 1/2	
7/8	1/4	1 1/16				2 1/4	2 1/2	2 3/4	3	3 1/2	4
1 1/4	3/8	1 1/4					2 1/2	2 3/4	3	3 1/2	4

Figure **7 · 16** Tentative ASTME standard heavy-duty perforating punch.

the head area should be redrawn to Rockwell C 40 to 50. In most cases, ideal hardness for the head area will be Rockwell C 43 to 47.

Finish. Quality of finish is important to the function and life of all perforators. Excepting the head diameter *D*, perforators should be smoothly finished by grinding all over. The working end, point length *E*, should be especially well finished. A nicely ground finish is considered adequate for most applications. For extra service requirements, points should be lapped or honed to produce an ultrasmooth finish.

Service requirements. The most prominent difference between a punch for normal service and a heavy-duty punch is the point length *E* (views *B* and *C*). For normal service the extra point length provides additional sharpening life. However, keep in mind that normal service covers a wide range from light duty to heavy duty. Point lengths should not be longer than necessary and should be proportionate for the individual application, as indicated by the dotted outline in view *B*. Heavy-duty punches require thicker heads *H*. Remember, though, that the rule for head thickness ($H = 2W$, min.) applies to all step-head punches.

Standard heavy-duty punches. The tentative ASTME standard for heavy-duty punches is given in Fig. 7 · 16. This is a generally practical proposal.

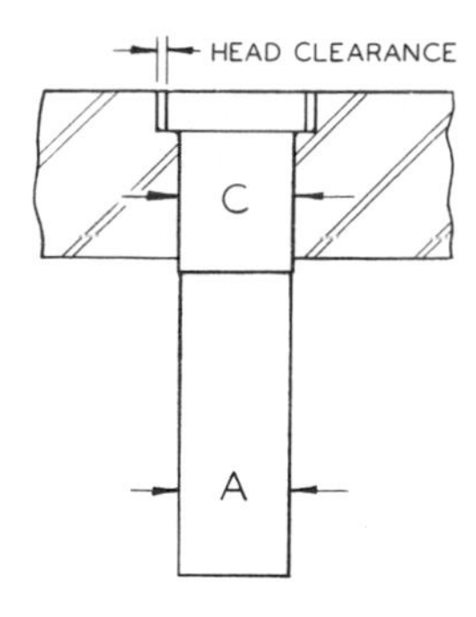

Figure 7·17 Step-head perforator, shankless type.

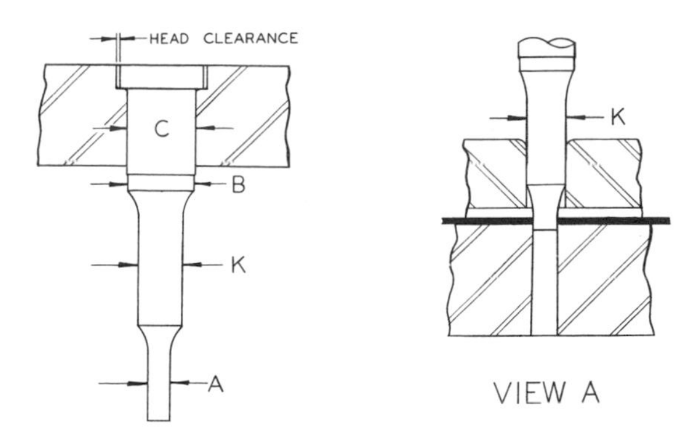

Figure 7·18 Pyramid perforator with step head.

It would, however, be better if the head diameter for the smallest group (3/16 shank diameter) were increased to 5/16 in. Also, it is more convenient for the diemaker to think of point length as the distance from the face to the radius tangent instead of to the shank diameter, as shown.

Off-standard shank diameters. In essence, this perforator (Fig. 7·17) is the same as the preceding perforator except for one feature. Here, the shank diameter *C* is simply made 0.001 in. larger than the required point diameter *A*. This is done in order to provide the necessary interference fit in the punch plate. At the same time, diameter *A* acts as a lead to facilitate assembling.

For punches which are a slip fit in the punch plate and for floating punches, *C* diameter can be eliminated. Point diameter *A* then extends to the head. For most perforators with this configuration, it will not be possible to use a standard-size reamer for sizing the punch-plate opening; it will be necessary to bore a special-size hole derived from the required punch size *A*. In spite of this, these punches are in common use because applications are encountered from time to time where they are desirable.

PYRAMID PERFORATORS

This configuration (Fig. 7·18) is often employed when for any reason there is a considerable disparity between the required point size *A* and the desired shank diameter *C*. It is particularly well suited to relatively long perforators with small point dimensions at *A*. Intermediate diameter *K* provides a more gradual transition from size *A* to diameter *C*. This is favorable for heat-treating as well as for providing a generally strong physical configuration.

The pyramid configuration is often handy for supported punches where point-guiding is impractical. These punches can be supported on the intermediate diameter *K* as indicated in view *A*. For such applications, the difference between *A* and *K* must be small enough to provide adequate stripping. Length proportions, too, are critical: The punch must be adequately supported before the point enters the stock material.

Step-head perforators lend themselves to floating applications. Be sure, however, that head diameters are adequate in such applications.

KEYED PERFORATORS

As mentioned earlier, if the point contour is other than round, some means of preventing rotation is necessary. One method is to key the punches to the punch plate. The arrangements shown in Figs. 7·19 to 7·21 are practical and efficient and as a result are in common use.

Figure 7·19. A slot is milled in the punch plate. Slot depth must suit the head height *H*, permitting flush grinding of the perforator head in assembly. An open slot permits the slot to be ground in a surface grinder, which is a desirable feature. Ideally, the slot should be finished as shown, making a step of 0.001 to 0.002 in. on the keying side and providing clearance on the opposite side.

Figure 7·19 Open slot keyway.

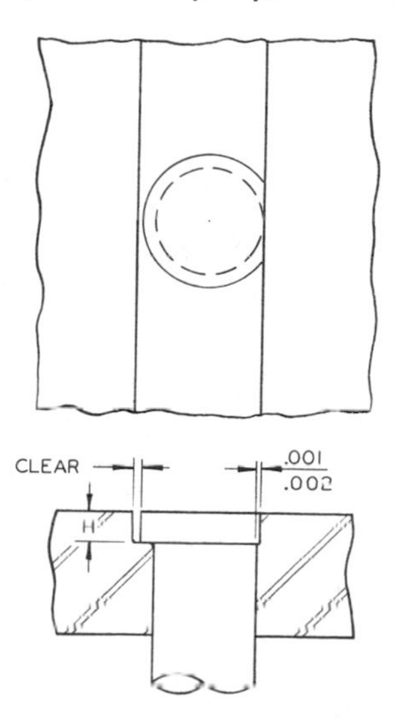

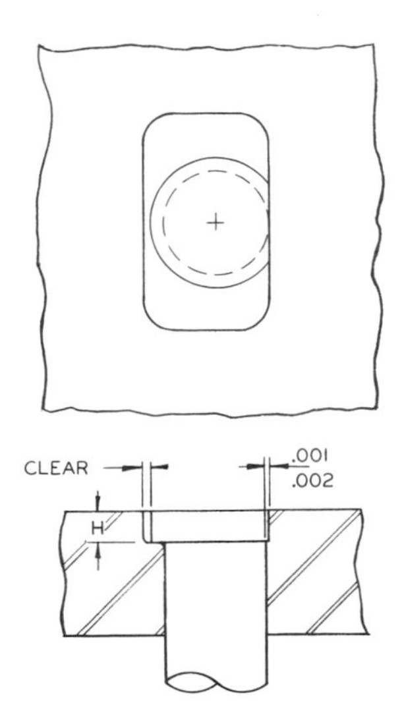

Figure 7·20 Pocket-type keyway.

The punch head is then ground to match on the proper key side.

Figure 7·20. It is not always feasible to provide an open slot as described above. In such cases, a pocket similar to that shown is a satisfactory method of keying. Essentially, the pocket is the same as the open slot except that it cannot be ground in a conventional surface grinder. If grinding is desired, the keyway can be jig-ground.

Figure 7·21. Mounting hole is prepared (including counterbore) as it would be for a round perforator. Key slot is then provided as illustrated. For foolproofing, it is best to use a key that is oblong in cross section, as shown. Dimension *S* should be 0.005 to 0.010 in. less than a standard stock size in order to facilitate grinding the key to fit. If an open slot is provided for the key, it is a good idea to supply some means of preventing the key from drifting.

Figure 7·21 Separate key.

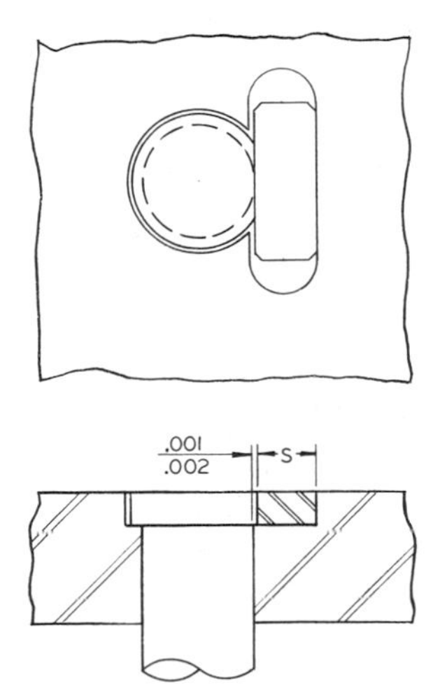

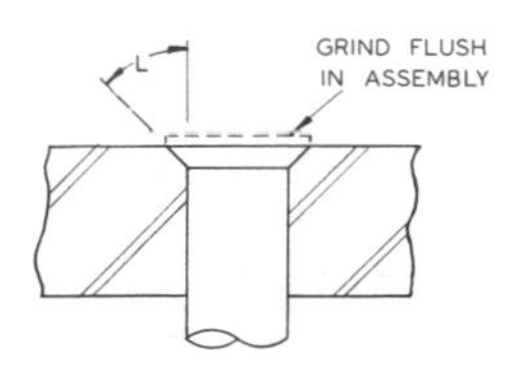

Figure 7·22 Bevel-head perforator.

Assembling keyed perforators. If the shank fits too tightly, it will make orientation more difficult. One assembly procedure is to start the shank in the opening with the key flat approximately aligned. Then twist the flat into exact alignment, checking with an indicator. Press the punch in the opening until the head is about to enter. Check again, restoring alignment if necessary. Then press the punch home.

Another procedure is to clamp a rectangular block to the punch plate with one edge exactly flush with the keying edge. The block then acts as a guide for the key flat on the perforator head. All keyed surfaces should fit nicely. If a punch is loosely keyed, it will not be oriented. If keying is too tight, it will exert a cocking action which may interfere with proper alignment.

There are other methods of keying perforators, and there are other assembly procedures. However, the methods described above are effective and proved. A clear understanding of the principles involved will enable the diemaker to estimate the desirability of various methods and to develop on-the-spot procedures for different situations where keying is required.

BEVEL-HEAD PERFORATORS

Heat-treatment and finish requirements are the same as for step-head punches. Possible general configurations (excluding the head profile) can also be the same. The head is beveled as indicated in Fig. 7·22. Bevel angle *L* may be any convenient angle from 30 to 45°. It is generally conceded that a 30° angle provides a stronger (therefore, better) head than larger angles do. Bevel heads are not as keyable as step heads. If keying is necessary, it will be better to switch to a step-head perforator. Bevel heads are not at all practical for floating perforator applications.

Bevel-head perforators produced by peening (see Fig. 7·8) used to be quite common, especially for light duty and/or low-production requirements. However, because of structural weakness resulting from cold-peening and because of the advent of commercially available high-quality, low-cost perforators, peened-head perforators have fallen into disfavor and are seldom seen.

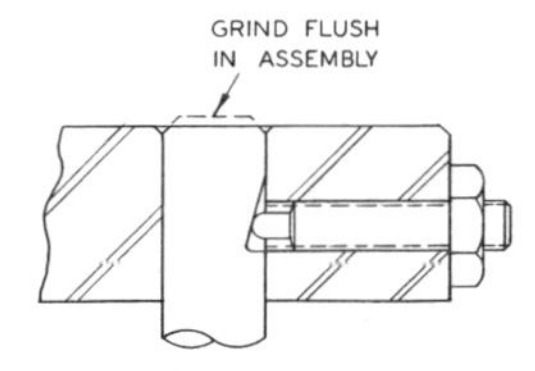

Figure 7 · 23 Headless perforator with whistle notch.

NOTCHED HEADLESS PERFORATORS

Hardness requirements and finish requirements are the same for this type except that, since the perforator is headless, the entire perforator may be hard. These perforators are assembled as shown in Fig. 7 · 23. The back surface should be ground flush in assembly, as indicated by the dotted outline. The whistle notch is essentially the same as that in Fig. 7 · 15. This type of perforator is not suitable where accurate keying is required (a round shank with a conventional whistle-type security notch is assumed).

SLUG EJECTION

To prevent slug-pulling, either air pressure or spring pins are commonly employed in the manner represented in Fig. 7 · 24. For air ejection the punch is provided with a suitable hole along its axis. Compressed air is led to and through the hole and tends to prevent the slugs from pulling out of the die opening. In some cases a constantly blowing air jet is satisfactory; in others the air blast may have to be timed. This can be determined by trial. The air-ejection method is considerably more complex when considered in the overall. Largely because of this, the spring-pin ejector (view *B*) is the predominantly preferred method.

Figure 7 · 24 Slug ejector punches, air or spring.

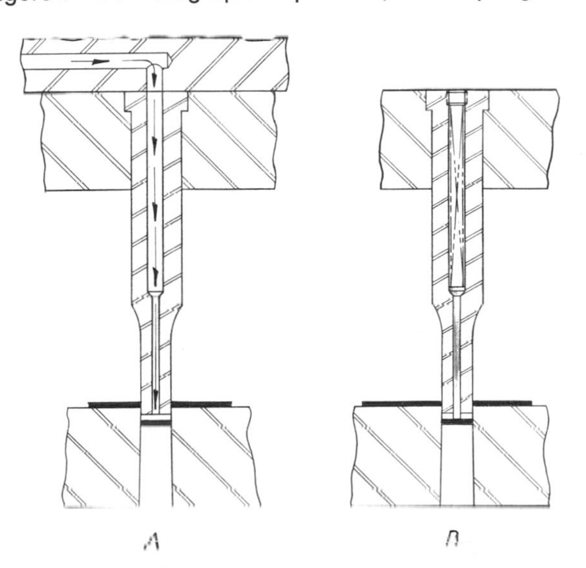

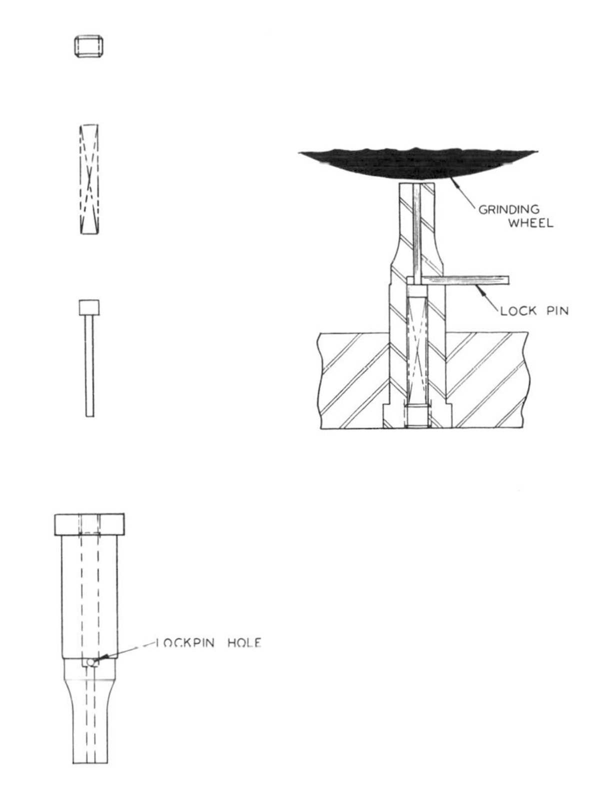

Figure 7 · 25 Lockout slug ejector pin.

Some commercially available perforators have spring pins which may be removed without dismounting the perforator or the punch plate. This can be a convenient feature for sharpening. Another commercially available type (Fig. 7 · 25) permits sharpening the punch in assembly without removing the spring pin.

QUILLED PERFORATORS

The purpose of quills is to provide support for relatively long, slender (consequently delicate) punches. A rather typical quilled-perforator assembly is pictured in Fig. 7 · 26. Constructionwise, the quill is a hollow step-head punch which contains and supports the actual perforator. The inside diameter of the quill should provide a close slip fit (no shake) for the perforator. Hardened thrust plates are normally required behind quilled

Figure 7 · 26 Quilled perforator.

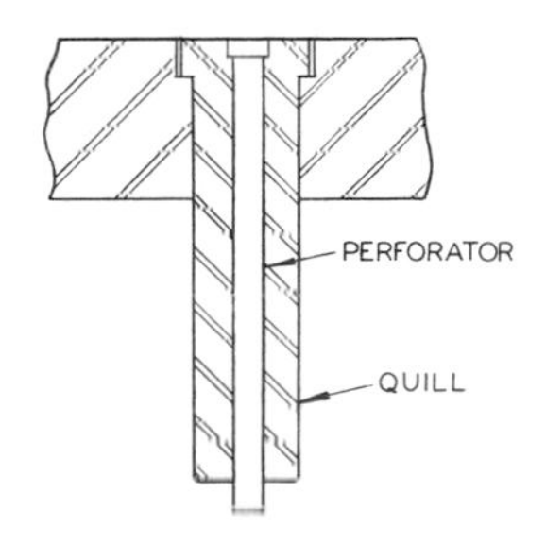

punches, because most quilled perforators can be considered heavy-duty (relative to perforator size) or severe-application punches. Since actual perforator-head area is comparatively small, there may be a decided tendency for this head to "set in" to the punch-holder face. The illustrated perforator has a step head, but bevel heads are also very commonly used for quilled perforators (not on the quill). When quill diameter is small in relation to perforator diameter, the bevel head will make possible a strong cross section in the quill.

Quilled-perforator assemblies can be quite expensive when made by the diemaker on an occasional basis. For this reason, the use of commercially available quilled assemblies is strongly recommended. The intermeshing sleeve-type quill is a prime example of the practicality of purchased punches.

PURCHASED PERFORATORS

High-quality, economical perforators are commercially available from many sources in any practical type, shape, and size. Just from the standpoint of education, it will be profitable for the diemaker to examine and study as many as possible of the commercial perforator catalogs. If a standardized perforator is not available exactly as required, one which will demand only a minimum of modification can in many instances be purchased. Keep in mind that this is true for die buttons (bushing-type die inserts) and pilots, as well as for perforators.

CHAPTER 8

PILOTS

PURPOSE OF PILOTS

It is the function of pilots to position workpieces or stock strips accurately for dieworking. When the work is brought into the required position by the pilots, it is said to be "registered."

REGISTRY ACTION

The piloting action, when registering a manually fed stock strip, is depicted in Fig. 8·1. The stop is located to permit overfeeding the strip a slight distance beyond the required registry position. For dies equipped with pilots, all stops except the first (primary) stop are positioned to permit overfeeding. Actual feeding distance is equal to the advance distance plus the overfeed.

SEQUENCE:

Operator feeds strip against stop. Strip is now at stop position. As press ram descends, pilot nose enters previously pierced opening in strip. This is the condition pictured in view *A*. As press ram continues to descend, bullet nose on pilot cams strip opening into alignment with pilot. Effective pilot diameter *P* engages pierced opening, registering the stock strip. Registered strip is shown in view *B*.

Overfeeding is a very important feature of the relationship of stops to pilots. Do not attempt to make the stop position coincide exactly with the registry position. All secondary and final stops in piloted dies should permit the strip to be overfed a slight amount.

Pilot action in displacing the strip from stopping position to registry position is a qualifying action. When the strip is registered, the feed is said to have been qualified.

When a stock strip is fed automatically by mechanical means, pilot action is the same, in principle. However, the direction in which the feed is qualified is normally reversed: Instead of being overfed, the stock strip is underfed. This is especially true for feeding mechanisms which incorporate unidirectional locking of the stock strip. For mechanical feeding, pilot action normally registers the stock strip by qualifying the feed in the same direction as the feeding motion. Figure 8·2 illustrates normal pilot function when operating in conjunction with a mechanical stock-strip feeding device.

In summary, hand-fed strips are overfed and backed into registry position by the pilots. Mechanically fed strips are normally underfed and pulled forward into registry position by the pilots.

PILOTING SIZE

The term piloting size refers to the effective gaging portion of the pilot. In Figs. 8·1 and 8·2

Figure **8·1** Pilot registry, hand-fed strip.

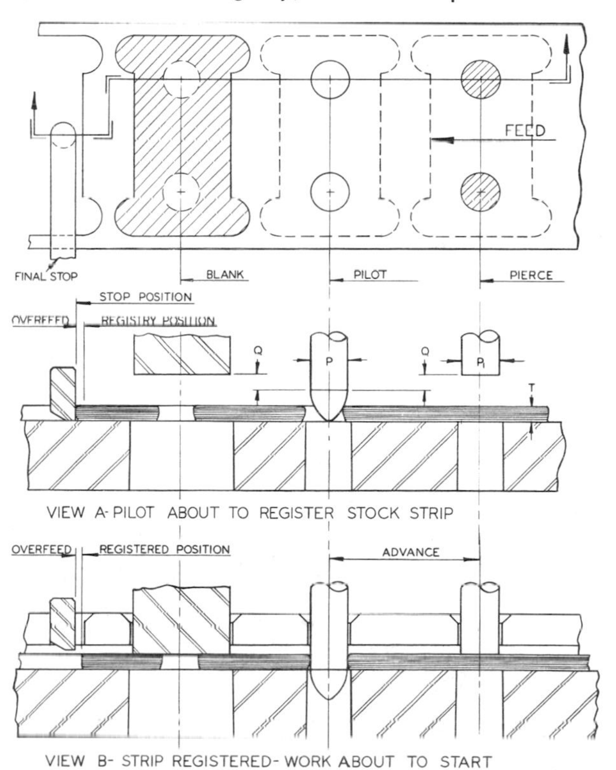

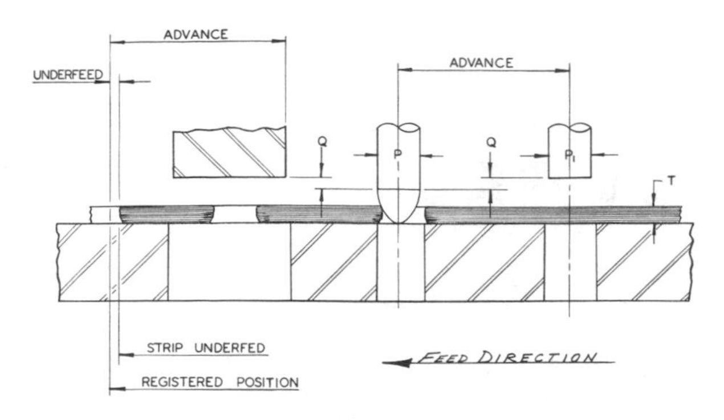

Figure **8·2** Pilot about to register mechanically fed strip.

the pilots are round. Here, piloting size is the diameter of the gaging portion *P*.

The accuracy with which the work can be registered is, of course, contingent upon proper pilot location. If the pilots are out of position, the work will likewise be out of position. If the pilots are precisely located, accuracy of registration then depends upon the relationship of the piloting size to the size of the respective openings in the stock material. The closer the pilots fit in the openings, the more accurate is the registry.

For very precise work, pilot fit is critical. It is desirable for the pilot to fit the opening exactly, permitting no lateral movement of the material when the pilot is entered. At the same time, the pilot should not fit tightly, because the resulting friction will be detrimental to both the pilot and the work. Excessive piloting friction causes undue pilot wear, which results in loss of accuracy and possible galling. This contributes to possible burring or distortion of the work in the area of the pilot opening. The piloting fit required for highly precise work is best described as a very close sliding fit. Where such a close fit is not necessary, piloting size should permit a slight clearance between the pilot and the work. This tends to alleviate piloting friction and to assimilate minor construction discrepancies. As implied above, the amount of clearance which may be provided is restricted by the accuracy requirements of the work. However, even for work which has wide tolerances, excessive piloting clearance is pointless and may be undesirable for die operation. For the majority of work, clearances will range from 0.0001 to 0.002 in. per side. Large, crude work may not only permit greater clearance but require it.

Since piloting size is derived from the corresponding opening size, piloting size is therefore related to the size of the punch P_1 which pierces the opening. In a general way, for average conditions the following may be used as a guide to determining piloting size (dimensions are inches):

Average work:

$$P = P_1 - 0.002 \text{ to } 0.004$$

Close work:

$$P = P_1 - 0.001 \text{ to } 0.002$$

High-precision work:

$$P = P_1 - 0.0005 \text{ to } 0.0007$$

This applies to piloting procedures approximating those illustrated in Figs. 8·1 and 8·2 and takes into account an average amount of compressive hole shrinkage, which is normal for pierced openings. Different procedures may necessitate a slightly greater differential between P_1 and *P*. Thicker stock material *T* will usually also require a greater differential. With some materials the pierced opening may close in more after piercing than with other materials. The amount of compressive hole shrinkage can cause variations in the normal relationship between piloting size *P* and punch size P_1.

PILOT LENGTH

Registry must be completed before any work begins, which makes it mandatory for the pilots to be longer than the punches in a given die. Pilot length must be such that the effective gaging portion of the pilot engages the stock material before the punches contact the stock material. If pilots are too short, they cannot perform their required function. This can have serious consequences, ranging from spoiled work to damaged dies. Pilots which are longer than necessary will, in most cases, perform their function of registry. One exception to this would be a progressive die mounted in a short-stroke press and requiring a relatively long feed stroke (advance). In such instances, pilot length can become quite critical, because overlong pilots may not permit an adequate feeding cycle (an automatic feed is assumed). In any die, however, it should be remembered that overlong pilots will increase piloting friction, which is objectionable.

In the application pictured in Fig. 8·1, the die is equipped with a fixed box-type stripper. For this and similar applications, piloting length is derived from the associated punch heights. Piloting size *P* should extend beyond the punch faces a distance *Q* which is equal to the stock material thickness *T* or to 1/16 in., whichever is greater.

This practice may be subject to slight modification because of circumstances peculiar to an individual application. It should be noted that piloting lengths derived this way apply only to unguided pilots (illustrated in Fig. 8·1) and to pilots which are guided in the stripper. They do not apply to pilots which are guided in the die block.

In many dies, the punches will not all be the same height. For such applications, keep in mind that at least one set of pilots must act to register before the longest (highest) punch contacts the work.

Many dies are equipped with pressure-pad strippers, which are, quite commonly, spring-actuated (Fig. 8·3). In addition to stripping the punches, these strippers may be employed to clamp the stock material against the die face, holding the stock flat during the working cycle in the manner indicated in view *B*. For this kind of application, be certain that the piloting length extends far enough beyond the stripper face to register the work before the stripper contacts the work. Thus, dimension *Q* applies when the stripper is extended as shown in view *A*. Slightly longer piloting lengths may be necessary, depending upon the amount of stripper overtravel *S*. Generally, any increase in piloting length will be quite small, since normal stripper overtravel is on the order of 0.005 to 1/64 in. Nevertheless, it is a factor which is related to piloting length. The diemaker must be aware of it and proceed accordingly, making certain that the pilots will accurately register the work without interference.

Figure **8·3** Pilot related to spring stripper which clamps against the stock strip.

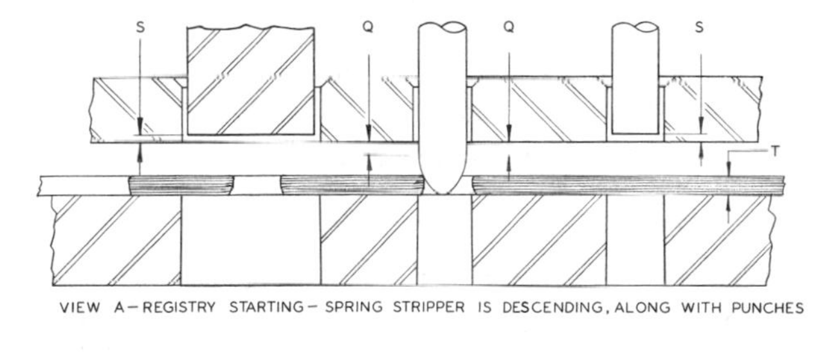

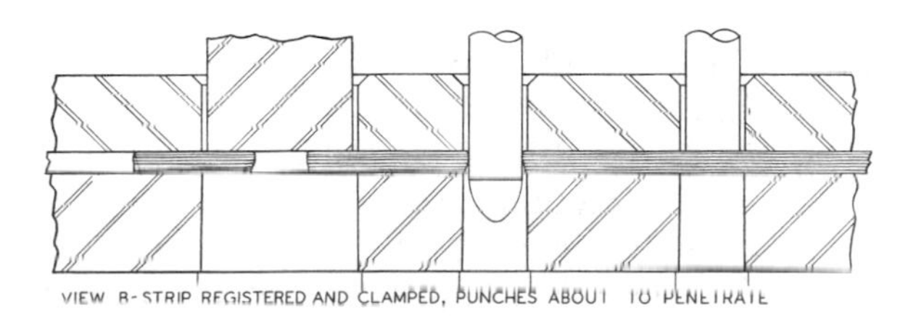

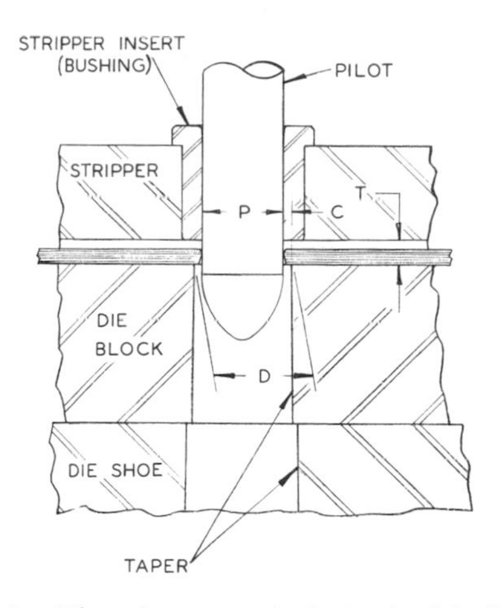

Figure **8·4** Pilot clearance hole in die block.

PILOT OPENINGS IN DIE BLOCKS

The most common type of pilot opening provided in die blocks is a clearance opening. This method applies to pilots which are stripper-guided and to pilots which are unguided. The pilot represented in Fig. 8·4 is guided in the stripper, but in relation to the pilot clearance hole, the condition would be the same for an unguided pilot.

To facilitate diemaking, it is usually desirable to make the opening *D* the largest practical size in order to permit as much discrepancy as possible in both its size and location. However, there are other circumstances which tend to curtail the amount of clearance permissible between the pilot and its clearance opening. If the opening is too large in proportion to the stock thickness *T*, the stock will tend to draw into the opening, distorting the material in the area of the pilot. This will cause objectionable pilot friction, in addition to possible work spoilage. Remember that the stock material is normally underfed or overfed. With thin, soft material, if the die-block opening is too large, the pilot may not displace the material into registry position but may instead draw the material on one side of the piloting opening into the clearance opening. The stock strip will then register out of position, resulting in work spoilage. With solid pilots, in the event of a misfeed the pilot should act as a perforator, cutting a slug into its mating die clearance opening. If the opening is too large, the material will tend to extrude and become jammed.

In view of the above considerations it is readily discernible that pilot clearance openings should be made some optimum size: neither too big nor too small. One generally satisfactory procedure for estimating optimum opening size *D* is to use twice the clearance *C* that would be used for the given stock material if the pilot were a perforator instead of a pilot. Then

$$D = P + 2C$$

where C = twice normal cutting clearance

In view of the fact that a pilot may pierce the material because of a misfeed, the clearance openings should in a sense be considered piercing-die openings. Thus the openings should extend through the die shoe to permit slug egress and should be tapered (as with a piercing operation) to eliminate the possibility of slug impaction.

Pilot support in die block. Occasionally a situation is encountered where it is desirable to support the pilots but not necessary to guide any of the punches. In such cases, it may be practical to guide the pilots in the die block as in Fig. 8·5, where a bushing is installed in the die block. One advantage of this procedure is that in the event of rapid wear in the opening, the bushing may be easily replaced. Another advantage is that as the die face is sharpened away in a cutting die, the bottom of the bushing may be ground to maintain its corresponding height. Thus, the bushing height is lowered without eliminating the straight guiding portion S of the inner diameter. When using this procedure, be sure to check the guiding inner diameter for size in order to ascertain whether or not it will provide suitable pilot support. Bushings of this kind are considered a necessity in high-production dies, if it is intended to support the pilots in the die block.

To provide support, the size of the opening which receives the pilot is made a slip fit for the pilot P. The opening should be made straight for a distance S of at least ⅛ in., or more, if deemed

Figure 8·5 Pilot opening in die block, fitted for pilot support.

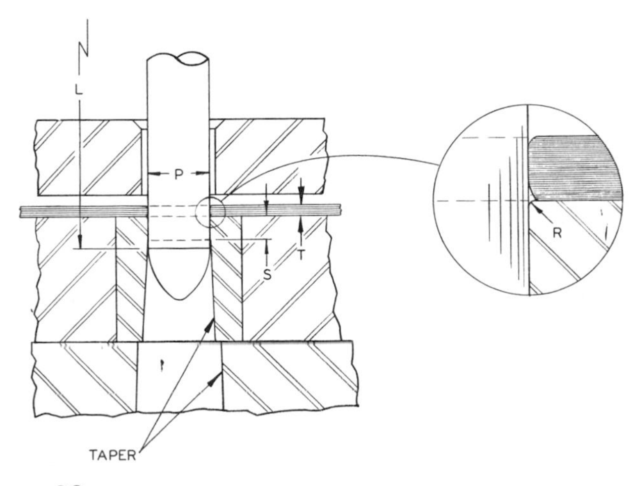

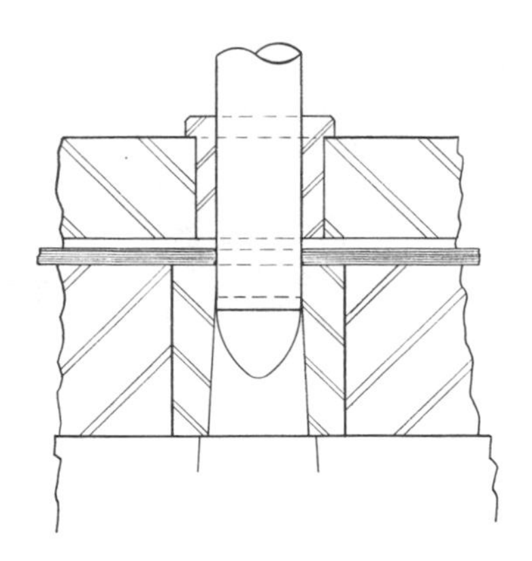

Figure 8·6 Pilot, doubly guided.

advisable. The balance of the opening should be tapered, larger at the bottom. The top corner of the opening should not be sharp. A small radius R should be provided as shown in the inset view. A radius of approximately 10 per cent of stock material thickness T with a minimum of 0.005 in. should be generally satisfactory. The radius alleviates the possibility of shearing the pilot. Piloting lengths L are necessarily longer for this method of piloting. Piloting size P must extend into the die opening far enough to assure adequate support before any punches, strippers, etc., effectively contact the stock material.

This method of providing support for pilots is also used in combination with stripper guidance for pilots. That is, the pilot is guided in the stripper and receives additional support from the die-block opening. This double-support method is used where pilots are relatively small and weak in proportion to the stock thickness and where relatively high lateral forces are imposed upon the pilots. Both the stripper opening and the die opening are made a slip fit for the pilot, as indicated in Fig. 8·6. The pilot is supported on both sides of the stock material and in close proximity to it as well. Normally, double support is superfluous, but for extreme piloting conditions it will achieve maximum pilot support.

PILOT PUNCHES

The general configuration of these pilots is very similar to that of perforators. The only difference is in tip configuration, as required by the difference in function. As far as die assembly is concerned, mounting methods and procedures for pilot punches are identical with those employed for perforators.

The similarity of pilot-punch construction to perforator construction is readily apparent in Fig. 8·7. Except for tip profiles (pilot noses), these pilots are identical in configuration with perforators of equivalent style. Therefore, the informa-

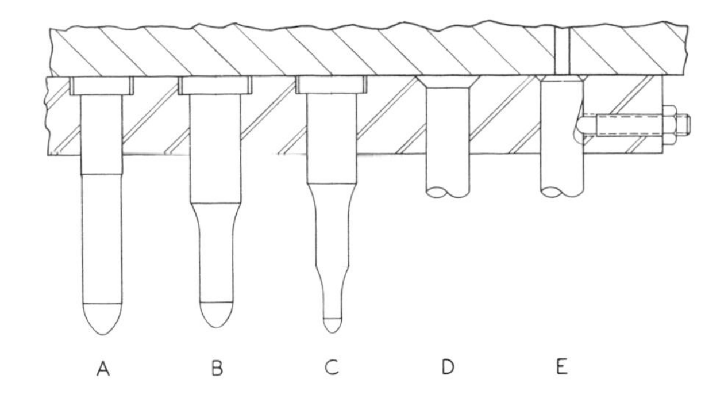

A- STEP HEAD, SHANKLESS
B- STEP HEAD, SHANK TYPE
C STEP HEAD, PYRAMID TYPE
D- BEVEL HEAD
E- HEADLESS, WHISTLE NOTCHED

Figure 8 · 7 Common styles of pilot punches.

tion and discussion applying to the various styles of perforators will also apply to the equivalent style of pilot. Always keep in mind, of course, the piloting-size and piloting-length relationships described earlier in this section.

It should be mentioned that, just as with perforators, pilots may be made floating or assembled in quills, etc. Also as with perforators, the use of commercially available pilot punches is recommended wherever feasible.

Pilot nose contours. The most prevalent nose profile is the bullet nose, illustrated in Fig. 8 · 8. The most frequently used profile proportion is pictured in view *A*. The bullet shape is formed by radius *R*, which is equal to piloting diameter *P*. Radius *R* is tangent to *P* at piloting length *L*. The tip is finished with a spherical radius *r* equal to ¼*P* and tangent to *R*.

For piloting sizes less than ¼ in. in diameter, it may be desirable to increase the length of radius *R* as an aid to reducing the lateral forces inherent in the piloting action. If this is done, the profile radius *R* must be kept tangent to *P* at *L*. For large pilots, however, an increase in the length of *R* may make the pilot nose objectionably long. The bullet-nose pilot profile is popular because it is strong, simple to make, and smooth in action. Pilot action is comparatively gentle because of the smoothly accelerated lateral motion imparted to the stock material during the piloting act. This is evident from examination of view *B*.

As with most generalized features in diemaking, pilot nose profiles are subject to variation in order to satisfy specialized circumstances peculiar to individual applications. Some nose-profile variations are pictured in Fig. 8 · 9. For comparison,

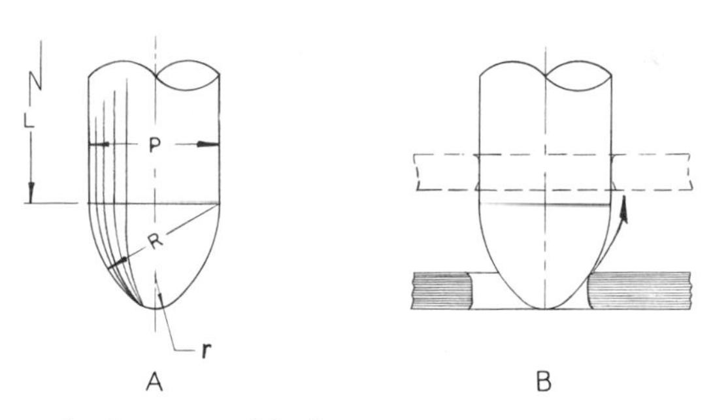

Figure 8 · 8 Typical bullet nose.

the conventional bullet-nose profile is indicated by dotted lines.

View A: 45° Conical Stub Nose. Radius *R* is made approximately ¼*P* and is tangent to *P* at *L*. Nose flank is conical at 45° and tangent to *R*. Tip is finished with spherical R_1 approximately equal to *R*. This profile is used where a shorter nose profile is desired. Mechanical advantage of the 45° cone is inferior to that of the bullet nose because the 45° cone increases the relative lateral forces involved in the piloting action. Therefore, speaking relatively, this configuration is not generally desirable for delicate pilots operating in easily distorted (soft, thin) stock material.

View B: 30° Conical Stub Nose. Procedure same as for previous nose, except cone angle is made 30° and radius R_1 is toroid instead of spherical. This is a compromise between the 45° stub nose and the conventional bullet nose.

View C: 15° Angular Long Nose. Procedure same as for 45° stub nose, except cone angle is 15°. The small angle provides good mechanical advantage, favorable for small pilots and/or soft, thin stock material. This contour is often used on small pilots in preference to a long bullet nose simply because it is easier to make. For practical purposes, any difference in efficiency between an angular long nose and a bullet-type long nose is hypothetical.

In the above discussions of nose profiles it has been assumed that the pilots were round. However, the principles involved are the same for all

Figure 8 · 9 Pilot nose profiles.

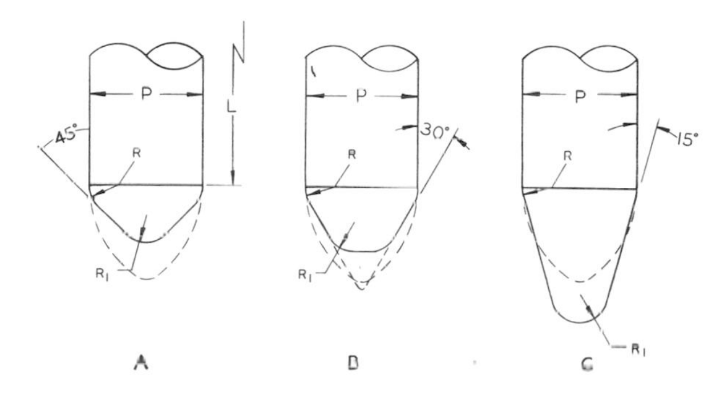

pilot contours (rectangular, etc.), except for one practical difference. For pilots which are not round, an arc-shaped nose flank (in the manner of a bullet nose) is seldom applied. It is usually simpler to make an angular nose flank on irregular pilots.

Removable pilot punches. As far as fits, proportions, etc., are concerned, mounting methods and procedures are generally the same as for perforators. This was evident in Fig. 8 • 7.

Note that only one of the illustrated pilots *E* can be considered removable in the sense that it can be removed from the punch plate while the punch plate remains mounted to the punch holder. Incidentally, a pilot is not referred to as removable unless this condition is satisfied.

Another kind of headless removable pilot is illustrated in Fig. 8 • 10, view *A*. An assembly variation is shown in view *B*, where a thrust plate acts as a head for the pilot in addition to backing it up. Whenever feasible, wrenching flats should be provided on the pilot to prevent rotation when tightening (or loosening) the pilot mounting screw. Because the mounting screw is aligned with the piloting axis, this construction can be more accurate than the whistle-notched shank, which may be affected laterally by pressure from the side screw. The headless styles are not generally well suited to applications for smaller pilots. Any of them may present accessibility problems for wrenching, etc., during assembly, especially if located within a cluster of punches. Caution! Retaining screws must be of adequate size.

Various methods for making step-head pilots removable are represented in Fig. 8 • 11.

View A. The pilot may be a tap fit or a close slip fit in the punch plate. A clearance hole for the pilot head is drilled through the punch holder. The hole is threaded from the top surface of the punch holder deep enough to receive the setscrew.

Figure **8 • 10** Headless removable pilots.

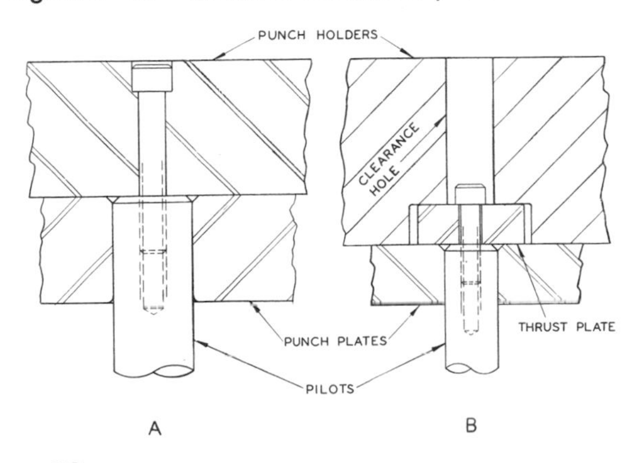

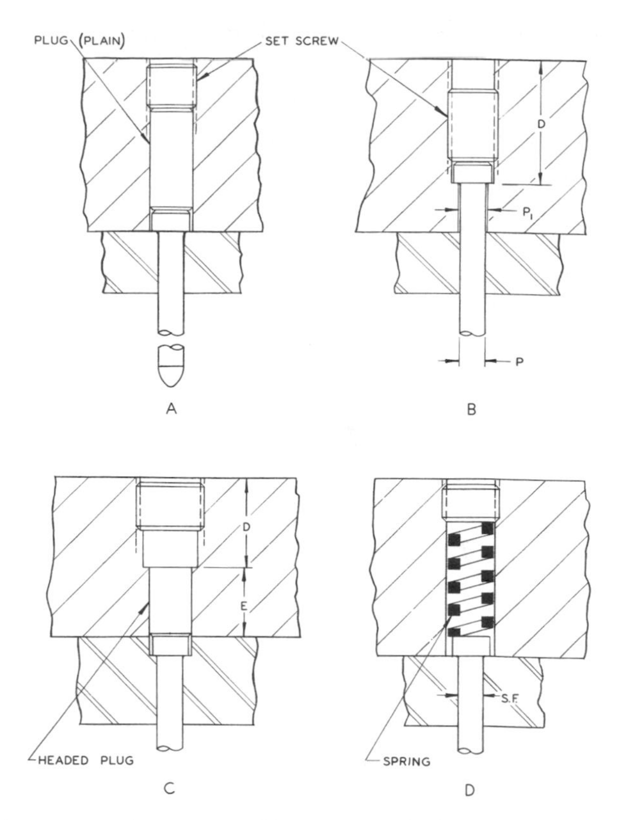

Figure **8 • 11** Removable step-head pilots.

A spacer plug of suitable length is inserted between the pilot and the setscrew.

This is a critical construction and can be a dangerous one. If too much pressure is applied to the pilot by the setscrew, the punch plate may be forced out of plane, causing misalignment of the pilot and any other components related to the punch plate. If this assembly method is used, be certain that it is adequately proportioned. The punch plate must be thick enough and the punch-plate mounting screws strategically located to compensate for the setscrew pressure.

View B. The punch-plate opening may provide a slip fit or a tap fit for the pilot diameter as required. To preclude interference, diameter P_1 should be slightly larger than pilot diameter P. To receive the pilot head, the hole is counterbored from the top of the punch holder to the required depth D. Counterbore diameter should provide clearance for the pilot head. The counterbore is then threaded in its top end to accept the setscrew.

This is a safe method and an excellent one. It is also quite expensive, and therefore its use seems to be confined to dies of very high quality.

View C. This is a compromise method using a spacer plug as in view *A* but providing the plug with a head to achieve the effect shown in view *B*. This construction, too, eliminates the inherent hazard of method *A*. It also is an expensive method, and in most circumstances it is not as efficient as method *B*.

View D. This is a semiretractable pilot. The punch-plate opening should be a slip fit to the pilot. Generally, the construction is identical to method *A* except that a heavy spring instead of a plain plug is incorporated to serve as a spacer. Purpose of the spring is to alleviate the tendency to bend the punch plate in method *A*. For the construction in view *D*, spring travel is not important. With light-gage stock material the spring may be strong enough to pierce the stock if a misfeed occurs. With stronger stock the pilot may retract, bottoming the spring before piercing. Either way, the pilot is expected to pierce through the stock material in the event of a misfeed, and therefore occasional piercing must be provided for. Just as with solid pilots, it is necessary to taper the pilot entry holes in the die block and extend these holes clear through the die shoe.

Retractable pilots. For a pilot to be considered retractable, it must be fully retractable. That is, in the event of a mishit, the pilot telescopes into the punch holder the distance necessary to preclude the possibility of piercing the stock material. As is evident in Fig. 8 · 12, construction is quite similar to a semiretractable construction. There is, though, an important functional difference in that spring action is critical with retractable pilots. The spring must be long enough and flexible enough to permit the required telescoping travel, but at the same time it must be strong enough to assure normal piloting action. The most common error encountered in retractable pilot constructions is selecting springs which are too short and/or too small in diameter.

These pilots are used for situations where it is desirable to eliminate the possibility of piercing with the pilots when a misfeed occurs. There may be a number of reasons for this—for example, die strength conditions or slug-disposal problems. Heavy-gage stock material may suggest the use of retractable pilots, especially if piloting diameter is relatively small in proportion to the stock thickness. Theoretically, a tapered pilot clearance opening in the die block is not necessary, but in practice it is a good idea to taper the die holes. Tapered holes will permit a changeover to solid or semiretractable pilots if this is found desirable after the die is in production. Also, a tapered opening may prevent damage to the pilot and/or other components in the event of pilot malfunction.

Figure 8 · 12 Retractable pilots.

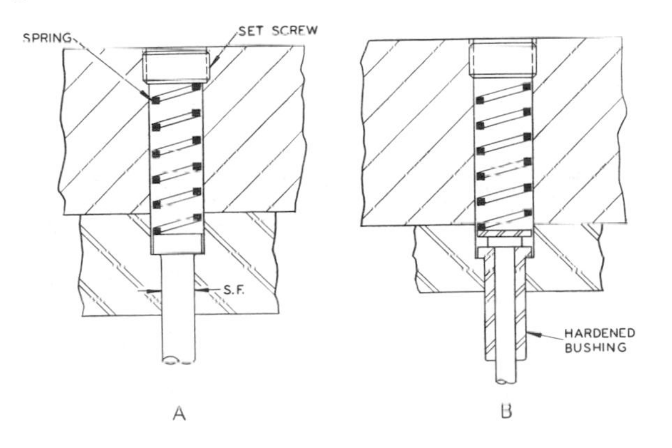

The illustrated pilots are removable, as well as retractable. With complex, closely fitted dies, it is a distinct asset to be able to perform maintenance sharpening with a minimum of disassembly. As described earlier, the pilots are longer than the punches, and in most cases they interfere with sharpening. This is the major reason for making pilots removable. Also, in the event that it is necessary to replace a pilot, a removable one can be replaced without disturbing the rest of the assembly. Removable pilots are not generally considered essential in simpler dies, especially if die clearances are greater than 0.001 in. per side. Removable pilot punches are neither necessary nor desirable for conventional low-production dies.

PILOTS IN PUNCHES

Pilots mounted in punches are, in trade parlance, conveniently called "punch pilots." A variety of different pilot types are commonly used in conjunction with the various kinds of punches. The function of the punch strongly influences the design and construction of a punch-pilot assembly, as is illustrated in Fig. 8 · 13.

View A: Pilot Mounted in Bending Punch. The pilot, in this case, is a "standard" step-head pilot. The pilot hole in the punch should be a light drive fit for the pilot *P*, because this will facilitate grinding the pilot head flush with the punch base in assembly. The hole should be relief-drilled slightly larger (P_1) to a depth *C*, which will leave sufficient fitted length *F* to provide adequate pilot location and support. Relief drilling is very practical: It eliminates unnecessary fitting. Relief diameter P_1 must not be too large, because it detracts from the area available for support under the pilot head. The following relief allowances are suggested as a guide for the diemaker to consider in conjunction with head diameter *D*:

$P < 1/4$ in.	$P_1 = P + 0.005$ to 0.010 in.
$P = 1/4$ to $3/4$ in.	$P_1 = P + 1/64$ in.
$P = 3/4$ to 1 in.	$P_1 = P + 1/32$ in.

For fitted length *F* (average conditions):

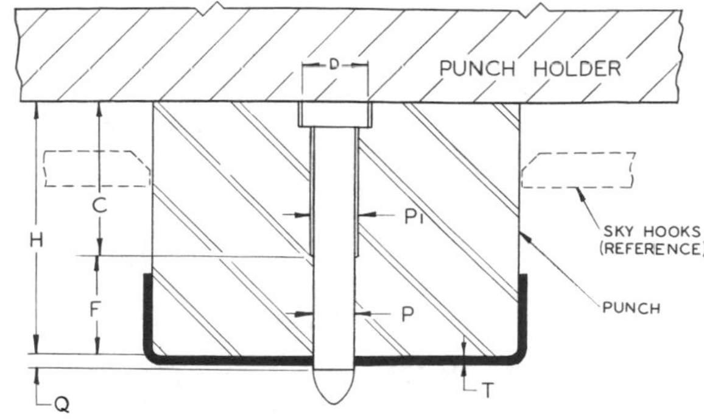

A- PILOT MOUNTED IN BENDING PUNCH

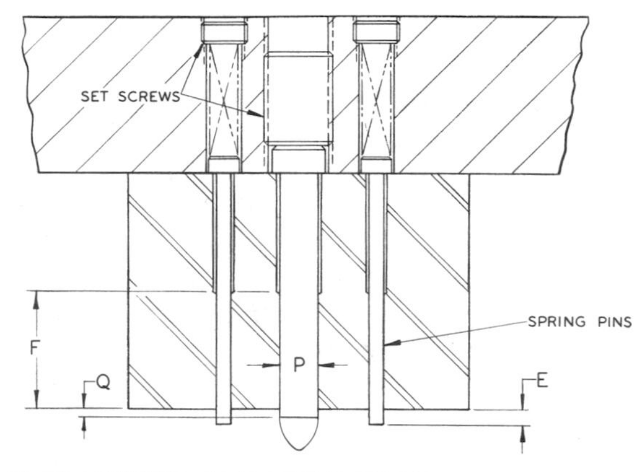

B- PILOT MOUNTED IN BLANKING PUNCH

Figure 8 · 13 **Influence of punch category on pilots mounted in punches.**

$P < \frac{1}{4}$ in.	$F = 2P$ to $5P$
$P = \frac{1}{4}$ to $\frac{1}{2}$ in.	$F = 2P$ to $3P$
$P = \frac{1}{2}$ to 1 in.	$F = 1\frac{1}{2}P$ to $2P$

Piloting size should extend beyond the punch face a distance Q of at least stock thickness T, but a minimum Q dimension of $\frac{1}{16}$ in. should be maintained for lighter-gage stock material. There is no stripping problem as far as the pilot is concerned; any of the conventional methods of stripping the work from forming punches will normally strip the work from the pilot also (see skyhooks, view A).

View B: Pilot Mounted in Blanking Punch. Allowing two exceptions, pilot-hole dimensions are the same as in view A.

Exception 1: Fitted diameter P should be a close slip fit or a tap fit.

Exception 2: Because this is a cutting punch, grinding life must be considered. Since fitted length F will become slightly shorter each time the punch is sharpened, be certain that F is long enough to permit the desired amount of sharpening life. Also in keeping with the fact that this is a cutting punch, the pilot is removable in order to facilitate sharpening. This may or may not be necessary, depending upon the individual application.

When pilots are installed in blanking punches, the hazard of blanks pulling becomes considerably greater. There exists a decided tendency for the pilots to lift the blanks out of the die opening on the return stroke. To overcome this tendency, spring pins are installed in the punch. Both the number of spring pins and their locations must provide a balanced stripping effect for the pilots. If the spring pins tend to tilt the blank, it may bind on the pilots, defeating the purpose of the spring pins.

Spring pins are not always needed. If, upon due consideration, they are deemed unnecessary, omit them. However, don't neglect to provide strategically located holes for them (in the punch only). The provisional holes will greatly facilitate adding spring pins later, if proved necessary. Spring-pin travel E must be sufficient to assure pilot stripping. Generally,

$$E = Q + \tfrac{1}{16} \text{ in., min.}$$

Spring-pin holes should be relief-drilled in a manner similar to the pilot holes. Dimension F applies to spring-pin holes as well as to pilot holes.

Pilots mounted in blanking punches require more clearance in the pierced opening than pilots which operate in the stock strip away from the blanking opening (see Fig. 8 · 1). This is one reason why pilots in blanking-type punches are somewhat less accurate than strip pilots.

Press-fit pilot. Refer to Fig. 8 · 14, view A. For retention, these pilots rely solely upon the press fit. When used for applications similar to the illustration, they are hazardous. They may pull and drop out of the punch (many have), causing damage and creating a hazard to the operator. Therefore, the construction shown is not recommended for "good" dies, although it is used for

Figure 8 · 14 Punch-mounted pilots.

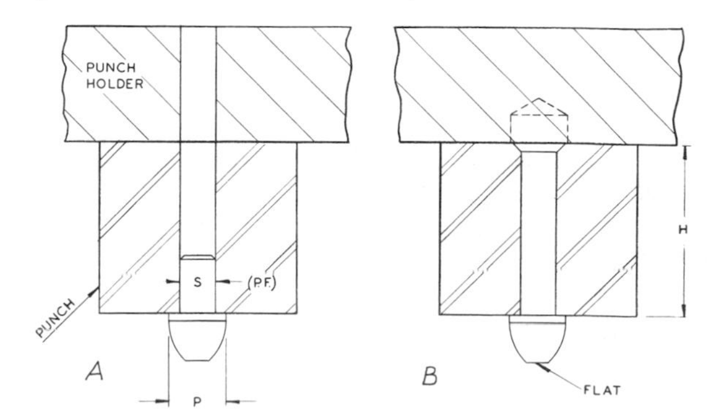

low-speed, low-production dies. However, it is best to avoid this method of construction entirely.

These pilots may be used for safer applications such as in inverted dies performing secondary operations, etc., where the pilot is not apt to pull and create a dangerous situation.

Peened-in pilot. The back end of the pilot shank is left soft and is peened into the countersink, forming a bevel head which retains the pilot in assembly (view *B*). The head is ground flush with the punch-base surface in assembly. If installed in a cutting punch, the pilot is driven out to permit sharpening the punch and then inserted again and repeened. Repeated peening may deteriorate the shank end. If deterioration occurs, grind the punch face to lower punch height *H*. This will expose the pilot shank a slightly greater distance at the back, providing new stock for peening. Of course, it will also detract from potential punch life; don't overdo it. Sometimes for cutting punches a clearance hole is provided behind the pilot, as indicated by the dotted outline. The idea is to eliminate the need for making the pilot flush with the punch base. A small flat on the pilot tip, as shown, will facilitate peening.

Threaded pilot. The pilot has a threaded shank and is secured by a nut (Fig. 8 · 15, view *A*).

Figure **8 · 15** Punch-mounted pilots.

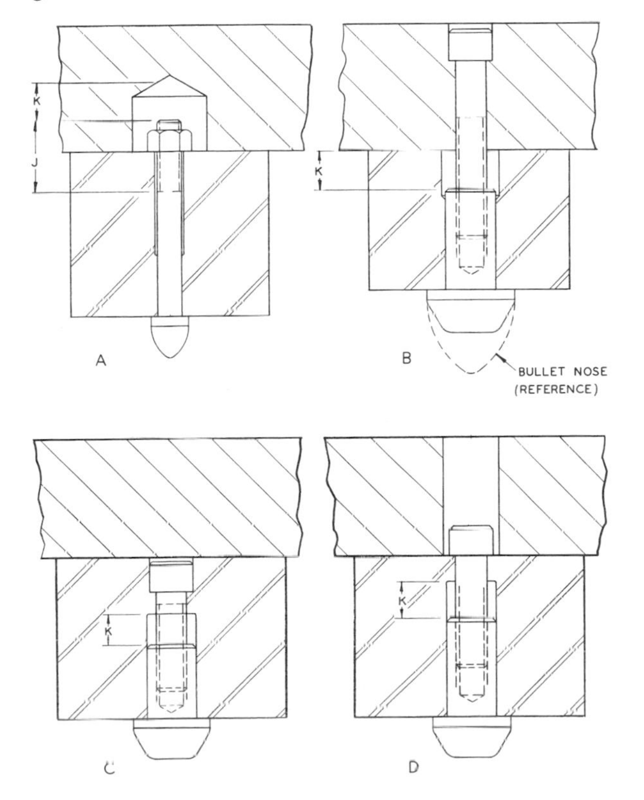

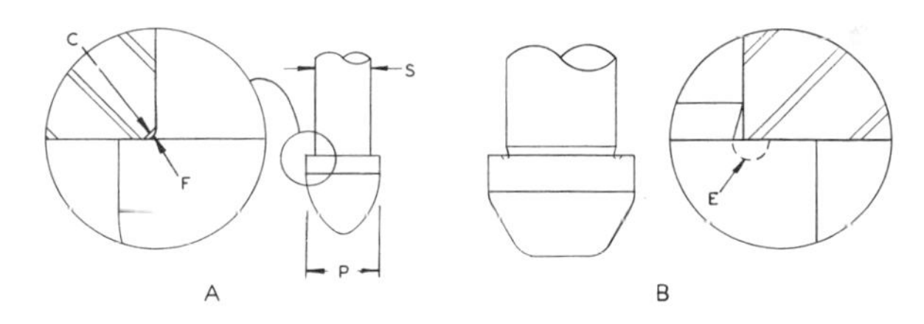

Figure **8 · 16** Stepped-down pilots.

Threaded length *J* is made long enough, and distance *K* deep enough, to allow for punch sharpening. This is generally considered a quality method applicable to high-production dies and suited to smaller pilots (⅜ in. or less shank diameter). The method may be used for larger shanks, but the relatively large nut required is often objectionable. For larger pilots (½ in. or more shank diameter), method *B* is commonly employed. In this view, a conventional bullet-nose outline is shown for reference to illustrate the disproportionately large nose which would be required if the bullet-nose configuration were applied to a large pilot. Methods *C* and *D* are, essentially, variations of *B*.

The pilots pictured in Figs. 8 · 14 and 8 · 15 are stepped-down pilots. That is, the piloting size *P* is larger than the shank size *S*, as shown in Fig. 8 · 16, view *A*. This view illustrates an important construction feature for smaller step-down pilots: A small fillet should be provided, as shown at *F* in the enlarged inset view. Fillet size should be 0.010 in. radius, min. In cases where the differential between dimensions *P* and *S* is proportionately large, the fillet radius should also be larger. To clear the fillet, countersink the mating member *C*. Countersink carefully: If countersink is too small, there will be interference; if too large, pilot support will be inadequate. Either extreme can cause the pilot to break. With cutting punches, the countersink will have to be renewed after sharpening. This can be done with a carbide burr (cutter) or by grinding.

With larger pilots, where the configuration is such that the pilot will be strong enough, an undercut is permissible. Such an undercut is shown, exaggerated, in view *B*, inset. Remember that undercuts should always be held to a minimum in size. Keep in mind also that undercutting is not practical on all large pilots but only on those with proportions which are strong enough to accept the undercut.

Flatted pilot. Pilots are, on occasion, provided with flats in some manner similar to Fig. 8 · 17. The purpose of pilot flats is to reduce the amount of piloting contact. Thus, flats are more apt to be

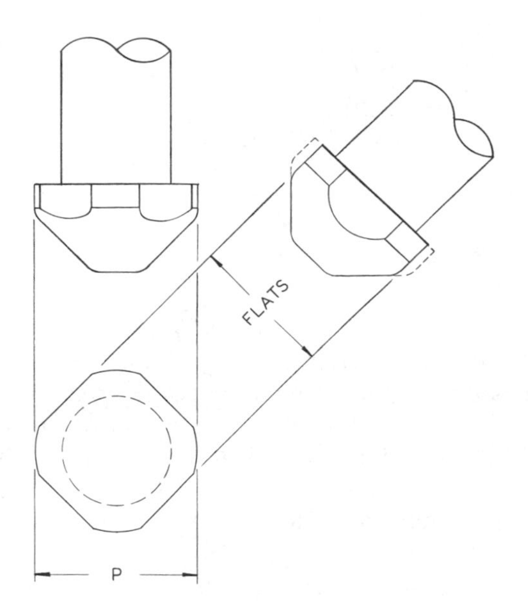

Figure 8 · 17 Flatted pilot.

appropriate on larger pilots. Since the reduced contact area will permit freer pilot-stripping, pilot flats can be desirable on punch pilots to help eliminate blank-pulling. A flatted pilot is functionally a round pilot intended to pilot in a round hole. The flats afford relief clearance. This kind of flatted pilot for this kind of application will not need to be keyed to prevent rotation.

MATERIALS AND HEAT TREATMENT FOR PILOTS

Pilots should be made of good tool steel, carefully heat-treated to conform to individual requirements. Normally, the piloting portion should be quite hard in order to resist wear. Heads on pilot punches should be drawn to a spring temper or softer. This applies also to shanks on step-down pilots.

For a given application, it is good practice to make the pilots of the same material that is used for the punches and/or perforators.

CHAPTER

9

DIE-BLOCK CONSTRUCTIONS

The term "die block" is used generically in reference to the complete unit component which mates with the punch (or punches) in the manner required to produce the desired effect upon the stock material. In the nomenclature of diemaking, the die-block unit may consist of a single piece, or it may be an assembled unit composed of a number of pieces. In the latter case, there are two predominant general constructions:

1. The die block serves as a yoke containing insert sections in addition to other necessary die contours.
2. The die block is sectioned into individual sections as required. In some instances, the sections may be secured solely by means of screws and dowels. Most applications, however, will require greater security. The die sections can then be mounted in a separate yoke or in a slot or a pocket, as required. The slot or pocket may be provided in the die set or in a special mounting plate, as determined for the specific application.

When the term "sectional die" is used in the general sense, it may refer to either of the specific constructions described above. Sectional and/or inserted die construction can afford the following major advantages:

1. Accessibility for the machining, grinding, checking, and inspection of die contours.
2. Greater accuracy for complex irregular contours. Sections and inserts can be finish-ground in their final hardened condition. This permits correction of any shrinkage or warpage which may have been induced because of heat treatment, etc. Also, any decarburized surface skin can be removed by grinding, which results in a better die life potential.
3. For heat treatment, unfavorable mass ratios can be eliminated or at least greatly alleviated.
4. For die maintenance, relatively fragile or perishable sections may be replaced individually, if required.

The foregoing discussion should not be construed to mean that sectional construction is always superior to one-piece construction. This just would not be true. Optimum die-block construction depends upon the set of circumstances peculiar to the specific application.

ONE-PIECE DIE RINGS

Among the simplest of die-block constructions is the one-piece ring pictured in Fig. 9·1. Dowel pins are used to locate and position this die ring. The screw and dowel locations shown are conventionally typical for smaller rings of this kind. Bolt-circle diameter C should be such that the screw and dowel positions will not be too close

Figure 9·1 One-piece ring-type die ring, dowel-mounted.

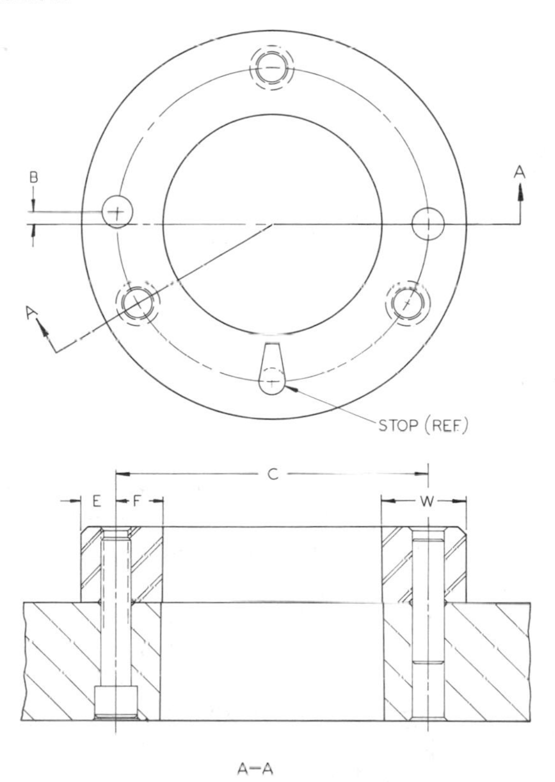

to the wall of the die-shoe clearance opening. For rings with minimal cross section W, this usually requires that dimension F be somewhat larger than E. For rings with wider cross sections W, the bolt-circle diameter may be made to divide W equally; that is, $E = F$.

The mounting screws are equally spaced on the bolt circle. For smaller rings, three mounting screws are generally appropriate. The number required increases as the diameter of the ring increases. For larger rings, it may be desirable to increase the number of dowels also. Three dowels for large rings and four for very large rings are not unusual. Smaller dowels than otherwise can then be used, and the strength of the ring in cross section W is correspondingly improved. This reasoning applies to the mounting screws as well as to the dowels.

Note that one dowel pin B is offset. This is a foolproofing procedure: If the dowel were not offset, it would be possible to mount this ring upside down. Foolproofing can, of course, be accomplished in other ways. For example, instead of offsetting, one dowel could be smaller, or one of the screw locations could be offset. The illustrated screw and dowel arrangement was chosen because:

1. It was deemed simpler to install two dowels of the same diameter, since the same drill and reamer combination can be used for both dowel holes.
2. Equidistant screw spacing is ideal from the standpoint of screw pressure distribution. Foolproofing procedures, however, vary to suit the specific situation: The optimum method for one application may not be optimum for another. It should be remembered that the configuration shown would be subject to modification or change according to the circumstances.

Another common construction for round work appears in Fig. 9·2. Here, the die ring is set in the die shoe. A pocket is bored in the die shoe, and the outer diameter of the ring is fitted to the pocket. This is a stronger, more stable construction. It is also more expensive. Dowels are not actually necessary. Foolproofing can be accomplished by offsetting one of the mounting screws. However, it is good practice to supply one dowel, as shown. The dowel not only foolproofs the mounting but also can be a convenience for the diemaker, since it is simpler to line up the mounting screw holes. The same effect can be achieved by identity marks on the ring, but the use of a dowel is the safer method.

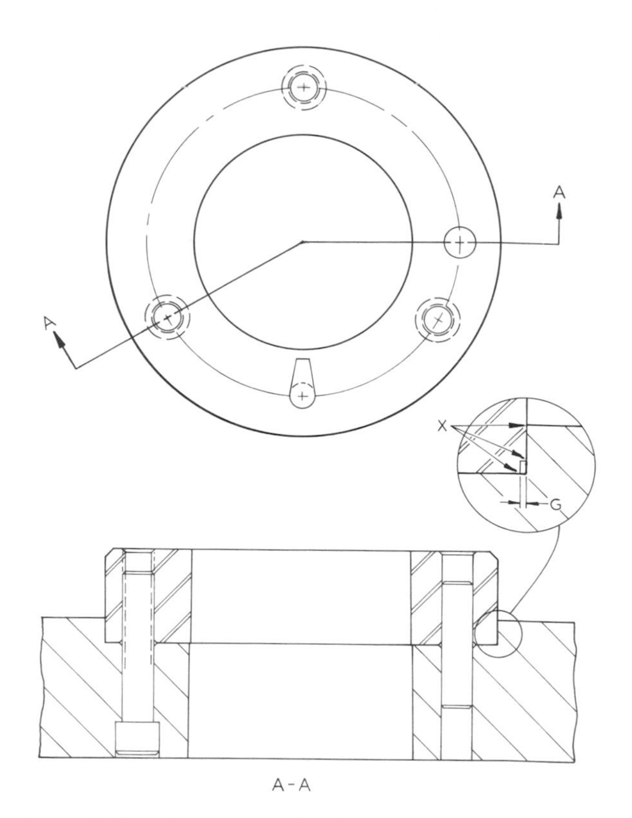

Figure **9·2** One-piece die ring, set in.

Normally, an assembly of this kind does not require a press fit between the ring and the shoe, and in fact, a press fit is usually undesirable. Fits ranging from a wring fit to a light drive fit are suitable. A small lead G, as shown in the inset, is of considerable aid for both fitting and assembling. Size of step G should be such that the lead portion is a slip fit in the bore. For small- to medium-sized rings, G will normally be 0.0002 to 0.001 in. Be sure to touch off the sharp corners at X before assembling. This is necessary to prevent shearing, which would adversely affect the fit and the positioning of the ring to the bore.

On occasion, circumstances may make it desirable to provide the lead step in the top of the bore instead of on the lower end of the ring. This could be true, for example, if a series of rings were meant to be interchangeably installed within the same bored pocket. As a general rule, however, it will be more expeditious to provide the lead on the ring, as shown.

Die rings are often assembled into yokes. There are a number of reasons for employing this construction. Among them are:

1. Material cost, which is in direct proportion to the size of the ring.
2. Replacement possibilities. Should replacement be required, this construction is generally more desirable than a one-piece ring.

3. Interchangeability factors. A die can be made using a series of interchangeable rings and corresponding punches to produce a number of different-sized piece parts.

Yoked die-ring assemblies can be made in various ways. The three most common methods are pictured in Fig. 9·3.

View A: Plain Ring Bushing and Yoke. Fit between bushing and yoke may range from a light drive fit to a shrink fit, depending upon the application. Mounting (hold-down) screws are generally required for the bushing when lighter fits are used, unless the bushing is held in assembly by another component, such as a stripper, etc. For heavy duty, a hardened thrust plate will be required under this type of assembly.

View B: Shouldered Ring Bushing. This type is normally used where lighter fits are desired between the bushing and the yoke. Mounting screws are not required for the bushing. Thrust plates are required for heavy service applications.

Views C and D: Tapered Assemblies. These assemblies are made by press-fitting or shrink-fitting. In view *C,* the bushing is tapered larger at the top. Optimum size for angle *L* depends upon the specific application. If maximum holding friction between bushing and yoke is desired, then angle *L* is generally made 1 to 1¾°. This angle range produces a self-holding taper. If high resistance to downward thrust is desired, *L* may be 8 to 10°, which would be a self-releasing taper. In the latter case, the fit alone should not be relied upon to hold the bushing in assembly. Thrust plates may or may not be required.

Figure 9·3 Cross sections through yoked die rings.

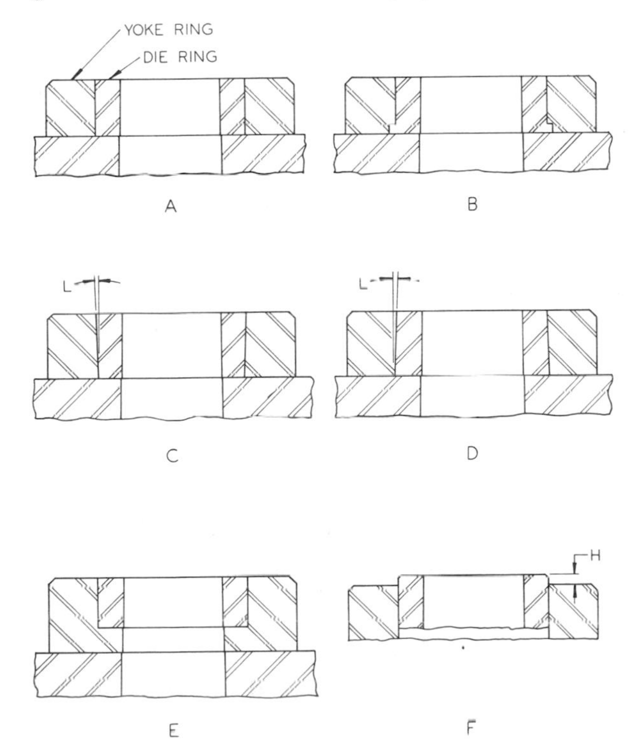

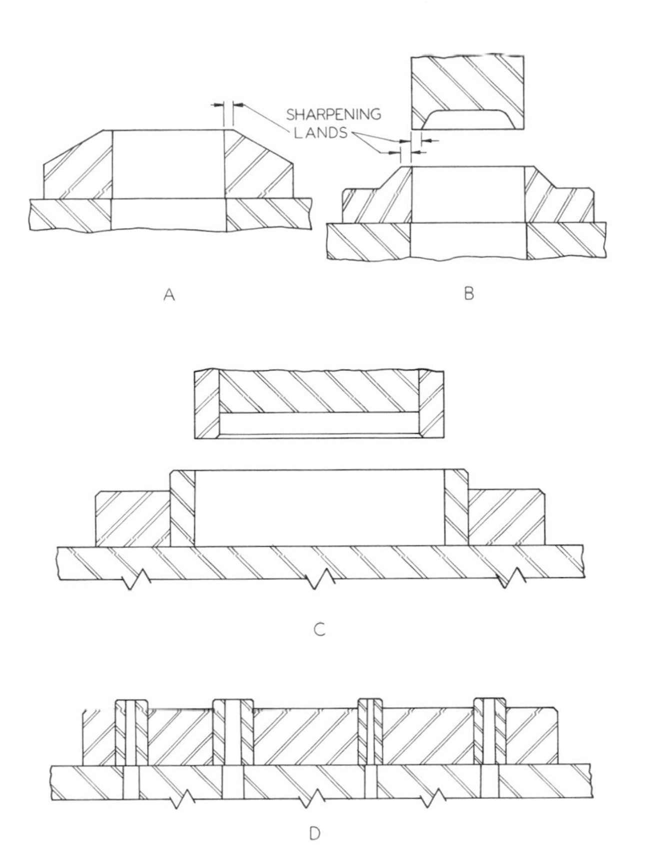

Figure 9·4 Sharpening lands.

In view *D,* the bushing is tapered in the opposite direction. With this method, angle *L* is conventionally made within the range ⅛ to 1¾°. Thrust plates are generally required for this type of assembly, but separate mounting screws (or the equivalent) are not required to retain the bushing in assembly.

View E: Plain Ring Bushing in Stepped Yoke. Fits are the same as for method *A.* The specific application determines whether or not mounting screws are needed for the bushing.

View F: Protruding Bushing. Face surface of bushing protrudes a distance *H* above the yoke. If feasible, this construction is employed when the application is such that a sharpening land is desirable.

SHARPENING LANDS

The cutting faces of die blocks or punches are sometimes relieved as typified in Fig. 9·4. This reduces the area of the cutting face to a comparatively narrow land which is contiguous to the cutting edges. The purpose of sharpening lands is to facilitate die maintenance by reducing the area which must be reground at each sharpening. Thus, the desirability of sharpening lands is associated with the size of the component, larger components tending to receive more maintenance benefits from sharpening lands.

Typical sharpening lands on one-piece components are depicted in views *A* and *B*. When making components of this kind, the diemaker must give especial advance consideration to his work sequence. In many instances it may become quite awkward to perform further machining operations after the component has been relieved to create the sharpening land. Indiscriminate use of sharpening lands should be avoided. The die designer should determine the feasibility of this construction in accordance with the particular application.

Views *C* and *D* indicate how insert-type construction can be used to create sharpening lands. Since composite constructions, such as these, are normally made with dissimilar materials, the land effect can provide a distinct sharpening advantage. For example, assume that the inserts are sintered carbide and the yoke blocks are tool steel. Diamond-impregnated grinding wheels are normally required for practical carbide grinding, but these wheels (or any type of wheel intended for carbide grinding) rapidly load up and glaze when grinding materials other than carbide. The constructions shown at *C* and *D* permit grinding only the carbide inserts when sharpening, which precludes the possibility of loading the grinding wheel with tool-steel particles. This principle applies in lesser degree to other dissimilar materials—for example, where the inserts are hardened tool steel and the yokes are soft machine steel. The sharpening benefit is, of course, proportionate to the size of the area which will require sharpening.

PEEN-AND-SHEAR SHARPENING

This is a practice applied in some industries to dies which cut soft and/or thin materials. Either the die block or the punch is hardened to the normal degree of hardness. Its mating component is not; it is merely toughened enough to provide optimum wear resistance while still remaining soft enough to permit peening (the terms peening, staking, and swaging are used interchangeably in regard to this procedure). Some steels may be peenable up to a hardness of Rockwell C 44. However, unless there is special justification for this, it will be more practical to hold the hardness down to Rockwell C 38 to 40. In some applications, the peenable member of the die is not even heat-treated but is merely left soft. In other applications a pretreated steel may be used. Such steels are generally machinable (Rockwell C 34, maximum hardness). Sharpening lands (refer to Fig. 9·4) are often desirable on the peenable member to facilitate the sharpening procedure.

If the die block is the softer member, use the following procedure:

1. Sharpen the punch, using normal grinding procedures.
2. Peen the land adjacent to the die opening, forcing the opening to become slightly smaller at the top.
3. Cleanup-grind the cutting face of the die block, keeping material removal to the minimum.
4. Assemble the complete die and shear the punch through the die opening.

As far as sequence is concerned, step 2 may precede step 1, if desirable. Also, in some instances it may be desirable to perform step 3 after step 4.

For applications where the punch is the softer member:

1. Sharpen the die block, using the necessary normal grinding procedures.
2. Peen the face of the cutting punch to expand the punch a slight amount at the cutting edges.
3. Cleanup-grind the face of the punch with minimum material removal.
4. Assemble the complete die and shear the punch through the die opening.

Step 1 may be alternated with step 2, and step 3 with step 4.

It is rather obvious that, opposed to conventional sharpening, the peen-and-shear method entails more work for die maintenance. The purpose of this method is to facilitate the making of very close-fitting punch and die combinations for cutting thin, soft materials. Percentagewise, the popularity of this procedure is declining. However, it is well suited to certain applications, and a great many dies will continue to be made this way. Most diemakers, at some time in their career, will find it advantageous to be familiar with the method.

SECTIONAL DIE BLOCKS

A sectional die block is one which is made up of two or more sections. This definition is illustrated by comparison in Fig. 9·5. View *A* depicts a die opening in a one-piece die block, which is often called a solid die block. Views *B*, *C*, and *D* are considered sectional die blocks.

Whether a die block should be sectioned, and how it should be sectioned, is determined by

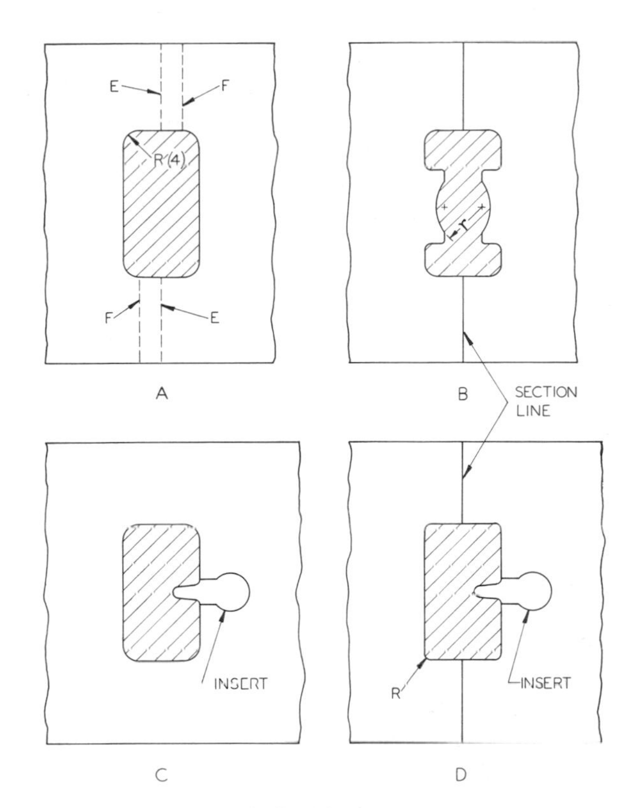

Figure 9·5 Typical die-block constructions in comparison (die openings shaded).

analysis of the specific application. However, the following generalities apply:

1. *Large die blocks.* As the size of the required die block increases, so do the advantages of sectional construction.
2. *Small die openings.* When the size of a die opening is too small to permit efficient internal working, the die should be sectioned.
3. *Complexity.* The advantages of sectional construction become greater as the complexity of the die contours increases.
4. *Perishability.* Sectional construction can simplify the making and the replacing of relatively perishable portions of the die.
5. *Ground openings.* Where the size and/or contour of the opening is such that internal grinding methods are not practical, the die must be sectioned if a ground opening is specified.

Using the above criteria, one can make the following analysis pertaining to the illustrated die contours:

View A. Ordinarily, this die opening would not require sectioning unless it were so large or so small that a one-piece die block would be impractical. If a filed die opening is acceptable, radius *R* can be any required size down to a sharp corner. If, however, a ground opening is specified, then *R* must be large enough to permit practical internal grinding of the contour. If *R* is too small, then the die block will have to be sectioned to provide access for grinding.

As simple as this opening is, it could be sectioned in a number of different ways. The simplest and most practical is indicated by the dotted lines shown at *E.* An alternate sectioning is indicated by the dotted lines shown at *F.* The alternate method would apply to wider openings, in order to eliminate side-grinding. Larger openings would, of course, require other approaches to sectioning (refer to Fig. 9·7).

View B. Here, the die contour is somewhat more complex, and the decision whether the die block should be sectioned is more complex also. If precision is relatively unimportant and production requirements are low, the most economical construction will probably be a filed die opening in a thin one-piece die block. If, however, a quality die is desired, the most efficient construction will be to section the die block, as shown.

View C. Here, an insert is provided in the die block because the small prong which intrudes into the die opening is relatively susceptible to breakage. If this occurs, only the insert needs to be replaced when effecting die repairs. Also, the insert permits the balance of the die opening to be internally ground. The insert can, of course, be finished by grinding before it is installed. This yields a completely ground die opening, if desired.

View D. Radius *R* is too small to permit practical internal grinding. Therefore, if a ground opening is desired, the die block is sectioned, as shown. Sectioning this die block also provides better accessibility for making the opening which is required for the insert.

Generally, when an insert is necessary, the need is obvious. Nevertheless, there are in-between situations where it is difficult to decide whether or not an insert should be provided. In such cases, it is often feasible to provide for an insert which can be made and installed later if breakage occurs. In Fig. 9·6, view *A*, hole *C* is provided in the die block when the die is made. If the prong breaks in service, the insert opening is then ground out as indicated by the dotted lines and the bulb-type insert is installed. Hole *C* can, of course, be hard-drilled or disintegrated after the die has failed, which will eliminate the need for the provisional hole. The decision as to which method is preferable depends upon circumstances pertaining to the specific situation. View *B* indicates an alternative procedure: the dovetailed insert. Note the desirability of the sectioned die blocks which give access to the die opening,

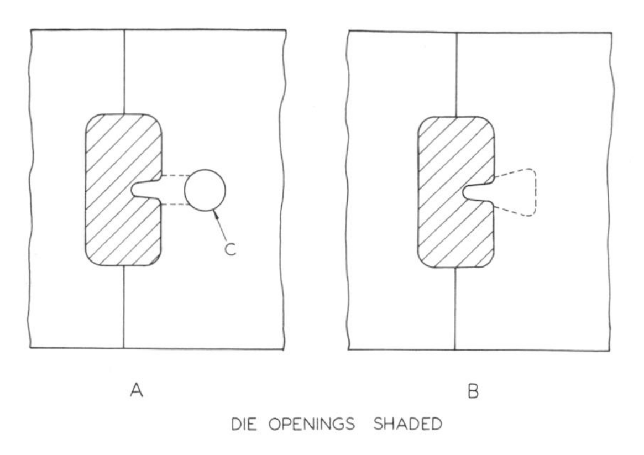

Figure 9·6 Provisional inserts.

thereby facilitating installation of the inserts, if required.

SECTIONING LARGE OPENINGS

Refer back to Fig. 9·5, view *A*, and consider the feasibility of a one-piece die block and the practicality of various ways of sectioning it. Then examine Fig. 9·7. When compared, these figures show how the size of a die opening can influence the decision whether to section a die block and the method followed when sectioning it.

STABILITY AND STRENGTH IN SECTIONAL CONSTRUCTION

During the operation of a die, the punching force is transmitted to the die assembly. A proportionate part of the force is transmitted to the individual die sections. High-impact conditions, rapidly and continuously repeated, are involved, and as a result, it is common for cutting die sections to be subjected between sharpenings to a hundred thousand continuously repeated hammerlike blows. During its total life, the section may be subjected to several million impacts. These estimates apply to die sections made of tool steel. A tungsten-carbide die may easily be subjected to ten times as many impacts, and such extra-long-life potential must be well considered in the design and construction of tungsten-carbide dies.

Figure 9·7 Sectioning for larger openings.

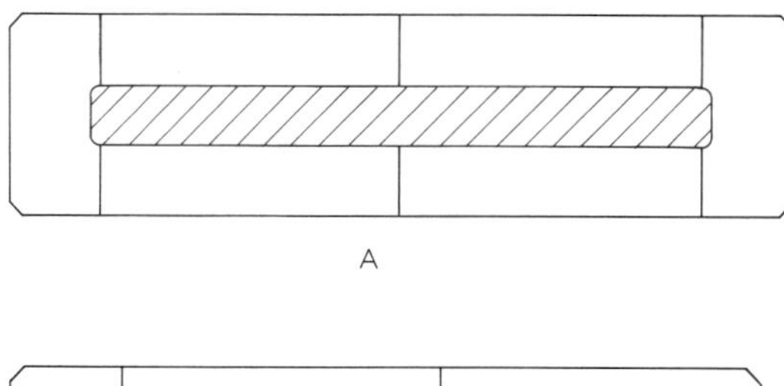

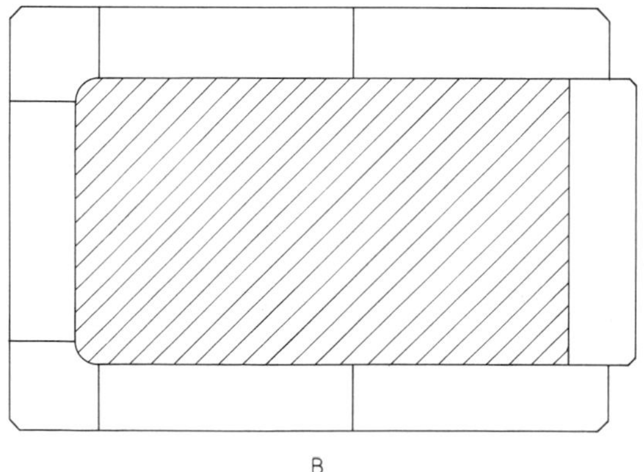

Punching force resolves into two major forces acting on the die block:

1. Direct thrust in the direction of the punch motion.
2. Lateral thrust in a direction which is at an angle to the punching direction. Most commonly, the lateral thrust radiates in a plane at right angles to the punching axis. With piercing dies and blank-through dies, lateral pressures are accumulative (see "Angular Clearance" in Chap. 1)—a factor which must be allowed for in the construction of these dies.

As a result of the forces involved, two special considerations are essential to the design and construction of sectional dies:

1. *Tilting.* Sections must be either self-supporting or supported by adjacent sections. There must be no possibility that a section will tilt when subjected to the punching impact.
2. *Spreading and separating.* Possible lateral displacement of die sections due to lateral thrust and/or pressure must be precluded.

Where die sections are relatively large, and where side thrust is self-canceling or minimal, security of the sections may depend only upon mounting screws and dowels. However, where greater forces are involved, some other means of assuring the strength and integrity of the assembly must be provided.

CROSSBOLTED DIE BLOCKS

Figure 9·8 illustrates the die-block assembly from a compound pierce-and-blank die. This die produces relay springs from 0.010-in.-thick nickel-silver strip. It is an inverted-type die. The die block is mounted to the so-called punch-holder member of the die set, which is attached to the press ram. The assembled die block is made up of two large die sections and two small insert sections. The small bulb-type inserts are accurately positioned by virtue of being fitted into the opening provided in the proper large die section. To

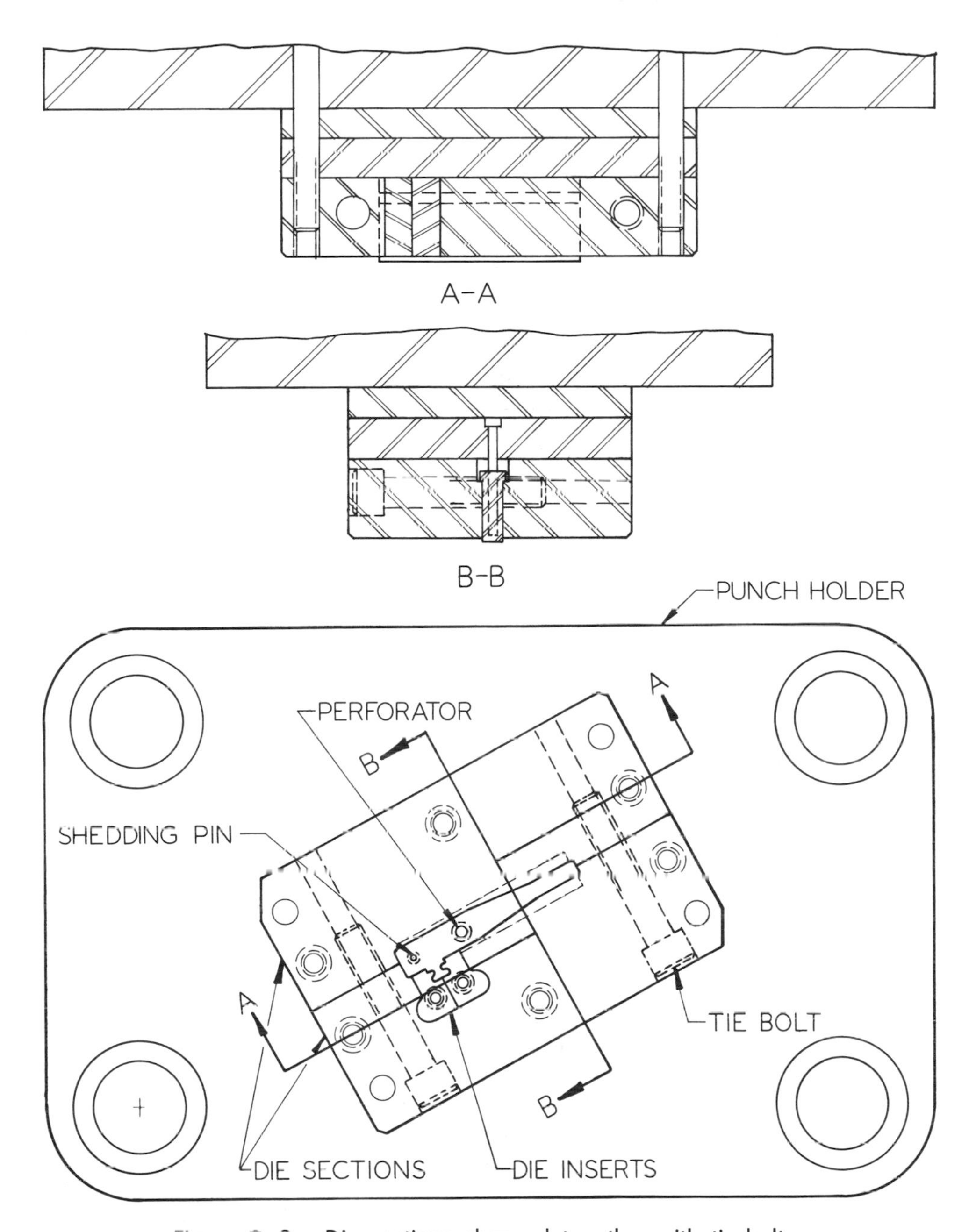

Figure 9·8 Die sections clamped together with tie bolts.

resist lateral blanking forces, cross screws (tie bolts) are installed through the large sections, clamping the sections tightly together.

Crossbolts are probably the most economical method of clamping sections together, but they tend to detract from die life. The tendency is not serious with shedder-equipped dies, which are designed as shown, with a pocket for the shedder head milled into the back of the die block. This creates a thicker die block, which permits installing crossbolts without appreciably detracting from die life. Also, lateral blanking pressures are not accumulative in shedder-equipped dies, so that crossbolts of a practical diameter can be used

Figure 9·9 Tie-bolt construction (bending die).

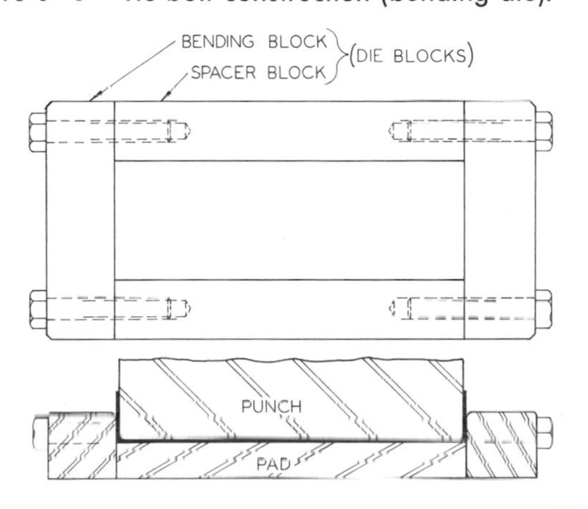

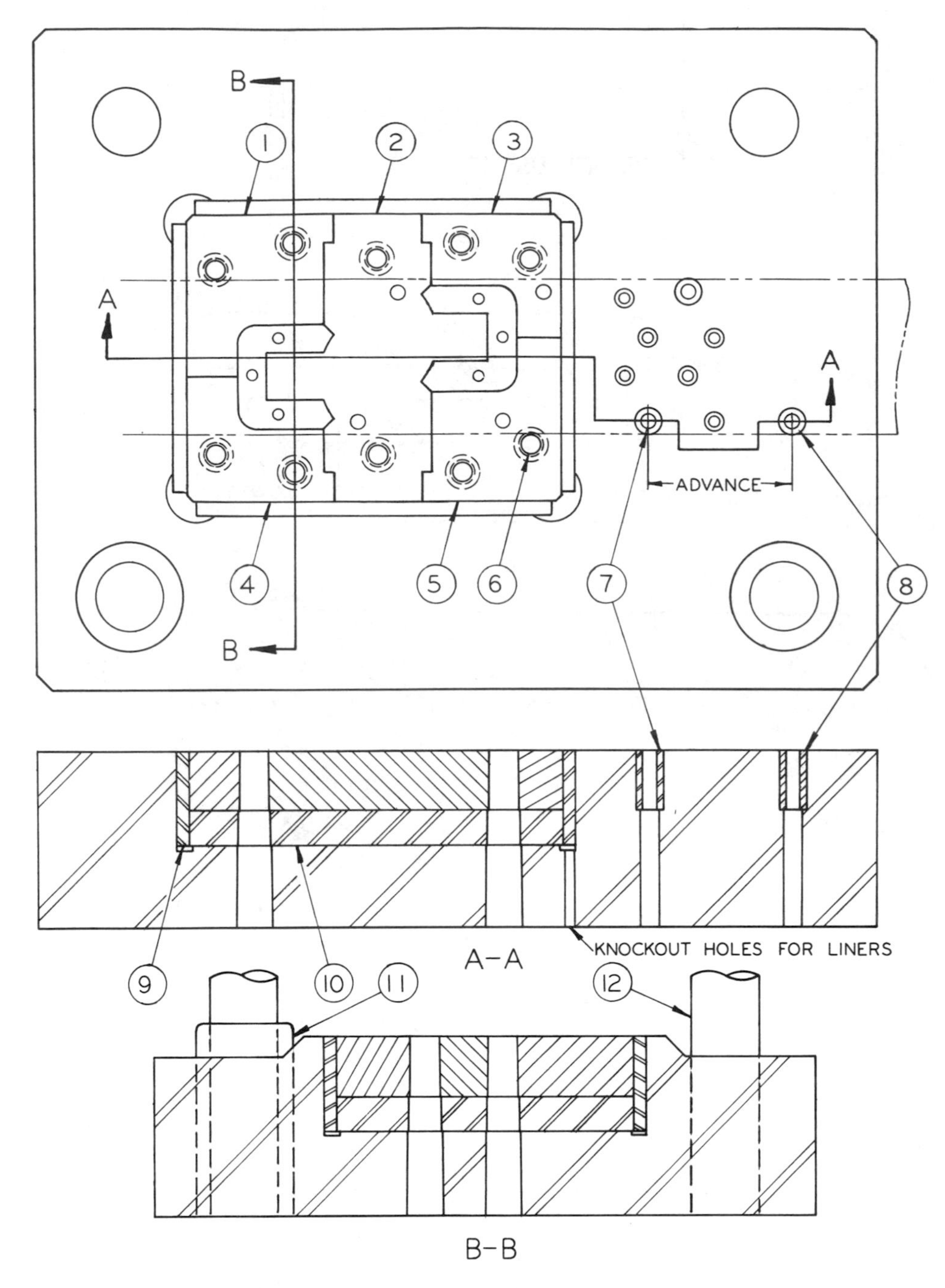

Figure 9·10 Die-block assembly with die sections pocketed in shoe.

Crossbolt construction is not generally practical for blank-through dies. In these and kindred dies, the crossbolts tend to detract seriously from die life. The condition is further aggravated because the lateral pressures are accumulative in such dies, and therefore larger crossbolts are required than would otherwise be necessary.

The crossbolt principle often lends itself to bending and other forming applications. A simple construction of this kind appears in Fig. 9·9. Because of the length of the die, short tie bolts are used which are simply threaded into the ends of the spacer blocks, as shown. When making dies of this sort, be certain the tie bolts are strong enough. For heavier applications, hexagonal head bolts are more desirable than socket heads.

In summary, the crossbolt principle is ordinarily the cheapest method for confining die sections, but it is limited in application.

DIE BLOCKS ASSEMBLED IN POCKETS

For a large number of applications, pocketing is an appropriate and effective method of confining die sections. The sections are generally set into a pocket which is provided in the die shoe, although on occasion it may be desirable to pocket the sections in a separate mounting block. If the latter is advisable, it will be obvious.

Figure 9·10 illustrates the die-block–die-shoe assembly for a progressive die which pierces and blanks field laminations from 0.014-in.-thick silicon steel, electric grade. The stripper, stock guides, etc., are not shown, in order to clarify the illustration. The die sections 1 to 5 are made of sintered carbide. As pictured, they are set into the rectangular pocket provided in the die shoe. Pocketing the sections does not eliminate the need for mounting screws. In this case, each section is bolted down with two socket head cap screws 6.

Heavy force fits should not be used to assemble the sections in the pocket. Die sections must fit into the retainer tightly, but assembly pressure must never be so great that it causes bowing or other distortion of the shoe. Fits for sectional die assemblies are best described as light force fits. Note how 1, 3, 4, and 5 are stepped and mated to 2. This is excellent construction. The sections are interlocked in assembly, which compensates for the somewhat unbalanced contours of some of the individual sections.

Liners (9) are an important feature where sections are assembled in pockets, yokes, etc. Liners are made of tool steel hardened to Rockwell C 54 to 58. They provide the following advantages:

1. Liners eliminate the possibility of shearing the walls of the pocket or yoke. This fact alone justifies the use of hardened liners.
2. Liners expedite accurate assembly of the sections. The liners are fitted and assembled last, which permits compensation to be made for discrepancies in size and/or position of the pocket.
3. Liners facilitate disassembly. The liners are pushed out first, which frees the die sections in the pocket. Therefore, push-out or knockout holes must be provided under the liners, as indicated. If liners are made too thin, their very real advantages will be nullified.

Figure 9·11 Up-and-down guideposts facilitate sharpening.

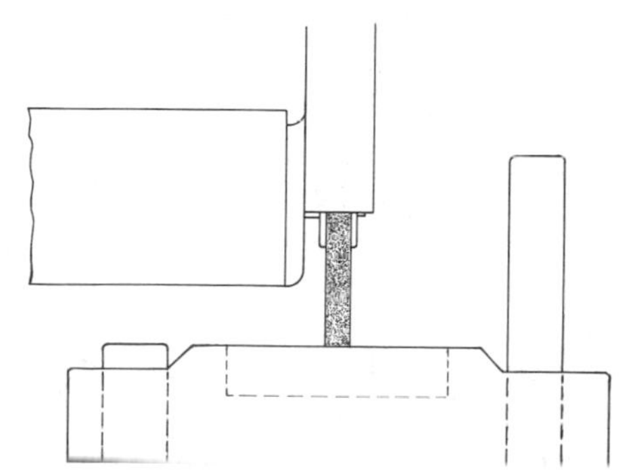

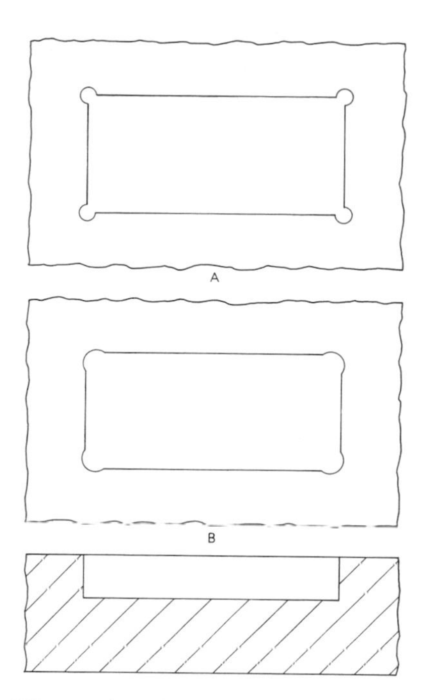

Figure 9·12 Pocket details.

Hardened thrust plates may or may not be necessary, depending upon the proportions of the sections. The hardness of the thrust plate should be selected from Rockwell range C 48 to 56, according to the configuration of the die contours. The thrust plate is provided in the case illustrated because of the projecting contours of 2. Thrust plates are also required behind sections whose area is proportionately small in relation to the forces applied. The thrust plate minimizes the possibility that such sections may become impacted into the bottom of the pocket. It also minimizes any creep tendencies due to the continuously repeated impact conditions which are inherent in die operation.

Die openings for the piercings are provided by the die buttons, of which 8 is typical. These bushings are made of sintered carbide and are pressed into the shoe, as pictured. It is not practical to enter pilots into soft blocks. Therefore, 7 is a hardened steel bushing which contains the pilot clearance hole.

Note the manner in which the guideposts and bushings are related to the die set: At the rear of the set, two of the four guideposts (12) are secured to the die shoe; at the front of the set, it is the bushings (11) which are mounted in the die shoe. This is a common feature of carbide dies. Its pur-

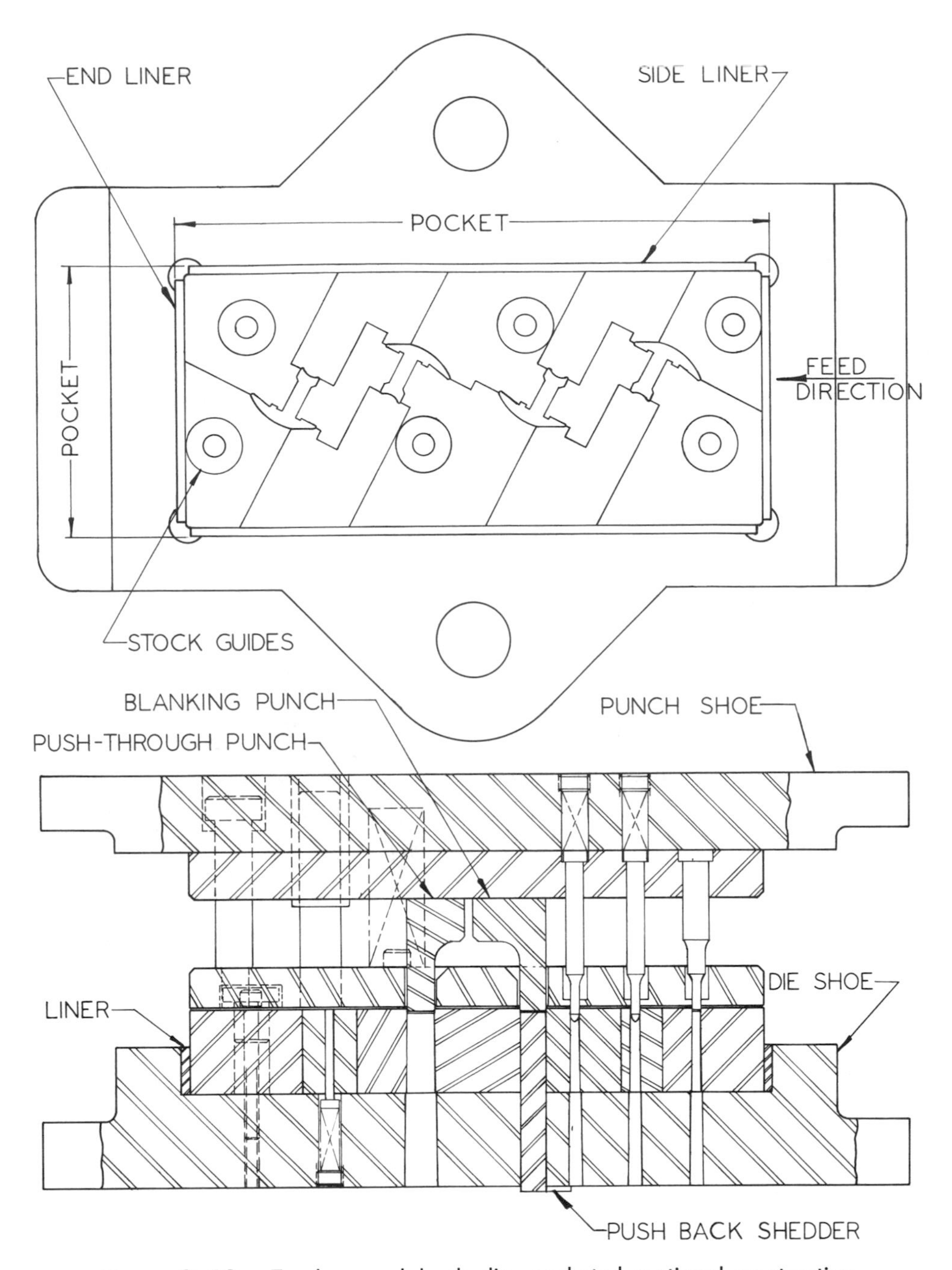

Figure 9·13 Tandem push-back die, pocketed sectional construction.

pose is to facilitate sharpening the die in assembly. The grinding wheels required for sharpening carbide dies are expensive, and the up-and-down guidepost configuration permits the use of smaller grinding wheels. This is illustrated in Fig. 9·11. Carbide dies are high-production dies. Their design should permit maintenance sharpening with an absolute minimum of disassembly.

Two pockets are pictured in Fig. 9·12. The pocket in view *A* is poorly planned. Its corner relief diameters are centered at the intersections of the pocket walls. This condition restricts the size of the milling cutter which can be used to finish the pocket, and the result is that more machining time is required. Also, even though the diameters are small, they detract from the strength of the yoking member. Both of these objections can be overcome by proceeding in the manner indicated in view *B*. This procedure obviously applies to yokes as well as to pockets.

In Fig. 9·13 the mounting screws, piercings, and pilot openings have been omitted from the plan view for purposes of illustration. This die makes two small stator lamination segments per stroke from 0.025-in.-thick silicon steel. The stock strip feeds from right to left. The piece parts are

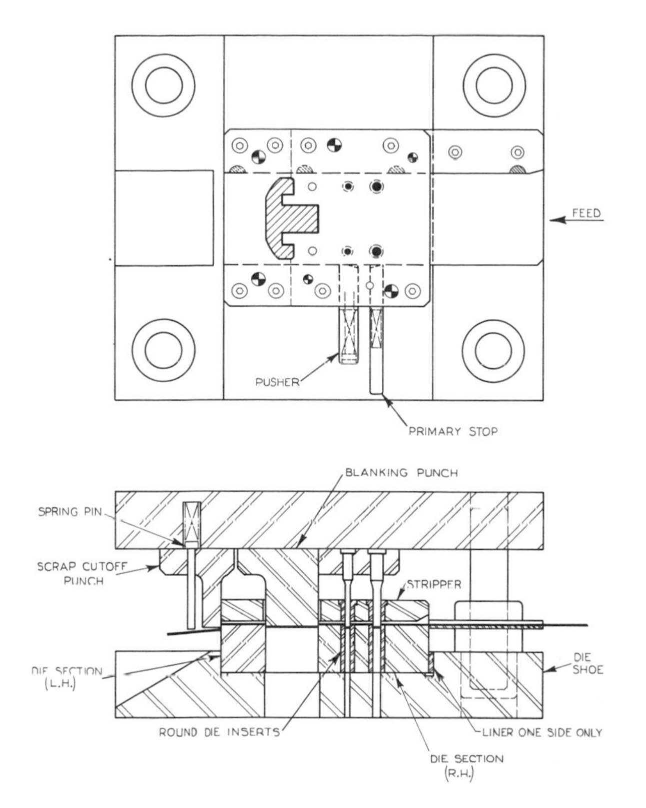

Figure 9·14 Sectional die block mounted in slotted die shoe.

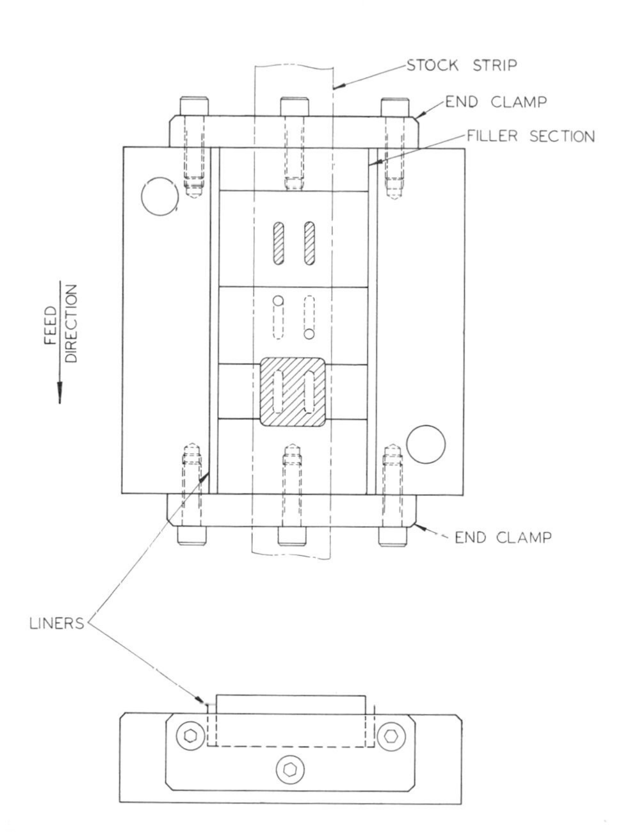

Figure 9·15 Die sections confined in slotted shoe with end clamps.

blanked into the first two openings. The push-back shedders which extend through these openings are actuated by the press cushion. The shedders push the blanks back into the stock strip, and the blanks are then carried along with the stock strip and pushed through the die at the next two openings.

The die is equipped with a spring stripper, as required for push-back applications. Spool-type stock guides are used in order to enhance the strength and rigidity of the stripper plate. This die is shown here because it is an interesting example of mechanically sound and well-planned die sectioning.

Because closed pockets are relatively difficult and exacting to machine, it is more desirable to mount die sections in open-end slots where feasible. The slots may be finished economically and accurately by grinding. Considered alone, such a slot provides confinement for the sections in one direction only.

DIE BLOCKS ASSEMBLED IN SLOTS

The die which appears in Fig. 9·14 has a die block which is composed of two large, heavy sections. The nature of the section pattern is such that they require strong confinement in the left-to-right direction only. The cutting forces are self-canceling from front to back, and therefore the mounting screws and dowels can adequately secure the sections in this direction. The die sections are mounted in a slot which extends through the die shoe from front to back. In this way strong confinement is provided to resist any tendency for the sections to spread apart. Since there are no section corners adjacent to the walls of the slot, liners are not needed to prevent shearing. Therefore, the die needs only one liner to aid in assembly fitting and in disassembly, when required.

This die has a primary stop, positioned to suit an automatic feed. It is also equipped with a scrap cutoff punch. A spring pin and a bridged trough are provided to assure and control the passage of the scrap from the die.

Where endwise confinement is necessary, in addition to the slot, the method illustrated in Fig. 9·15 is often satisfactory. In a sense, this could be interpreted as a sectionalized pocket which incorporates the tie-bolt principle for the end plates.

Figure 9·16 Slot detail.

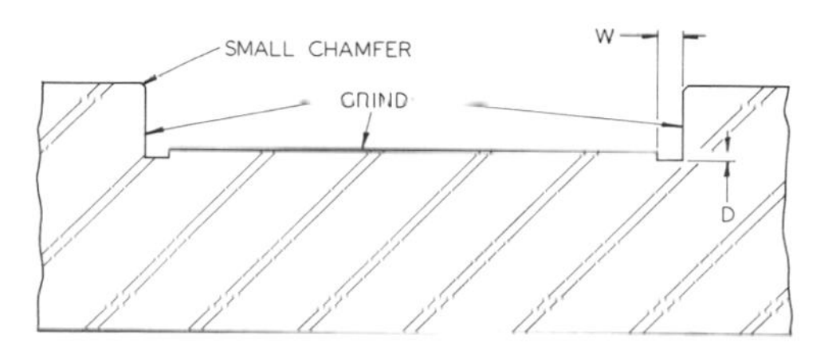

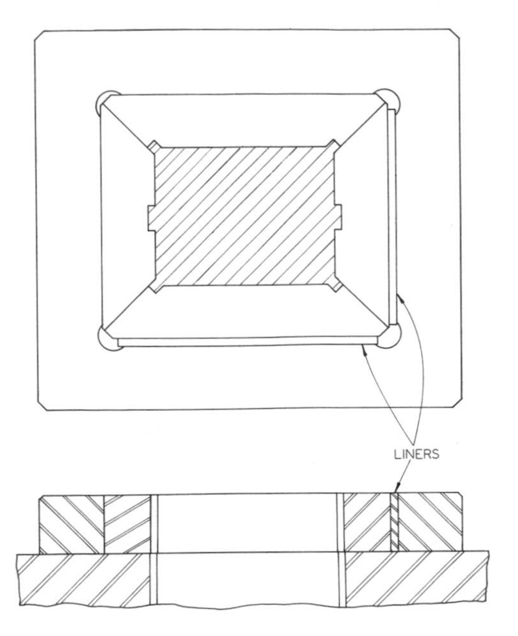

Figure **9 · 17** Yoked die sections.

It is an economical and precise method for confining die sections. Be certain that the clamp plates and the tie bolts are strong enough for the application. Check the tie-bolt locations to be sure that they do not detract from the required die life.

Open-end slots can be both precise and economical because they are easily finished by grinding, as indicated in Fig. 9·16. To facilitate grinding, a small relief step should be provided, as shown. A practical step depth *D* is 1/32 in. Step widths *W* may vary to suit the specific application: Most will be on the order of 1/4 in.—in small dies slightly less and in large dies possibly more.

YOKED DIE BLOCKS

In Fig. 9·17 the die sections are encompassed by a one-piece yoke. Yokes may be made of machine steel, alloy steels, or tool steels. They may be left soft, toughened, or fully hardened. The optimum material and heat-treat procedure depends entirely upon the requirements of the specific application. In a sense, a yoke can be considered a pocket which does not have a bottom. The similarity between yokes and pockets is obvious in the plan view. The earlier discussion pertaining to pocketed sections can be generally applied to yoked sections also.

CHAPTER 10

STRIPPERS AND STOCK GUIDES

Stripping is the act of removing the work from the punch or punches. A stripper is a device for stripping.

STRIPPING ACTION

There are two basic stripping actions, depending upon whether the punch is motivated or stationary. With a motivated punch, the stripper acts to arrest the stock material, permitting the moving punch to withdraw from it. This action is depicted in Figs. 10·1 and 10·2. Figure 10·1 illustrates the arresting action of a fixed stripper.

View A. Die is in closed position, with punch entered, at the bottom of the press stroke.

View B. Punch is ascending on the return stroke. The stock material (work) clings to the punch, moving upward with it until it contracts the intervening stripper, as shown.

View C. The punch as it continues upward is withdrawn from the stock material, permitting the stock material to drop toward the die surface.

View D. Working cycle completed. Punch has returned to its up position. Workpiece has dropped back to the die face and is ready to be removed.

Figure 10·2 depicts the arresting action of a pressure-pad-type stripper. The stripper shown is spring-actuated. Gages, spring pins, etc., have been omitted for purposes of illustration.

View A. Die is in closed position, punch entered, bottom of press stroke. Potential stripper travel for the stripping action is indicated at *M*.

View B. Punch assembly is ascending on the return stroke. The stripper is still in the position shown in view *A*. The springs have maintained this stripper position, forcing the ascending punch to withdraw from the stock material as shown. Stripper travel *M* relative to the punch is noted.

View C. Working cycle completed, top of press stroke. The punch holder has carried the punch and stripper assembly with it to the up position. The workpiece has been left behind, lying on the die surface to be retrieved or ejected as required.

In dies with stationary punches, the stripping action pushes the stock material from the punch. A fixed stripper is not appropriate for stripping a stationary punch: If the punch is stationary, the stripper must travel. Stationary punches are a characteristic of inverted type dies. Stripping action with such a die is pictured in Fig. 10·3.

View A. Die is in closed position, punch entered, bottom of press stroke. Potential stripper travel *M* is shown.

View B. Die block is ascending on the return stroke. The stripper springs have forced the stripper to ascend also. The ascending stripper motion has stripped the stock material from the

Figure **10·1** Stripping action: fixed stripper, punch moving with ram.

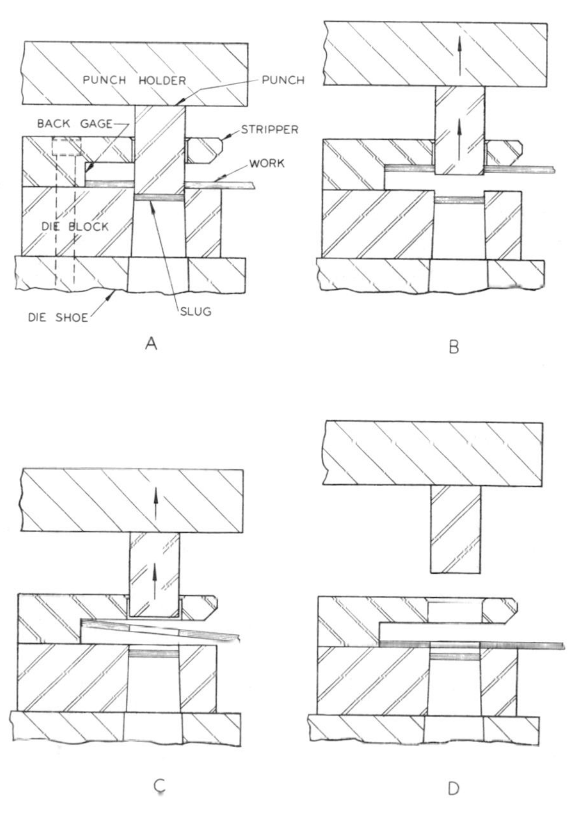

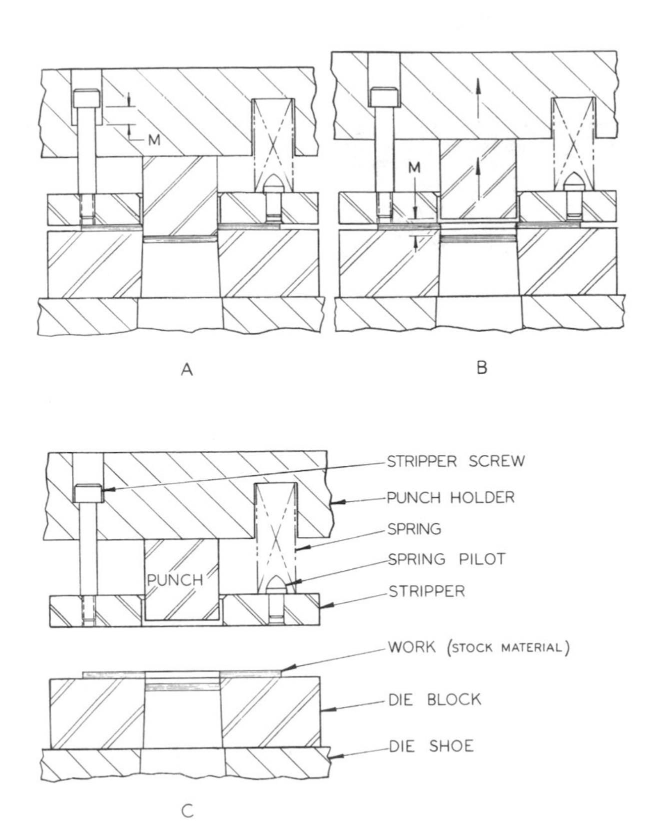

Figure 10 · 2 Stripping action: pressure-pad stripper, punch moving with ram.

punch. At this point the material is still sandwiched between the die face and the stripper face.

View C. Working cycle completed, top of press stroke. The shedder, driven by the knockout, has ejected the slug from the die opening. The workpiece has been left behind, lying on the stripper face to be retrieved or ejected as required.

STRIPPER CATEGORIES AND TYPES

Essentially, strippers can be classified in two categories: fixed strippers and traveling strippers. Each category is composed of numerous types and variations. For example, the stripper illustrated in Fig. 10 · 1 belongs to the category of fixed strippers and is also an open-gap-type stripper. Specifically, therefore, it is an open-gap fixed stripper. An open-gap stripper is distinguished by its single cantilevered configuration when viewed in cross section. The strippers shown in Figs. 10 · 2 and 10 · 3 are in the traveling category and of the pressure-pad type. Specifically, they are spring-actuated. Custom has decreed that these and similar strippers be referred to as spring strippers.

Generally, fixed strippers are simpler and easier to make than traveling strippers. Fewer components are required in the construction of fixed strippers when compared to the equivalent traveling strippers. Therefore, fixed strippers are economically desirable as far as die construction costs are concerned. Mechanically, a fixed stripper is solid (positive) in performance of the stripping act. This is an advantage where strong stripping forces are necessary. Thus, where practical for the application, fixed strippers are usually preferable. However, it must be kept in mind that there can be many situations where a fixed stripper may be impractical. For example, it may be necessary for the stripper to act as a hold-down in addition to performing the stripping function. It may be desirable or necessary to keep the punches engaged in the stripper during the entire press cycle. Visibility may be a factor. A traveling stripper can permit the operator to observe the work while the die is operating. A traveling stripper can make the die and/or the work more accessible for operating efficiency.

The above discussion should convey some idea of the major factors which influence any choice of strippers. The actual choosing is a matter of determining the most suitable stripper type for the given application. A complex die may require the incorporation of a number of different stripper types. Such applications compel careful consider-

Figure 10 · 3 Stripping action: pressure-pad stripper, punch stationary.

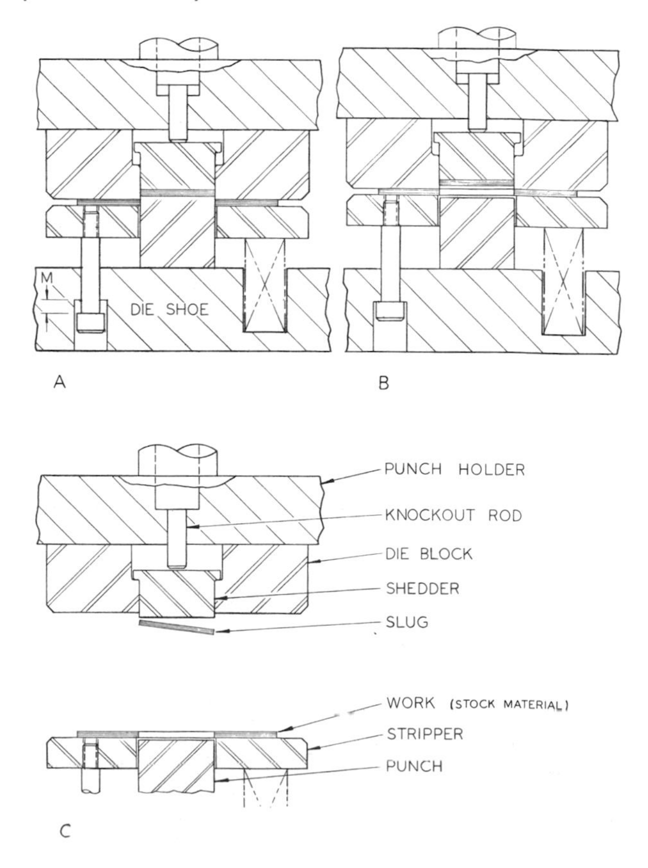

ation of the action of each stripper both singly and in relation to the others.

HOOK-PIN STRIPPERS

A typical hook pin is pictured in Fig. 10·4. Hook pins may be made from cold-drawn steel rod or drill rod to suit the application. The pins are heated and forged to produce the rough contour, then machined and/or filed to a shape approximating that of the illustration. Hook pins may function as stop pins for applications which require an offset fixed-pin stop. The smaller hook pin in Fig. 10·5 is a stop pin; the larger pins are the stripper hooks.

An arrangement of this kind creates a fixed stripper. In addition to stripping, the hook pins serve as stock guides. The back hooks are located with their shanks tangent to the desired back gage line ($B + S$, S being the desired back scrap dimension). Front hooks are located with their shanks tangent to the front gage line ($W + F$, where W is the stock-strip width and F is the desired feeding clearance).

Because of the small and isolated stripper contact areas, there is a tendency to distort the scrap skeleton during stripping. This can become objectionable, especially with soft stock material. The distorting tendency increases as the blank size increases. However, hook-pin strippers have functioned successfully on dies producing 2-in.-diam blanks from soft cold-rolled steel with a thickness T of 0.012 in. and a scrap allowance S of 1/32 in.

For low-production dies, the hooks may be made of cold-drawn steel and left soft. For higher production they may be pack-hardened (1/64 to 1/32 in. case depth). If they are made of tool steel, a hardness range of Rockwell C 45 to 55 is generally appropriate. Stripper hooks should be provided with a flat to prevent rotation and should be secured with setscrews, as shown.

Hook-pin strippers are not as popular as they once were, probably because forging is required. However, they do make a nice open stripper, with excellent operator visibility and access to the stock strip. They are used on combination dies, as well as on blank-through dies. In consequence, many diemakers may find themselves in situations where it is an asset to be familiar with this type of construction.

Figure 10·4 Typical hook pin.

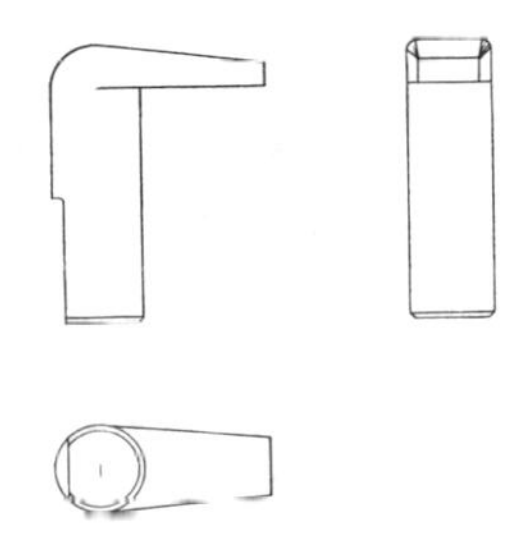

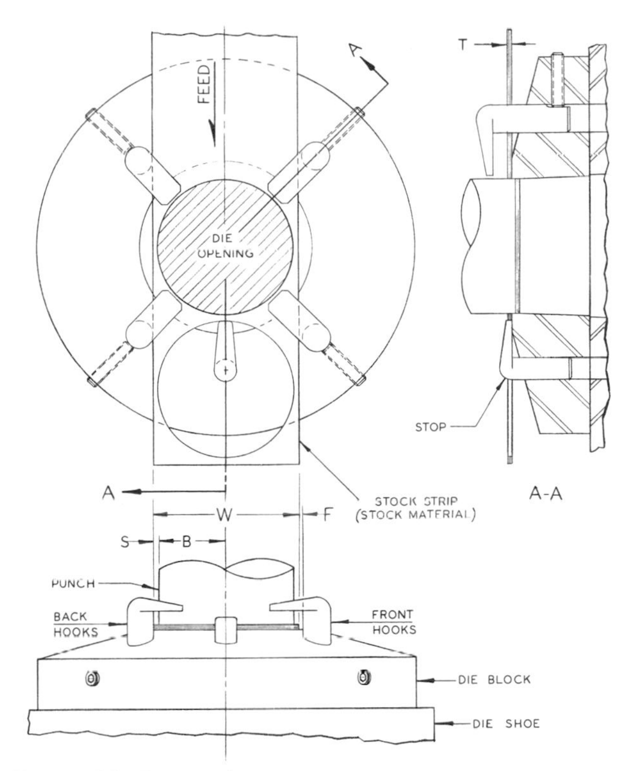

Figure 10·5 Hook-pin stripper.

BOX STRIPPERS

The box stripper illustrated in Fig. 10·6 is sometimes referred to as a solid stripper. This name stems from the fact that the stock guides are an integral part of the stripper plate.

Conventionally, overall dimensions E and D are made the same as the die block. Although not necessary, this procedure is practical, because in many cases making and mounting the stripper will be facilitated if the outer stripper edges are flush with the outer edges of the die block.

Before machining the tunnel slot in the stripper plate, analyze the situation to determine the most efficient machining sequence. In the illustrated case, for example, it would be advantageous to machine the clearance opening for the blanking punch before machining the tunnel slot.

The back edge of the tunnel acts as the back gage and must be located accordingly ($B + S$, S being the desired back scrap width; see Fig. 15·18). Tunnel width X is then made equal to $W + F$, where W is the stock-strip width and F is the desired horizontal feeding clearance. Tunnel height H is made equal to $T + G$, where T

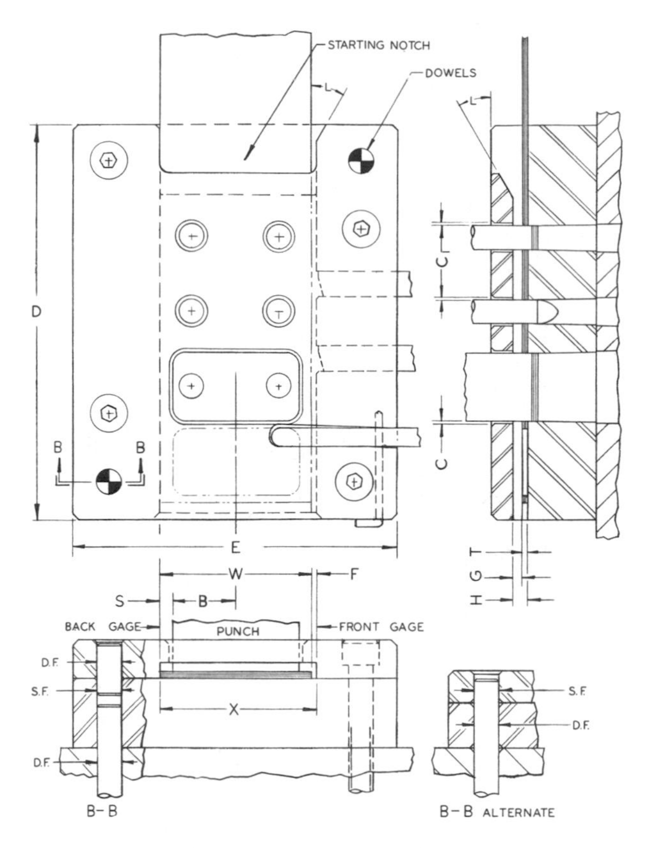

Figure 10·6 Box stripper (channeled).

is the stock material thickness and *G* is the desired vertical feeding clearance.

These strippers are frequently doweled as indicated at *B-B*. The stripper dowels are installed in the same holes as the die-block dowels. The dowels are fitted as called out (*D.F.* and *S.F.*). A light drive fit is made where indicated at *D.F.* A close slip fit is used where indicated at *S.F.* This procedure will permit easy removal of the stripper, if desired, without disturbing the die-block doweling. An alternate method, used primarily on low-production dies, is shown in alternate view *B-B*. This method is not as convenient for die maintenance.

There are two reasons for providing clearance C and C_1 in stripper plate openings.

1. Interference with punch action is prevented.
2. Stripper is easier to make.

When providing clearance, the idea is to avoid making the task any more exacting than necessary. Stripper-opening clearance should not be less than the cutting clearance at any given site around the periphery. Maximum clearance should normally be restricted to ½T at any given peripheral checkpoint. Do not work to either the maximum or minimum clearance dimension but to an approximate mean dimension somewhere between them. Think of the permissible plus or minus clearance variation as a tolerance which facilitates making the stripper opening. Since stripper-opening clearance is related to the type of stock material associated with the application, larger clearances may be used when punching harder material than when punching softer. For conditions similar to the illustrated instance, clearance C_1 can (and often should) be less than clearance C. There are two reasons for this:

1. Punch Contour. These perforators (and pilots) are round. It is a simple matter to locate and drill the required clearance holes in the stripper. Therefore, these stripper openings can be made in an expeditious manner, using less clearance than for an irregularly shaped opening.

2. Effect of Possible Stripping Distortion. Excessive clearance will encourage much distortion of the stock material. If this occurs in the area surrounding the blanking punch (see clearance C), the distortion will show up in the scrap skeleton. A slight amount of distortion here is usually tolerable. However, if the material is distorted in the area of the pierced openings (clearance C_1), the quality of the piece part will be affected.

Suitable lead angles L must be provided at the tunnel entrance. The lead angles facilitate starting the stock strip into the tunnel and are very necessary for practical die operation. In addition to the lead angles, a starting notch, as shown, will greatly enhance die operation.

On occasion it may be desirable to heat-treat strippers of this kind, but in most cases they are left soft. The resulting soft stock-guide edges may not be adequately wear-resistant. For example, the back gage would wear very rapidly in an application employing a pusher in this kind of stripper. To overcome this condition, hard inserts may be installed in the stripper to provide wear resistance along the back gage line. Hard inserts may, of course, also be installed at the front gage line, if needed. The inserts may be made of tool steel hardened to Rockwell C 60 to 64. Or if maximum wear resistance is required, they may be made of sintered carbide.

A simple method of providing stock-guide inserts is illustrated in Fig. 10·7. The inserts are cylindrical. They may be tool steel or carbide, as required. Carbide inserts ground to standard diameters are commercially available at nominal cost. The inserts are generally installed tangent to the gage lines. However, to provide a larger wear surface, they are sometimes made with a flat and installed as shown in view *B*. An alternate construction is pictured in view *C*, where a longer

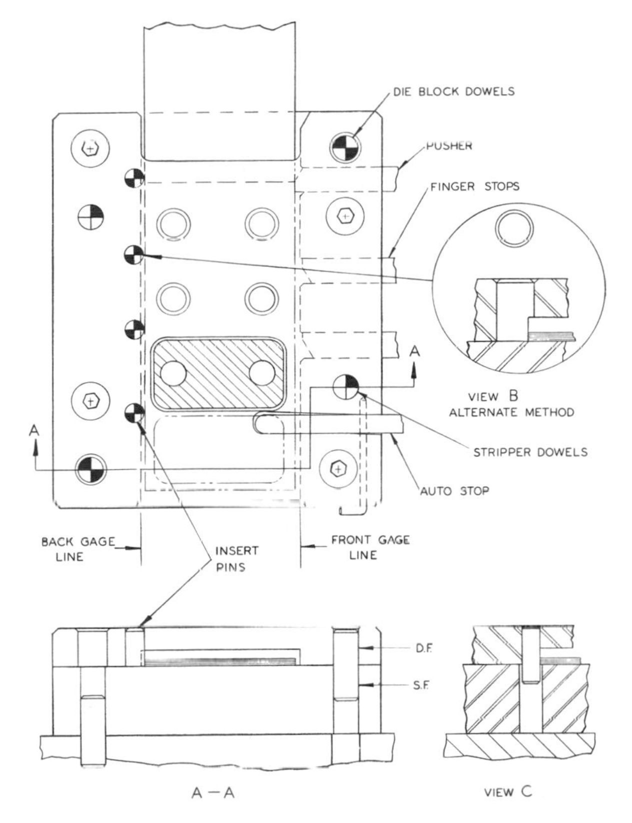

Figure 10·7 Box stripper (channeled) with back-gage insert pins.

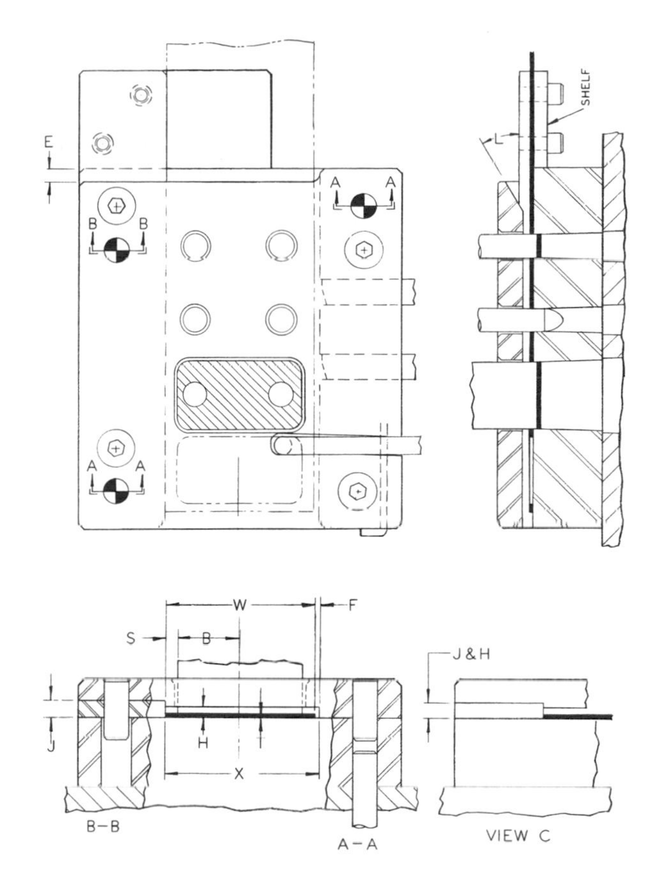

Figure 10·8 Box stripper with separate back gage.

insert pin is used. The insert projects into a clearance hole provided in the die block. This method is often used with carbide pins to avoid the bother of shortening a standard pin length. It also lends itself nicely to applications where the stock material is very thin.

This illustration depicts another stripper doweling procedure, where separate dowel holes are provided for the stripper dowels. Separate stripper dowel holes are customarily employed for larger, more complex dies and are used more often for guiding strippers than for nonguiding strippers. It is often advantageous to provide separate stripper dowel holes when die construction is such that the die block is composed of a number of sections.

Separate back gage. In Fig. 10·8, the rear stock guide (back gage) is not an integral part of the stripper plate. It is a separate piece assembled to the stripper plate and the die block.

The stripper plate is made with a double step to permit adequate back-gage thickness *J* and suitable tunnel height *H*. For applications which permit an adequate tunnel height, the double step may be eliminated and the stripper plate made with a single step, as shown in view *C*.

Doweling procedure for dowels shown at *A-A* may be the same as that described for Fig. 10·6. An additional dowel is installed to assure back gage alignment. This dowel pins the back gage rail to the stripper plate, as indicated at *B-B*. A clearance hole is provided in the die block to accept the intruding dowel. If the back gage *J* is thick enough, a shorter dowel may be installed, eliminating the need for a clearance hole.

Notch *E* is a variation of the starting notch previously described. It may or may not be worthwhile, depending upon the specific application. The shelf attached to the extended back gage is often all that is required to facilitate starting the strip. Lead angle *L* is, of course, always necessary at the tunnel entrance.

Separate guides (front and back). Another box-stripper assembly variation is pictured in Fig. 10·9. In this instance, both the front and back gages are separate units assembled to the stripper plate and the die block. If dimension *H* is large enough, the stripper plate may be made flat (no offset), as shown in view *C*. If the tunnel height is limited, the stripper may be offset, as indicated at section *B-B*. Note that the finger-stop construction would be different in each case. Doweling

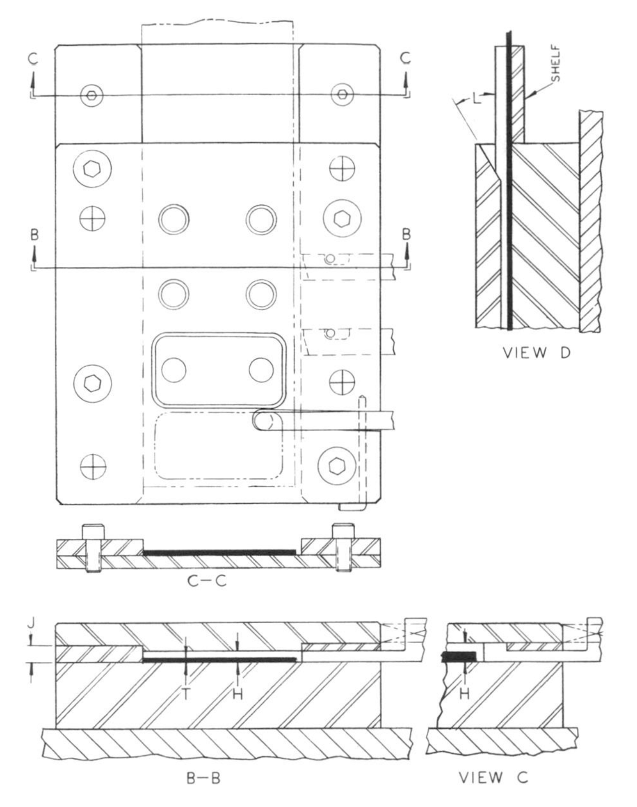

Figure **10·9** Box stripper with separate stock guides (back gage and front gage).

procedure is generally the same as for Fig. 10·8, except that an additional dowel is required for the front gage.

With this arrangement, both of the guide rails may be extended beyond the stripper on the infeed side. A shelf may be attached, as shown at view *D*. This makes a very nice arrangement for

Figure **10·10** Box stripper with extended back gage.

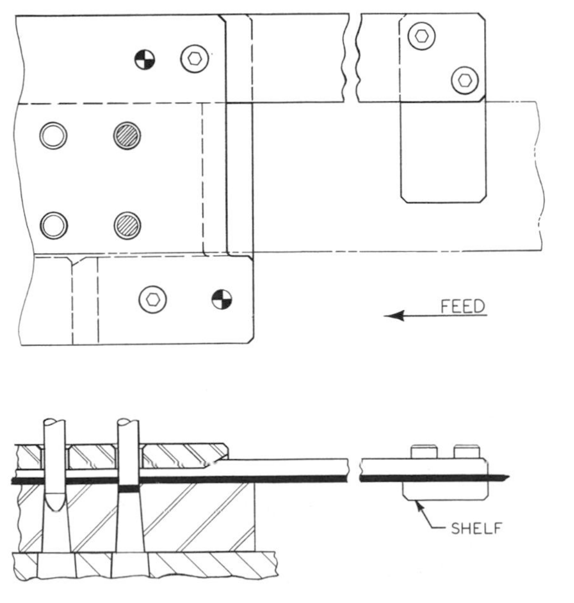

starting the stock strip. Lead angle *L* is, of course, present.

Extended back gages. Back gages are extended as represented in Fig. 10·10 when it is felt necessary to aid in aligning the stock strip for starting and feeding. In other respects, construction can be the same as that shown in Figs. 10·8 and 10·9. The shelf (sometimes called stock rest) provides support under the stock strip. Shelves are commonly made from cold-rolled steel.

MATERIALS FOR STRIPPER PLATES

Choice of material for stripper plates depends upon the cost and quality factors pertaining to each specific instance. In the majority of box-stripper applications, the stripper plate is made of low-carbon steel, left soft. Both cold-drawn and hot-rolled steel are used. Hot-rolled steel in bars finished by grinding to standard sizes is commercially available under the designation "low-carbon ground stock." Where a stronger stripper is desired, the stripper plate, even though it is left soft, may be made from a hardenable alloy steel or a tool steel. Another method, desirable on occasion, is to use a pretreated alloy steel. This is heat-treated in advance for greater strength and wear resistance, but it is still machinable.

Plain stripper plates (Fig. 10·11, view *A*) are often made from cold-drawn steel, because it is well suited to this type of stripper construction and because its initial cost is low. Also, machining time is minimized. These factors make cold-drawn steel economically desirable.

Channeled plates and offset plates (views *B* and *C*) are also often made from cold-drawn steel. However, since cold-drawn steel may warp considerably when the channels or offsets are machined, it is not employed for these strippers as frequently as it is for plain stripper plates. The warping tendencies of cold-drawn steel will be even more noticeable with stepped stripper plates (view *D*).

Figure **10·11** Cross sections through typical box-stripper assemblies.

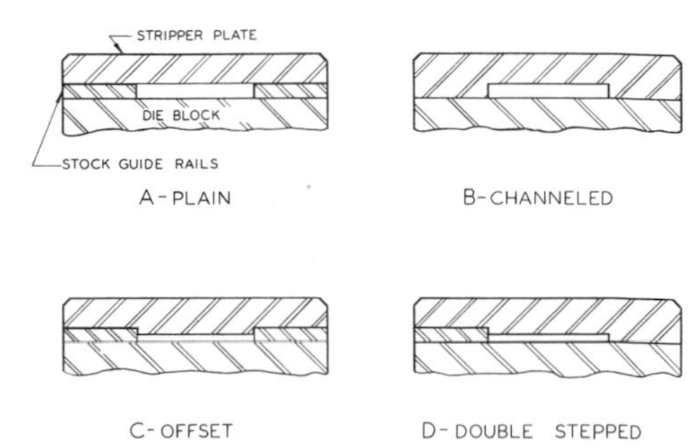

If considerable time is required to straighten a stripper plate after machining, the original economic advantage of cold-drawn steel may be lost. To alleviate warpage, cold-drawn plates may be annealed or stress-relieved before machining, but this may cancel the original economic advantage. It is best to avoid the use of cold-drawn steel for precision strippers which are intended to guide and/or support the punches.

The following materials, listed in descending order of quality and durability, are commonly employed for making stripper plates.

1. Tool steel, hardened
2. Pretreated alloy steel
3. Tool steel, left soft
4. High-strength, low-alloy steels
5. Hot-rolled, low-carbon steel, including low-carbon ground stock
6. Cold-drawn, low-carbon steel bar stock, commonly called cold-rolled steel

This list is arranged in order of quality and durability, not of desirability. Optimum material choice varies in accordance with circumstances peculiar to the specific application. After all, it would not be good diemaking practice to use a hardened tool-steel stripper for an application which requires only a soft cold-drawn steel plate.

MATERIALS FOR GUIDE RAILS

For low-production requirements, stock guides are commonly made from cold-drawn, low-carbon steel bar stock (cold-rolled steel). Otherwise, they can be made of tool steel and hardened, or they can be provided with hard inserts. Or instead of hardening the entire rail, the diemaker can use localized hardening techniques to harden a series of spots along the gaging edge in the manner shown in Fig. 10·12(*A*).

Hard inserts may be brazed or silver-soldered along the guiding edge. They also may be mechanically assembled in the guide rail. Inserts

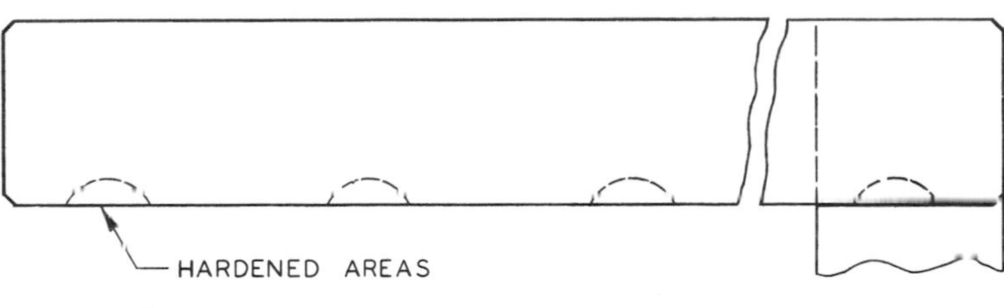

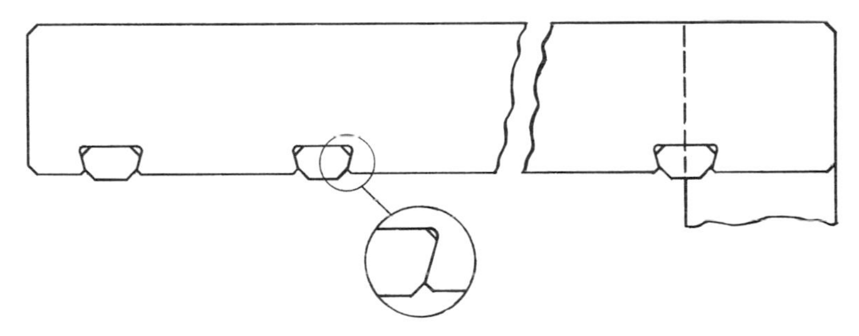

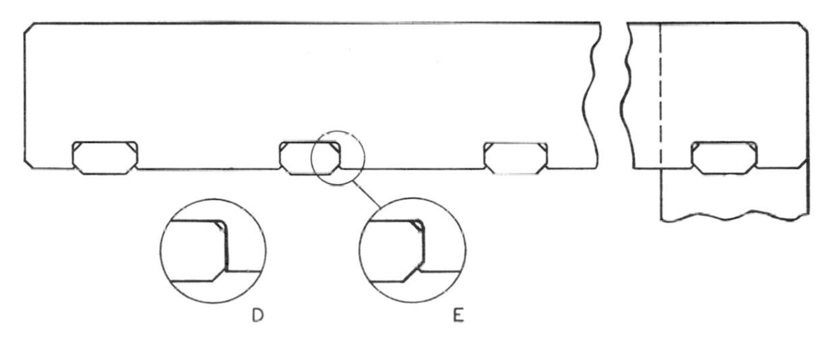

Figure **10·12** Stock guides (back gages shown).

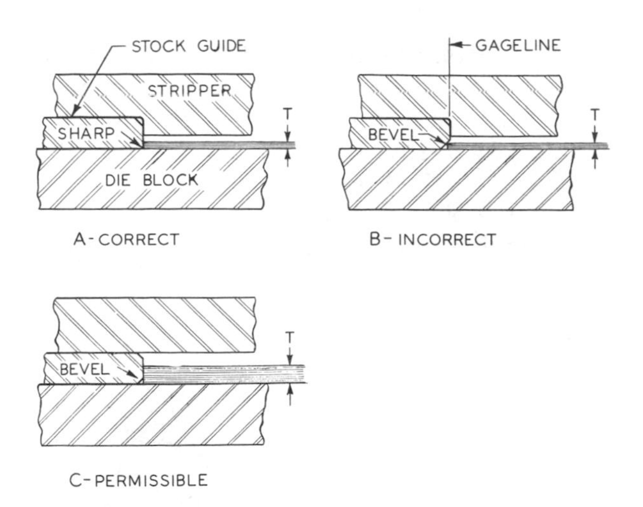

Figure 10·13 Stock guide edges.

are commonly high-speed steel, stellite, or carbide. Dovetail inserts are pictured in view *B*, where they are pressed into matching dovetailed notches provided in the rail. Plain inserts are shown in view *C*, where corners of the inserts are beveled and the edge of the rail is staked against the bevels, locking the inserts into the rail. When inserts are installed, be certain that a lead is provided on them. Sharp corners, which will interfere with feeding, must be eliminated.

When pushers are required along the front gaging line, they may eliminate the need for a hard-surfaced front guide rail. They will, however, exaggerate the need for hard gaging surfaces at the back gage line. In fact, it is not generally practical to use a soft back gage in conjunction with pushers.

GUIDING-EDGE CONDITION

Generally, the guiding edge of stock guides should be finished with a sharp corner in the manner shown in Fig. 10·13, view *A*. This is especially so for light-gage stock material *T*; the condition pictured in view *B* must be avoided. When cold-drawn steel guides are used in conjunction with thin stock, be sure to grind the guiding edge in order to provide a sharp-cornered gaging surface. Otherwise the natural edge on cold-drawn steel bar will have an effect similar to that shown, and the stock strip will wedge under the gaging corner, causing misfeeds, etc. Of course, if the stock material thickness is heavy enough (view *C*), a very slight bevel or rounding is permissible.

BULGE CLEARANCE

Work performed close to the edges of the stock material will cause bulges in these areas. The bulges will affect subsequent gaging unless bulge

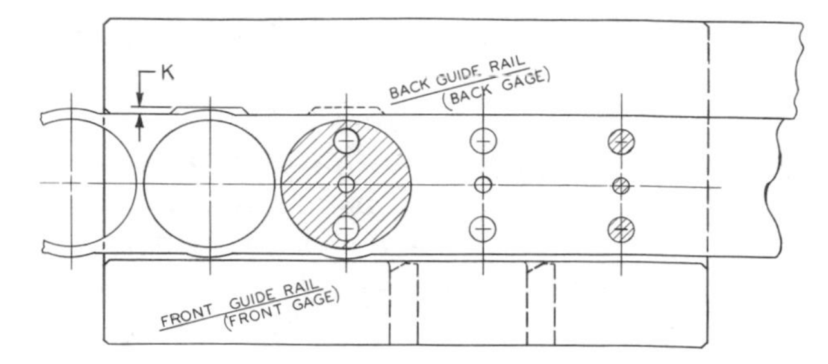

Figure 10·14 Clearance in guide rail to accommodate blanking bulge.

clearance is provided in the subsequent operations. One method of providing it is pictured in Fig. 10·14. The blanking punch causes the stock strip to bulge, but clearance *K* in the back gage permits the stock strip to assume its proper position against the back gage as the strip is advanced after each press stroke. Depending upon the individual application, bulge clearance may or may not be necessary in the front gage. It may also be necessary or desirable at the blanking station, as indicated by the dotted notch outline.

Another example of bulge clearance is represented in Fig. 10·15. Here, the pierced holes are close to the strip edges, and bulge clearance must be provided early in the progression. Since bulge clearance shortens the effective guiding length of the rail, it may be necessary to compensate by making the rail longer on the infeed side.

TUNNEL DIMENSIONS

Tunnel width *X* (Fig. 10·16) can be determined as

$$X = W + F$$

where W = stock-strip width at max. tolerance
F = desired horizontal feeding clearance

For the average progressive blanking die, assuming there are no other specific requirements, clearance F may be 1/32 in. per foot of tunnel length. This figure will accommodate average strip camber. Badly cambered stock may require more

Figure 10·15 Clearance in guide rails for piercing and blanking bulges.

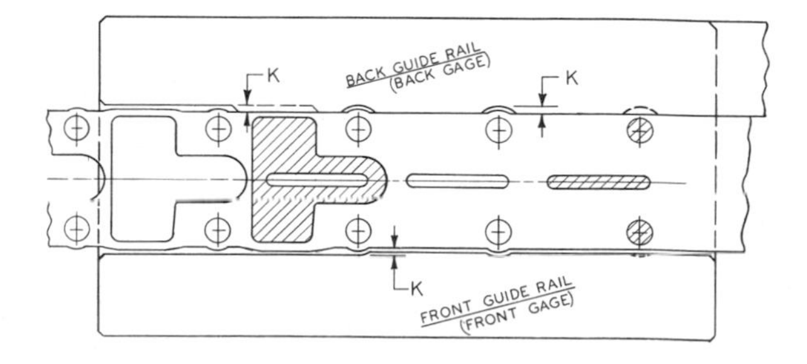

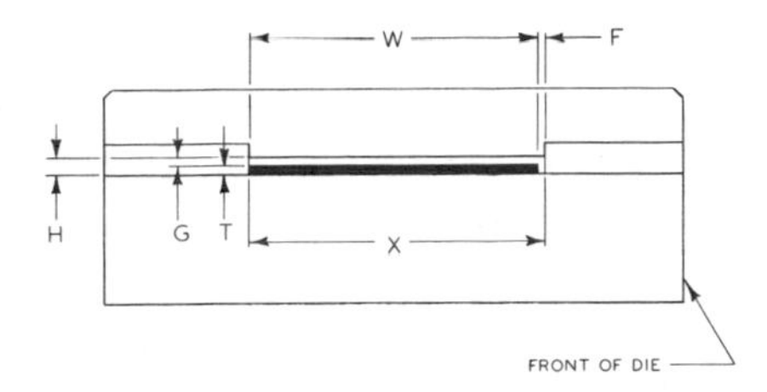

Figure 10·16 Tunnel dimensions.

clearance. Strip stock with restricted camber tolerance will not require as much clearance. Precision slit stock applications (usually scrapless dies) may operate successfully with F clearances as small as 0.004 to 0.005 in. per foot of tunnel length. Keep in mind that special conditions may require special allowances.

Tunnel height $H = T + G$, where T is the stock material thickness and G is the desired vertical feeding clearance. Clearance G cannot be specified in a broad and general way appropriate for a wide range of applications. G may be as little as ½T in a flatwork cutting die with a relatively short tunnel length, or it may be several times larger than T when required. Progressive dies involving forming operations can function successfully with a surprising amount of vertical stock-strip movement, providing this has been well considered in the design of the dies.

PRESSURE-PAD-TYPE STRIPPERS

A pressure pad may be defined as a die component which delivers holding pressure to the material being worked. Pressure-pad strippers are so called because they are capable of acting as true pressure pads if required. They may be actuated pneumatically, hydraulically, or by springs or rubber cushions. The use of pneumatic and hydraulic strippers tends to be limited to situations where:

1. Very high pressures are required.
2. Special timing or control of stripper motion and/or pressure is required.

The above conditions often apply to strippers which act as draw pads, forming pads, etc., in addition to performing the stripping function. Such components are not, as a rule, called strippers. They are referred to in accordance with their dominant function—that is, draw pad, forming pad, etc.

SPRING STRIPPERS

A spring stripper is a pressure-pad-type stripper actuated by springs. Spring strippers are employed where it is necessary or desirable to hold the stock material flat (or very nearly flat) during

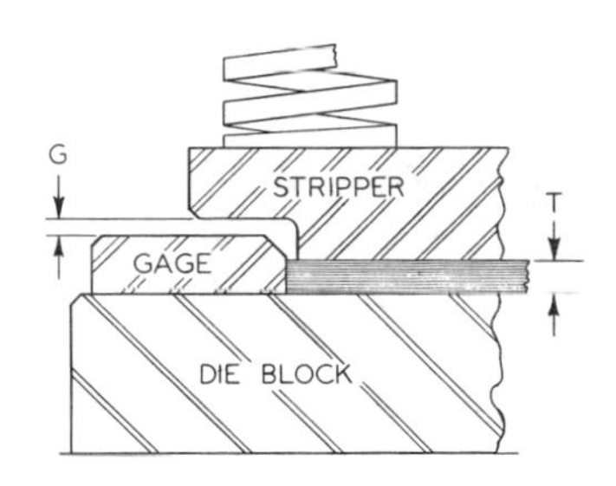

Figure 10·17 Clamping.

the working cycle. In such applications they can be important in producing flatter work. Another advantage of spring strippers is that they provide better visibility and access to the lower die assembly while the die is mounted in the press. Inverted dies have stationary punches and therefore require traveling pressure-pad-type strippers. The great majority of these are spring-actuated. Pressure-pad strippers are also required for push-back applications, where the blank is pushed back into the stock strip after blanking. Pressure-pad strippers for push-back dies are very often spring-actuated.

Clamping spring strippers. The spring-stripper group includes a number of types, variations, and combinations. Clamping strippers (Fig. 10·17) are true pressure pads: They bear against the stock material, applying pressure to it. Thus, the stock material is clamped between the stripper and the die block. Clearance G must be large enough to assure clamping the minimum contemplated stock material thickness T. Spring strippers which clamp the stock material are the predominant type; they are generally made nonclamping only when necessary.

Nonclamping spring strippers. Applications are encountered where it is not practical to permit the stripper to clamp the stock material. For such applications the stripper is made in some fashion which produces the result shown in Fig. 10·18. Here, the stripper contacts the gage, causing a clearance gap G between the stripper and the

Figure 10·18 Nonclamping.

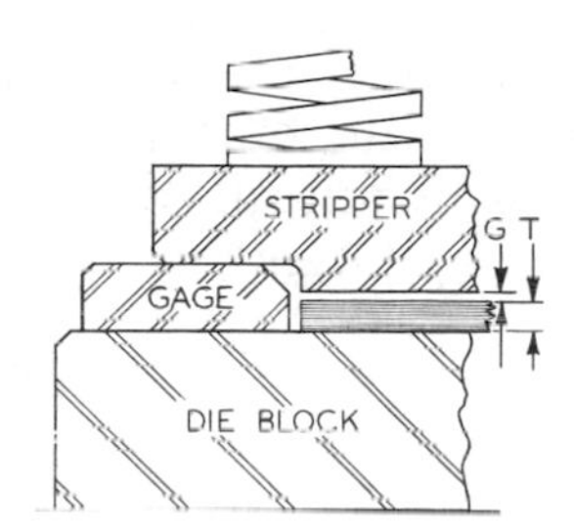

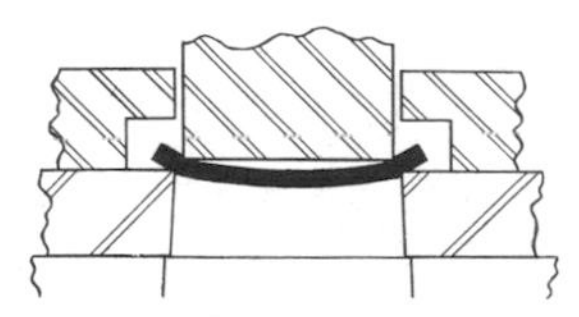

Figure 10·19 Twist when cutting close to edges of stock material (exaggerated).

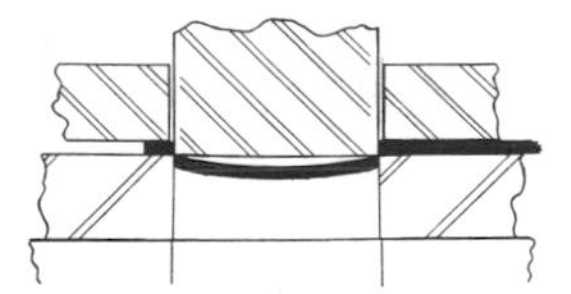

Figure 10·20 Dish when blanking (exaggerated).

stock material. Size of gap G depends upon the application. For light-gage stock material T, gaps of only 0.002 to 0.005 in. are used successfully. These minimal gaps are employed for applications where it is desired to hold the work as flat as possible without clamping. It should be mentioned that surprisingly flat work can be produced this way. In fact, the work is often comparable to that produced with clamping strippers. Other applications may require gaps G which are comparable to those used for box strippers, and these, of course, do not hold the stock material flat.

Spring strippers and flatness. The normal spring stripper (or other pressure-pad stripper) is not a flattening device in the sense that, of itself, it permanently flattens the stock material. A clamping stripper (or a closely fitted nonclamping stripper) will hold the stock material flat while the work is performed. Any improvement in the flatness condition of the stock material is the result of work straining the material while it is held flat by the stripper. Such improvement is related to the degree of working and the kind of work.

This is not meant to imply that there will always be improvement in the flatness condition. Some work may be detrimental to flatness in spite of the fact that the material is constrained by the stripper. However, for most applications, a spring stripper is an effective aid to producing piece parts which have rigorous flatness requirements.

A spring stripper can prevent distortions similar to that indicated in Fig. 10·19, but it will not prevent dishing from blanking pressure (Fig. 10·20). The prevention of dishing would require a pressure pad within the die opening to hold the material flat during the cutting process.

PILOTING THROUGH SPRING STRIPPERS

It is often necessary to employ pilots in conjunction with spring strippers in the manner illustrated in Fig. 10·21. If the stripper is a clamping stripper, it cannot be used to strip the pilots completely. This is because they must register the stock strip (or workpiece, in a secondary-operation die) before the stripper contacts the stock strip. Thus, when the stripper is at its maximum extended position (die open), the stripper-to-pilot relationship is that shown at view B. The effective piloting size projects beyond the stripper a distance Q. Normally, $Q = T$, with a minimum of $Q = 1/16$ in. The stock material clings to the pilots and will ascend with them on the up stroke unless some additional stripping means is provided. When the pilots are located favorably in proximity to the guide rails, the hooked portion of the guide rails will act as fixed strippers, causing the pilots to withdraw from the stock strip on the return stroke of the press (see Fig. 10·1). Total pilot stripping involves two stripping actions:

1. Pilot is stripped to distance Q by the spring stripper. This is a pressure-pad action. The stock material is then lifted by the pilots until it contacts the guide-rail ledge H.
2. Balance of pilot stripping is performed by the ledge. This is a positive stripping action.

Figure 10·21 Piloting through clamping stripper.

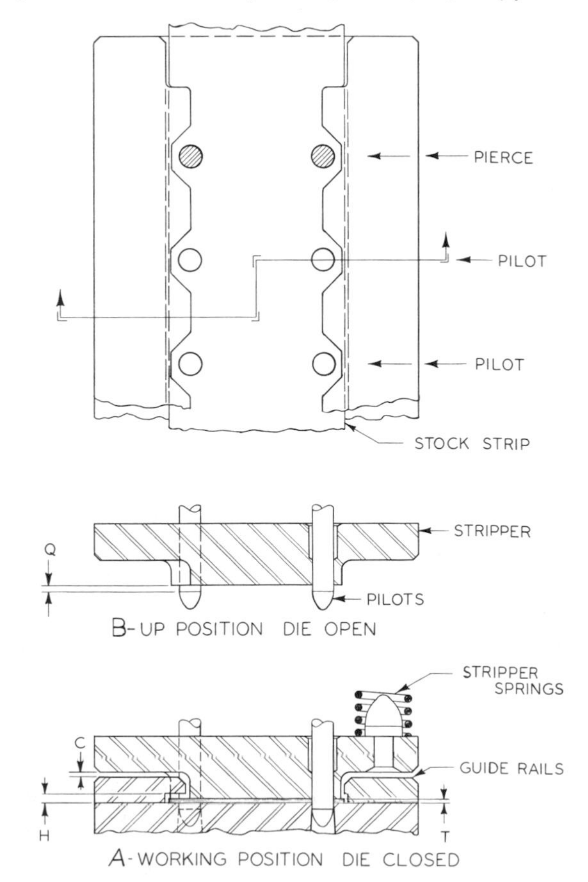

If the pilot locations are too far removed from the hooking action of the guide-rail ledges, the stock material may pull up, bowing in a manner similar to Fig. 10 • 22. It is possible for the stock strip to pull all the way out of the guide rails. As the stock strip pulls higher at the center, the radius of the bow decreases, which tends to tighten the clinging effect at the pilot. Even if the stock strip does not pull out of the rail confinement, the bowing action is deleterious. It can cause excessive pilot wear, seriously deteriorate the quality of the pierced opening, and adversely affect the ultimate flatness of the piece part.

When possible, a nonclamping stripper (Fig. 10 • 23) is used for applications where the pilot locations are too far removed from the guide rails. The pilots are not required to lead a nonclamping stripper. Therefore it is practical for the stripper to strip the stock material completely from the pilot. This is obvious in view *B*. It is equally obvious in view *A* that clearance *C* permits the pilots to register the stock strip without interference from the stripper. Most applications of this kind will require that clearance *C* be held to a practical minimum in order to assure the flatness of the stock strip.

In practice, clearance *C* will not necessarily exist in the ideal manner illustrated. The stripper will come into contact with the stock strip. However, the stock strip will not be clamped by the stripper. The pilots will overcome friction engendered by a normal amount of undulation in the stock strip. If the out-of-flat condition of the strip is excessive, it will have to be straightened before it is run. This, of course, applies to all applications, not merely those which employ nonclamping strippers.

In some instances a clamping stripper must be incorporated in a die in spite of the fact that pilot locations are unfavorable. One solution for this problem appears in Fig. 10 • 24, where shedder pins are installed in the stripper. In progressive dies, the shedder pins operate in conjunction with the stock-guide hooks to prevent the stock material from bowing and locking on the pilots. In secondary-operation dies, the shedder pins usually operate without help from hooked gages. In any case, the shedder-pin springs must be strong enough to strip the projecting portion of the pilot without causing undue resistance to the registry action.

A variation of the general stripper type shown in Figs. 10•21 to 10 • 24 is the laminated stripper in Fig. 10 • 25. This is a very common construction for guiding strippers when the application is such that stripper support is required for punches with irregular cutting contours. The facer plates can be made up of hardened sections assembled to the master plate, which is a distinct construction advantage. With this variation, the most common error is an inadequate master plate. Be certain that the plate is thick enough for the application.

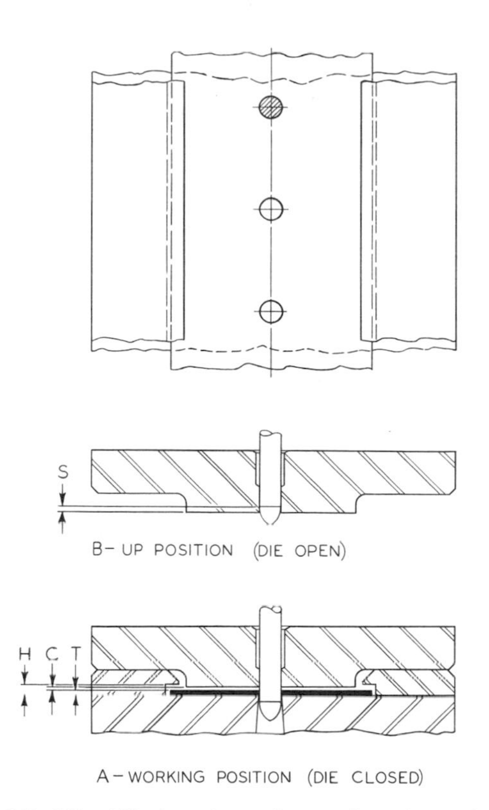

Figure **10 • 23** Piloting through nonclamping stripper.

STOCK GUIDES USED WITH SPRING STRIPPERS

When spring strippers are employed in conjunction with guide rails, they are typically similar

Figure **10 • 22** Stock-strip bowing because of poor pilot stripping condition.

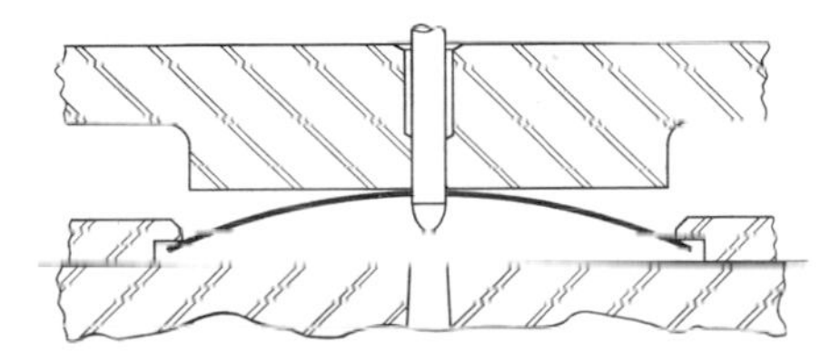

Figure **10 • 24** Shedder pins to strip pilots (clamping-type spring stripper).

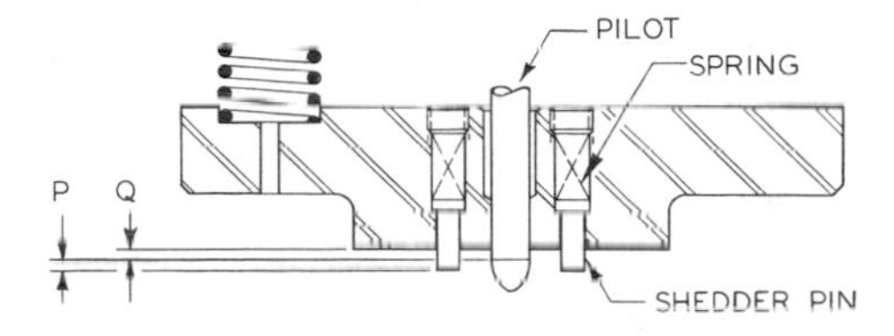

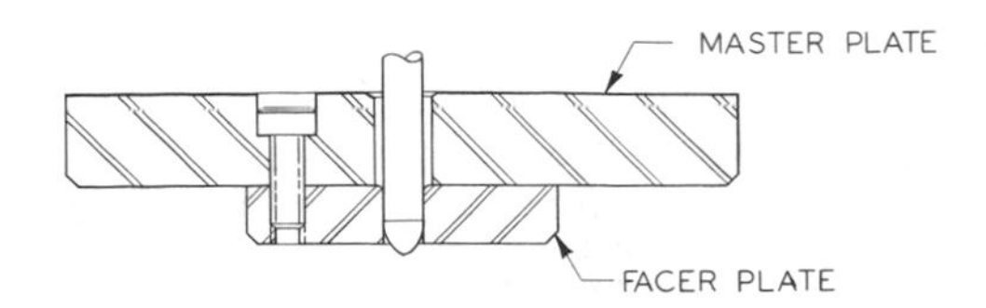

Figure **10·25** Laminated stripper.

in cross section to those pictured in Figs. 10·21 through 10·25. A different stripper and stock-guide combination appears in Fig. 10·26. Spool-type guides are not as efficient as rails. With spools it is usually somewhat more difficult to start the stock strip through the die. Also, spools are not generally as desirable as rails for hook-stripping. For applications where the spools are expected to act as hook strippers, be sure they are strategically located in proximity to the required stripping sites. In circumstances where vertical die space is restricted, spool-type stock guides will often permit the use of a thinner stripper plate. Keep in mind, however, that the most common spring-stripper design fault is strippers which are too thin. The use of spool-type stock guides can often result in a more economical stripper and stock-guide combination. Spool hardness may range from Rockwell C 50 to 60. For heavy stock material, spool hardness should approach the low end of the above range; for light work, the high end.

Lifter guides (Fig. 10·27) perform the dual function of guiding the stock strip and lifting it a required distance L above the face of the die block (4). They are used in progressive dies where forming operations necessitate lifting the stock strip in the manner illustrated. Lifter guides can be made as shown detailed. Radius R_1 accepts the washer (3), which serves as both a key and a retainer. R_1 is sunk to depth D required for the application. Radius R_2, which also originates at the washer center, provides a lead and, in addition, will absorb minor discrepancies in angular location. Lifting action is derived from the spring (5). If the spring is backed by a setscrew (6), it will be decidedly more practical than a blind spring pocket. Counterbore depth E in the die shoe (7) should include an allowance for die life. The stripper (1) is pocketed to depth F required to properly position the lifter guide (2) when the die is closed (view A). Lifter guides similar in function to the kind shown here are commercially available at moderate cost.

The lifter guide in Fig. 10·28 is a simpler design that is not as strong as the preceding one. Nevertheless, it is in common use and gives sat-

Figure **10·26** Spool-type stock-strip guides.

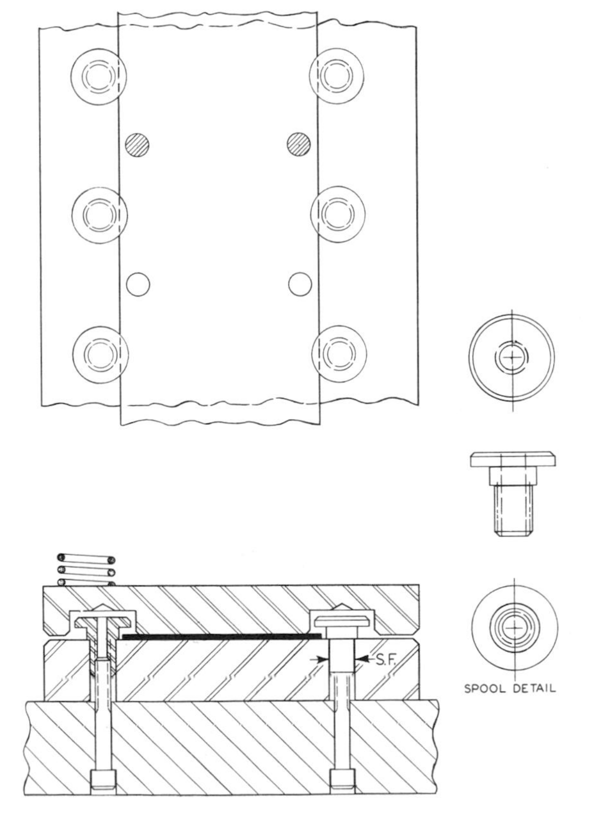

Figure **10·27** Lifter guides for stock strip.

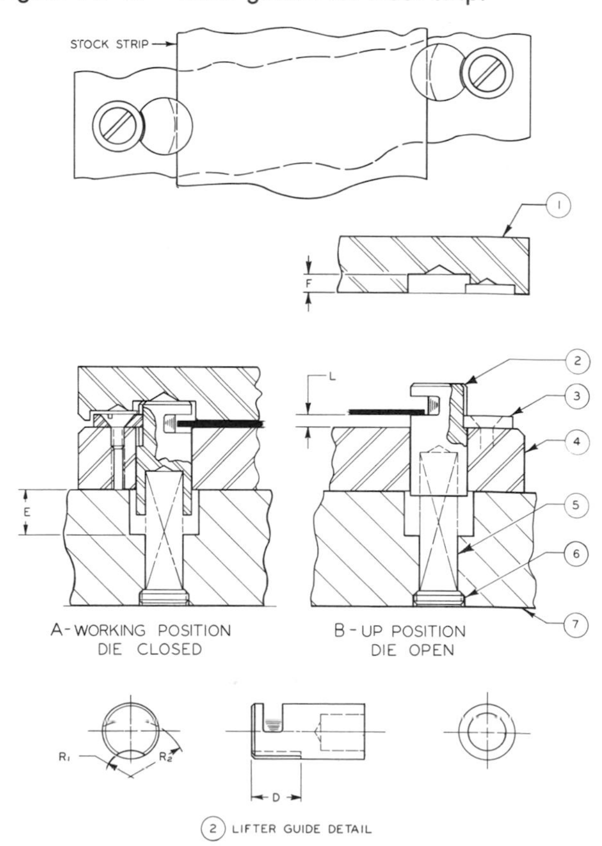

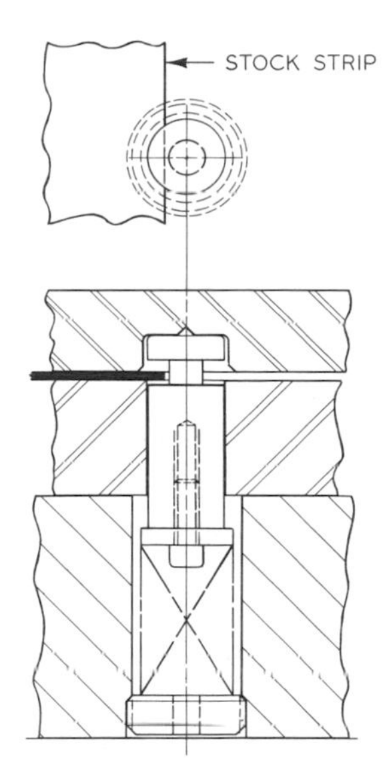

Figure **10·28** Simplified lifter guide.

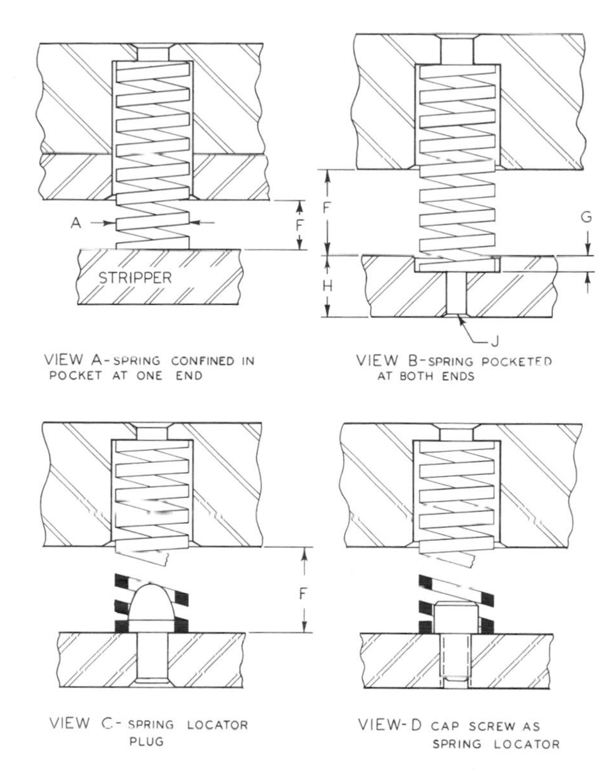

Figure **10·29** Spring confinement.

isfactory service. Being simpler, it is more economical to make. Remember, though, that a standard commercial lifter guide also offers economy advantages.

SPRING CONFINEMENT

It is essential for springs to be confined to their required locations. Typical stripper spring confinement methods appear in Fig. 10·29. Spring pockets are a natural solution for spring confinement, because they make use of available vertical space and/or limit vertical space requirements to reasonable amounts. It is possible to use standard-length springs for virtually all stripper applications.

In view *A*, the spring is shown pocketed at one end only. This is generally acceptable practice where gap *F* is not greater than the spring diameter *A*. For larger gaps *F*, possible lateral spring movement should be restricted at both ends. One conventional method of achieving this is depicted in view *B*, where a shallow spring pocket is provided in the stripper plate. Permissible pocket depth *G* is generally proportionate to stripper thickness *H*; the pocket must not seriously weaken the stripper. A minimum pocket depth *G* of ⅛ in. is necessary to stabilize springs up to 1 in. diam. Pocket *G* should be proportionately deeper for larger springs. A spotting hole *J* should be provided with this method of construction. The spotting hole is used to transfer the pocket location from the stripper to the punch holder (or other mating component).

Spring locator plugs (view *C*) are another conventional method for confining stripper springs. These plugs are sometimes merely press-fitted in the stripper, but it is safer to peen them in assembly, as shown. They can be purchased in suitable size combinations as standard commercial items. An effective spring locator can be made as indicated in view *D*. It is simply a short cap screw of suitable diameter assembled into the stripper, as shown, at the required spring sites.

Double spring pockets are often required even though gap *F* is less than the spring diameter. A typical instance, shown in Fig. 10·30, is where double pockets are necessary in order to provide adequate spring length.

Pertinent conventional spring-pocket data are depicted in Fig. 10·31. For springs ranging from ½ to 1¼ in. outer diameter *A*, pocket diameter *B* may be

$$B = A + \tfrac{1}{16} \text{ in.}$$

For larger springs

$$B = A + \tfrac{1}{8} \text{ in.}$$

Figure **10·30** Double spring pockets for restricted vertical space situations.

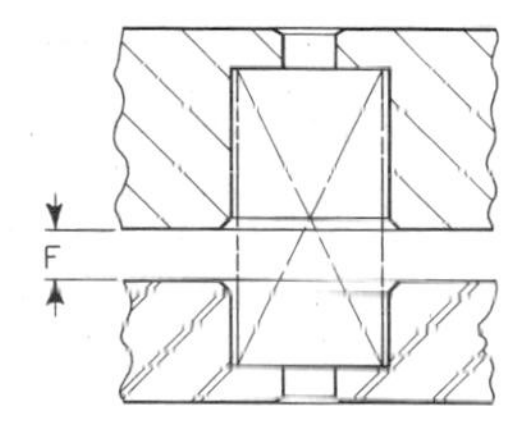

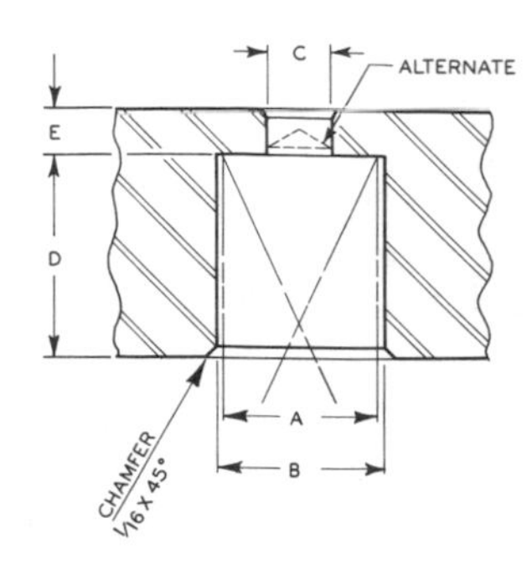

Figure **10·31** Spring pocket details.

Spring manufacturers customarily furnish specification charts indicating the minimum hole diameter in which the spring will work and the maximum rod diameter over which the spring will work. Pocket depth *D* should permit using standard-length springs, initially compressed to the required preload condition. Dimension *E* will vary according to *D*. However, the following minimums are generally appropriate.

$A < 1\frac{1}{2}$ in.	$E = \frac{1}{4}$ in. min.
$A = 1\frac{1}{2}$ to 2 in.	$E = \frac{3}{8}$ in. min.
$A = 2$ to $2\frac{1}{2}$ in.	$E = \frac{1}{2}$ in. min.

It will facilitate drilling the spring pocket if a lead hole *C*, as shown, is drilled first. Lead-hole diameter *C* may be any convenient diameter smaller than the inside diameter of the spring. An alternate method, if desired, is indicated by the dotted outline. Spring pockets should be chamfered, as shown.

Compensator slugs. When cutting punches are sharpened, they become shorter. In many applications, the springs are simply compressed a little more after each sharpening. However, this is not always desirable. A practical method of eliminating the compression is illustrated in Fig. 10·32. A cylindrical slug is installed behind the spring, as pictured. Each time the punches are sharpened, an equal amount is ground from the slug height *H*. Thus, the compressed length of the spring remains constant throughout the life of the slug. For economic reasons these compensator slugs should be made from cold-drawn steel and left soft.

Figure **10·32** Compensator.

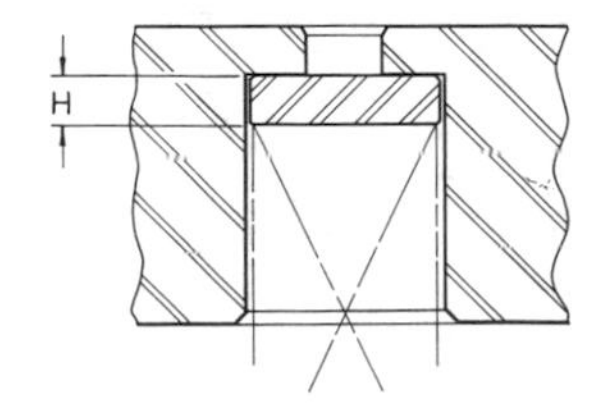

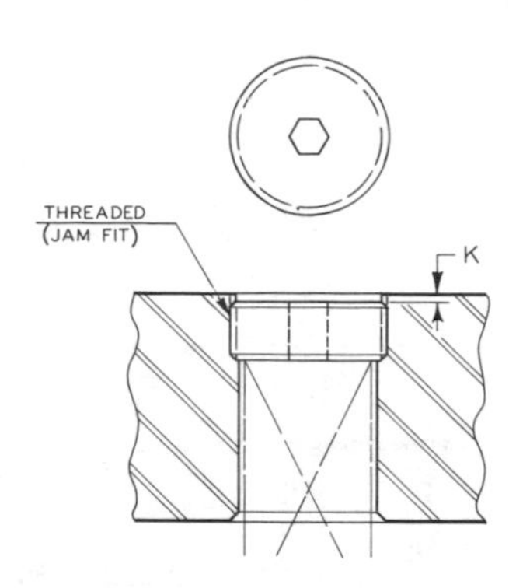

Figure **10·33** Screw cap behind spring.

Screw caps. On occasion, it may be desirable to install screw caps (Fig. 10·33) behind stripper springs. Standard screw caps should be used. These are provided with hexagonal wrench openings, as pictured. Spring hole must be tapped a jam fit for the screw cap. When the screw cap is in locked position, clearance *K* should be a minimum of 1/32 in.

Springs around stripper bolts. The construction indicated in Fig. 10·34 has desirable features and undesirable features. Desirable ones are:

1. The bolt retains the stripper at center of spring pressure.
2. The bolt acts to confine the spring in location, so that for applications where space *F* is adequate, spring pockets may be eliminated.

Undesirable features are:

1. An assembly of this kind requires considerable vertical space, often more than is available.
2. With this construction, springs are often made too short, with no life provision. (This is not actually a separate reason; it is cited as such because it is an error which occurs all too often with this construction.)

Figure **10·34** Stripper bolt within spring.

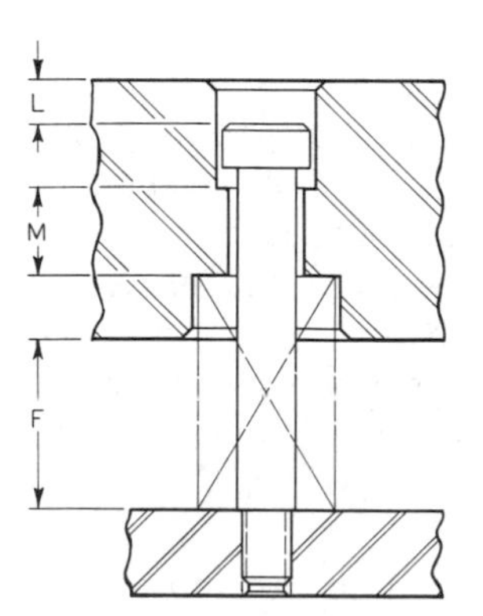

When making the installation, be certain that dimensions *L* and *M* are adequate.

STRIPPER-BOLT SUSPENSION

Most commonly, spring strippers are retained in suspension by socket head stripper bolts. Stripper bolts, a standard form of shoulder screw, are commercially available as standard items conforming to the specifications in Fig. 10·35.

Stripper-bolt installation is typified in Fig. 10·36. Bolt hole *B* is drilled larger than shoulder diameter *A*. Clearance should be the same as a clearance hole for a regular cap screw. Counterbore diameter *D* is larger than head diameter *C*, and this difference too should be the same as for a regular cap screw. When the die is fully closed and the stripper bolt is at its maximum travel position *F*, distance *E* must be sufficient to assure adequate punch grinding life. Minimum *E* dimension is normally 5/16 in. This will permit replacement of stripper bolts in 1/4 in. shorter increments when and if required.

The majority of spring-stripper applications will not permit the end of the stripper bolt to protrude beyond the stripper. Check for this condition. Normally, a space *G* should exist between the end of the stripper bolt and the face plane of

Figure **10·35** Standard stripper bolts.

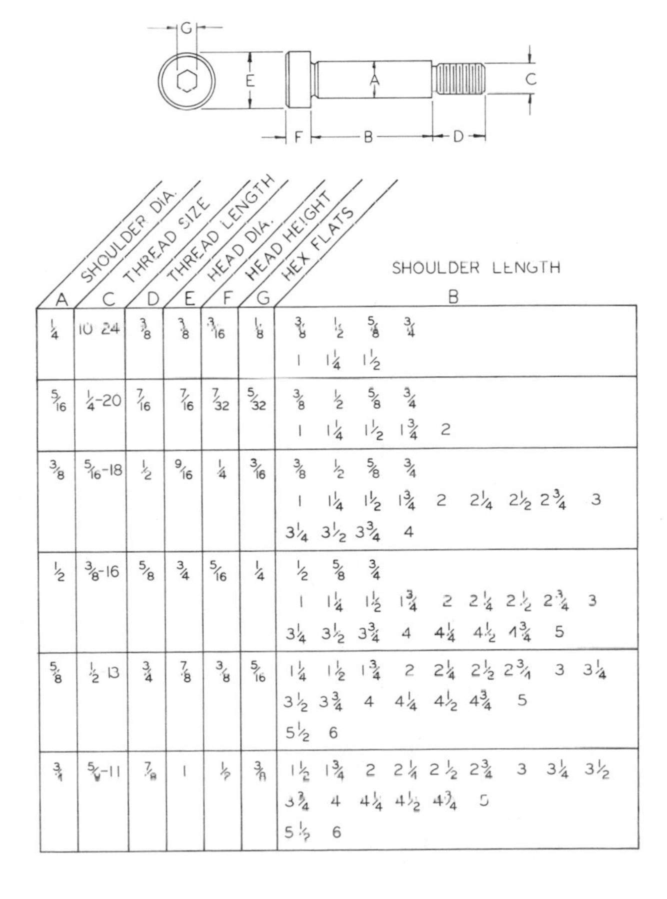

A SHOULDER DIA.	C THREAD SIZE	D THREAD LENGTH	E HEAD DIA.	F HEAD HEIGHT	G HEX FLATS	B SHOULDER LENGTH
1/4	10 24	3/8	3/8	3/16	1/8	3/8 1/2 5/8 3/4 1 1 1/4 1 1/2
5/16	1/4-20	7/16	7/16	7/32	5/32	3/8 1/2 5/8 3/4 1 1 1/4 1 1/2 1 3/4 2
3/8	5/16-18	1/2	9/16	1/4	3/16	3/8 1/2 5/8 3/4 1 1 1/4 1 1/2 1 3/4 2 2 1/4 2 1/2 2 3/4 3 3 1/4 3 1/2 3 3/4 4
1/2	3/8-16	5/8	3/4	5/16	1/4	1/2 5/8 3/4 1 1 1/4 1 1/2 1 3/4 2 2 1/4 2 1/2 2 3/4 3 3 1/4 3 1/2 3 3/4 4 4 1/4 4 1/2 4 3/4 5
5/8	1/2 13	3/4	7/8	3/8	5/16	1 1/4 1 1/2 1 3/4 2 2 1/4 2 1/2 2 3/4 3 3 1/4 3 1/2 3 3/4 4 4 1/4 4 1/2 4 3/4 5 5 1/2 6
3/4	5/8-11	7/8	1	1/2	3/8	1 1/2 1 3/4 2 2 1/4 2 1/2 2 3/4 3 3 1/4 3 1/2 3 3/4 4 4 1/4 4 1/2 4 3/4 5 5 1/2 6

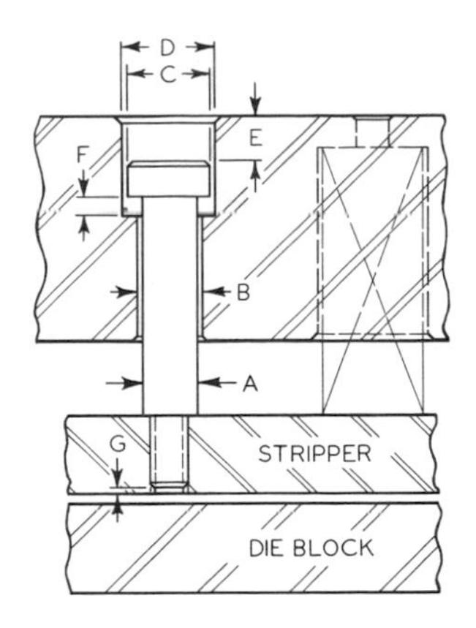

Figure **10·36** Stripper-bolt installation (die in closed position).

the stripper. The customary minimum for *G* is 1/64 in. In many cases it may be cheaper, as well as better, to use a thicker stripper plate than to alter the threaded length of the stripper bolts.

To assure stripping, a spring stripper should overtravel a slight distance *S* (Fig. 10·37) when the stripper is at its extended position. Normal overtravel may be as little as 0.005 in. for small- to moderate-sized precision work. For large, heavy work, overtravel may be on the order of 1/16 in. or occasionally more. In any case, each time the punch is sharpened, the overtravel increases by an amount equal to the stock removed from the punch. An accumulated increase in overtravel may or may not be permissible, depending upon the specific application. In situations where increased overtravel may not be tolerable, compensating washers can be installed under the stripper bolt head, as shown. Heat-treated compensating washers in appropriate thicknesses may be purchased as standard items.

At times, special stripper bolts are deemed advisable. Figure 10·38 illustrates one manner in which a special stripper bolt can be made up. A standard socket head cap screw 1 is assembled into a hardened tube 3. A hardened washer 2 is required also. The washer serves the same

Figure **10·37** Overtravel (die open, stripper extended).

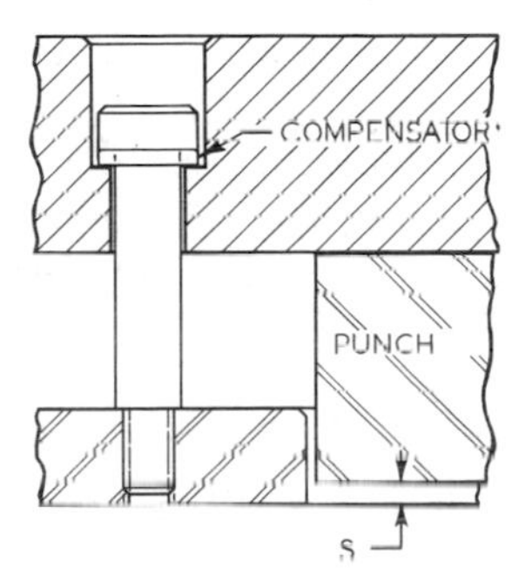

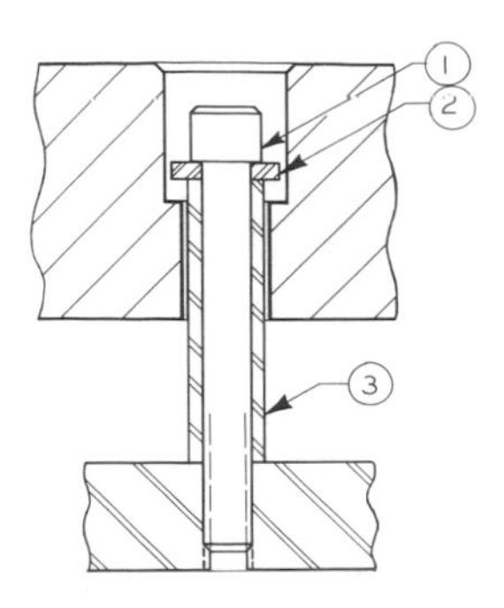

Figure **10·38** Special stripper bolt.

function as the head of a standard stripper bolt. Properly designed and made, this assembly can be superior to a standard stripper bolt. It is also considerably more expensive. Tubing should be reamed to a slip fit for the screw. Wall thickness should range from 1/16 in. when used with a 1/4-in. diam screw, to 3/32 in. for a 1/2-in. screw. Washer thickness should range from 3/32 in. for a 1/4-in. screw, to 3/16 in. for a 1/2-in. screw. Procedure for assembly in the die is generally the same as for a standard stripper bolt.

Flanged strippers. Spring strippers may be retained by flanges called shoulders. Since spring strippers belong to the family of pressure pads, the description and discussion pertaining to shouldered pressure pads (see Fig. 3·24 and accompanying text) is generally applicable to shoulder-retained spring strippers.

LEVELERS

When the stock strip is started into a progressive die, it introduces a tilting effect on the spring stripper. Tilting a spring stripper can have serious consequences. One method of counteracting the tilting effect is illustrated in Fig. 10·39, where leveling buttons are installed. For clamping strippers (view *A*), the height of the levelers is made to provide a small clearance *C* between the leveler and the die face when the stripper is contacting the stock strip. When the stripper is in contact with a stock strip of minimum thickness (tolerance-wise), a clearance of 0.002 in. should exist at *C*. For nonclamping strippers (view *B*), the leveler contacts the die face, providing the desired clearance *D* between the stripper and the stock material. Levelers in strippers should be installed in such a manner that they will not fall out into the die.

Levelers may also be required on inverted-type dies. Small and/or fragile punches are very susceptible to breakage resulting from any tilting of the stripper.

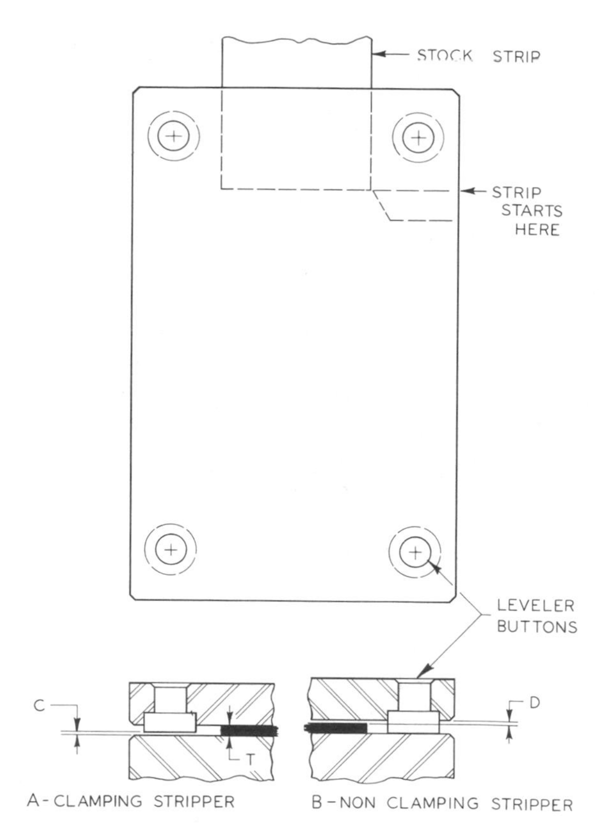

Figure **10·39** Levelers.

Numerous different ways of providing stripper leveling are possible and practical. They will all be similar in principle. For example, in Fig. 10·21 clearance *C* could be minimized to provide a leveling effect. In Fig. 10·23 the stripper is leveled by the guide rails.

GUIDED SPRING STRIPPERS

Stripper bolts do not provide precision stripper alignment. For applications requiring this, special provision must be made to guide the stripper. In Fig. 10·40, for example, the stripper is guided between blocks which assure that it will be maintained precisely in position laterally.

Although guide blocks are a conventional construction, stripper alignment is more commonly secured and maintained by guide pins. Two typical stripper guide-pin arrangements are pictured in Figs. 10·41 and 10·41(*A*). The drawings are self-explanatory. Fits are as indicated. Do not omit air vents. A minimum of two guide pins is necessary to prevent rotation. However, to assure stability, a minimum of three is generally installed in round strippers and four, if possible, in rectangular strippers.

Numerous variations of the illustrated guided-stripper constructions are possible and practical. There are various reasons for using guided strippers. Some are:

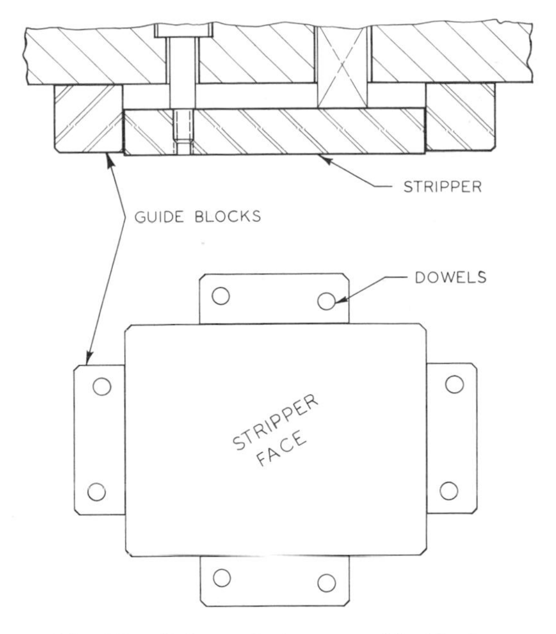

Figure **10·40** Stripper between guide blocks.

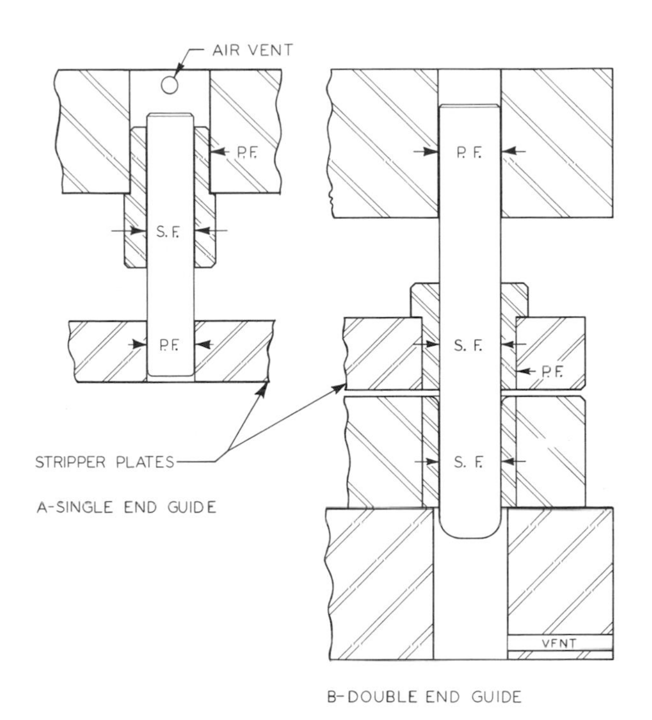

Figure **10·41** Guided strippers.

1. Normal lateral "float" of a nonguided stripper may permit objectionable contact between stripper and punches.
2. Stripper may fit closely between guide rails or nest plates.
3. Stripper is a "guiding" stripper: That is, the stripper is intended to guide and/or support punches. Guiding strippers must be guided in some manner similar to the strippers described above.

GUIDING STRIPPERS

The commonly known demonstration of driving a needle through a coin is illustrated by the sketch in Fig. 10·42. The coin is placed on a suitable surface. The needle is inserted in a cork and placed, as shown, on top of the coin. The needle, supported by the cork, can be driven through the coin by a smart hammer blow. This demonstration is not recommended as a parlor trick: It can be hazardous. It is cited here because it is an appropriate illustration of the basic principle of guiding strippers. A guiding stripper provides punch support in much the same manner that the cork supports the needle.

The spring-actuated stripper in Fig. 10·43 is a basic guiding stripper. It is called a guiding stripper because it guides and supports the punches. In consequence, it also is guided. Clearance S between punch and stripper may range from as little as 0.00005 in. to as much as 0.001 in. One practice is to make clearance S equal to one-half of the cutting clearance C but to restrict S to a maximum of 0.001 in. where C is greater than 0.002 in. That is,

where $C < 0.002$ in. $\quad S = \frac{C}{2}$

$C > 0.002$ in. $\quad S = 0.001$ in.

Another practice is simply to make the guiding openings a close slip fit to the punches for all applications. Piece-part tolerances pertaining to center distance between holes may require a closer guiding fit than would otherwise be necessary.

Figure **10·41A** Guided strippers.

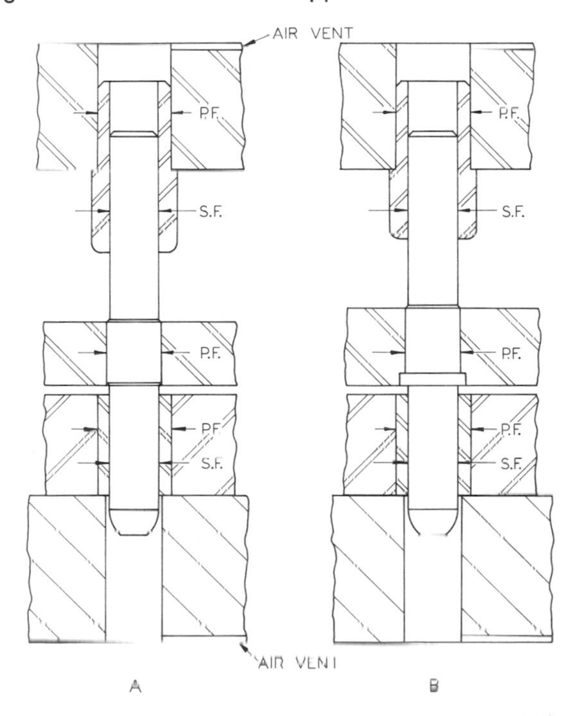

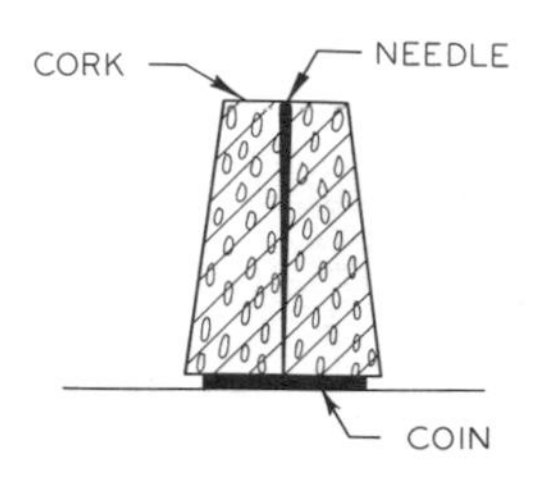

Figure **10·42** Needle and coin demonstration.

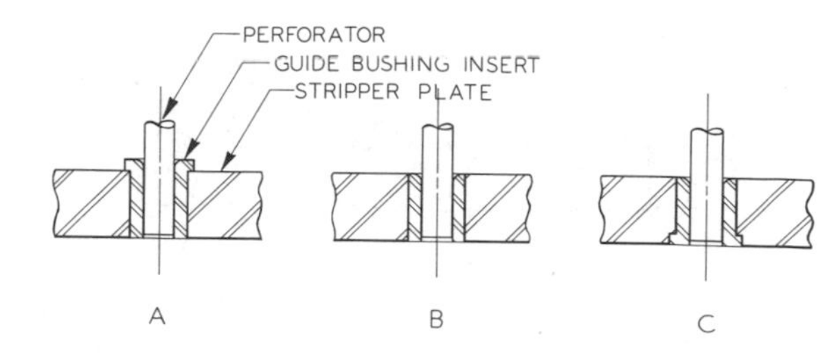

Figure **10·44** Hardened bushing inserts in guiding strippers.

This stripper should be hardened because of friction with the punches. Optimum degree of hardness may range widely, from Rockwell C 48 to 60, depending upon the application. Ordinarily, stripper hardness is directly in proportion to production requirements: Higher production necessitates harder strippers. However, stripper proportions may strongly influence the degree of hardness. If a stripper has relatively weak, or fragile areas, the hardness must be correspondingly reduced. For low-production dies, strippers of this kind may sometimes be made from pretreated alloy steel, as furnished.

Hardened inserts and/or bushings are commonly incorporated in guiding strippers. When this is done, the stripper plate may at times be left soft. However, this, like many other conditions, must be in accord with the specific requirements of each application. A hardened or a toughened stripper plate may often be required in conjunction with hardened inserts.

Typical bushing inserts are depicted in Fig. 10·44. The method shown in view *A* is generally the most practical. Bushings are sometimes assembled in spring-stripper plates in the manner shown in views *B* and *C*, but these constructions can be hazardous and should be avoided. Standard drill-jig bushings are the logical choice for most bushing-insert applications in guiding strippers. They are commercially available as a precision quality item at nominal cost, they are interchangeable, and they are easily replaced, if necessary.

Figure 10·45 indicates two possible constructions which would apply to small punches operating through relatively thick stripper plates. In view *A*, the stripper is counterbored to receive a shorter bushing. This construction permits a shorter (and stronger) perforator point. In view *B*, the punch is guided on its shank instead of its point. This method, too, provides a stronger perforator. The difference between the inside diameter of the bushing and the perforator point cannot be too large, or it may result in distortion of the stock material during the stripping act. This can cause perforator breakage in addition to possible detrimental effects on piece-part quality.

An intermeshing quill is pictured in Fig. 10·46.

Figure **10·43** Basic spring-actuated guiding stripper.

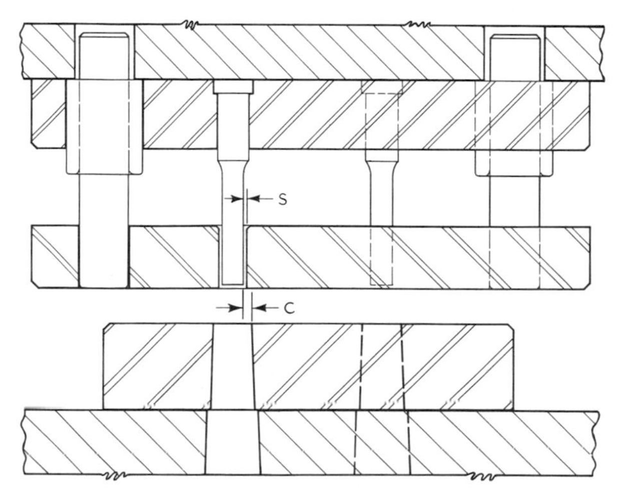

Figure **10·45** Guiding of small perforators.

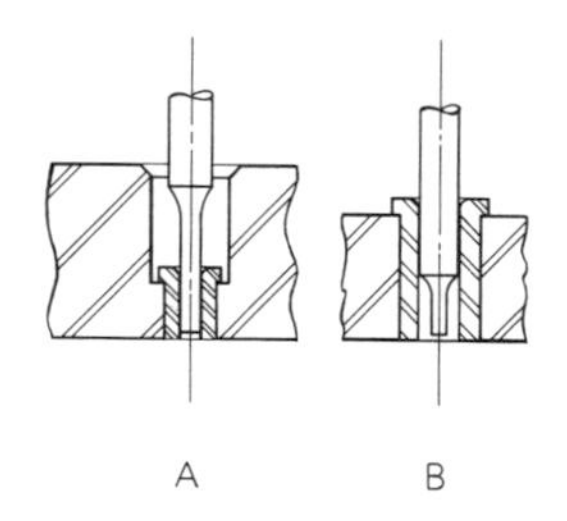

Figure **10·46** Intermeshing quill sleeve.

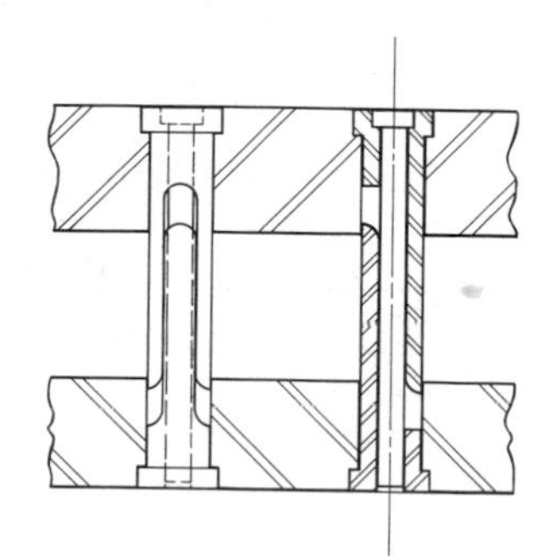

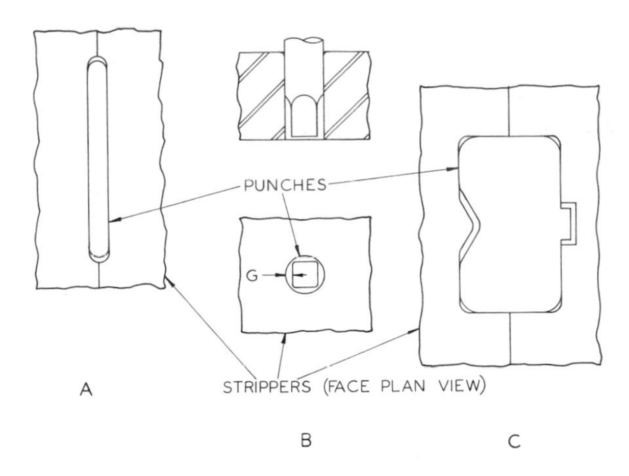

Figure **10·47** Guiding and supporting of contoured punches.

These are commercially available as standard die components. They are especially desirable for piercing heavy stock material.

A guiding fit around the entire punch contour is not normally required. Wherever practical, the stripper openings should be relieved if this will save fitting time. In the relieved areas, clearance between punch and stripper opening should be the same as for a nonguiding stripper. Typical guiding-stripper relief is pictured in Fig. 10·47. In many cases, long and narrow punches may not require stripper support at the ends, as shown in view *A*. Square punches made with round shanks (view *B*) can often be supported on the shank instead of at the cutting contour. This simplifies stripper construction, since only a round hole is required in the stripper. The punch

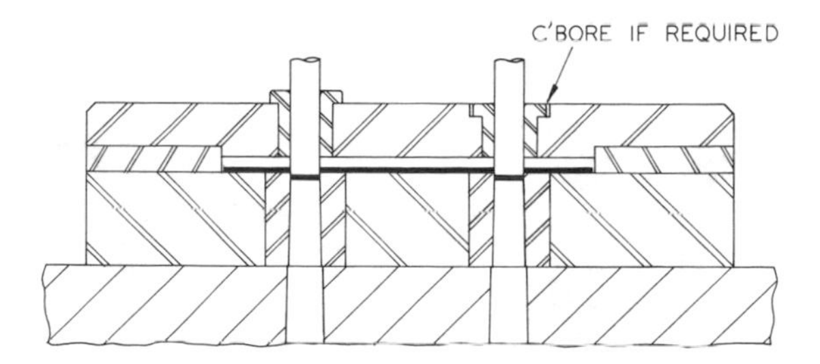

Figure **10·48** Guiding fixed stripper.

should, of course, be keyed in the punch plate to prevent rotation. Gap *G* must be small enough to strip the specific stock material without causing distortion. Wherever practical, provide the relief at sites where fitting is most difficult (view *C*).

GUIDING FIXED STRIPPERS

Fixed strippers, too, may act as guiding strippers. In fact, this is very common. The same general fitting, etc., applies to both spring and fixed strippers, but since a fixed stripper is secured in position on the die-block assembly, it does not need to be guided in the sense that a traveling spring stripper does. A guiding fixed stripper is typified in Fig. 10·48. Note that because of tunnel height *H*, a fixed stripper cannot provide punch guidance as close to the stock material as a spring stripper can. In this respect a spring stripper is more efficient for guiding very small punches.

Keep in mind that a fixed stripper is almost always more economical to make than an equivalent spring stripper.

CHAPTER 11

SHEDDERS AND KNOCKOUTS

MULTIFUNCTIONAL DIE COMPONENTS

Before considering shedders as specific die members, it might be well to consider briefly the multifunctional aspects of many die components. In keeping with specific requirements dictated by differing individual applications, most die components can be multifunctional. In such instances, the component is usually called by its dominant or more obvious function. For example, the forming-pad component in Fig. 11·1 acts as a shedder, stripping the piece part from the die cavity. Its function as a shedder is vital to the successful operation of the die. However, in this and similar applications, it is customary to refer specifically to this component as a forming pad.

Figures 11·2 and 11·3 illustrate a combination blank and draw die. The piece part is a round cup which has a depressed bottom, as shown. This die incorporates a number of multifunctional components. For example, the blanking punch 3 contains the draw-die cavity. Thus, the blanking punch acts as a draw-die opening in addition to its function of blanking. The face of the shedder 4 is contoured as shown. The shedder bottoms against the thrust plate 2 and acts as a punch to deboss the bottom of the cup (piece part). The top surface of the draw punch 9 is contoured to act as a die cavity for the debossment. The draw pad 7 serves as a shedder in the sense that it acts to lift the cup out of the blanking-die cavity. The draw pad also acts as a stripper, assuring stripping of the cup from the draw punch.

It is essential for the diemaker to analyze the die design and determine the complete functional requirements of each component. This is necessary because the less obvious functions of multifunctional components may be just as vital as the more obvious.

PURPOSE AND ACTUATION OF SHEDDERS

Specifically, a shedder acts to expel the workpiece from the die cavity. Shedder actuation may be achieved by:

1. Pneumatics, usually air cylinders or cushions
2. Hydraulics, generally hydraulic cylinders or cushions (hydropneumatics)
3. Rubber pads (cushions)
4. Springs
5. Some kind of positive mechanical knockout

POSITIVE SHEDDERS

A positive shedder can be described as one which is not actuated by springs or other compression media. Figures 11·4 to 11·6 illustrate a very basic positive shedder actuated by means of a simple knockout rod. The shedder and knockout-rod assembly shown is functionally typical

Figure 11·1 Multiple function: forming pad and shedder.

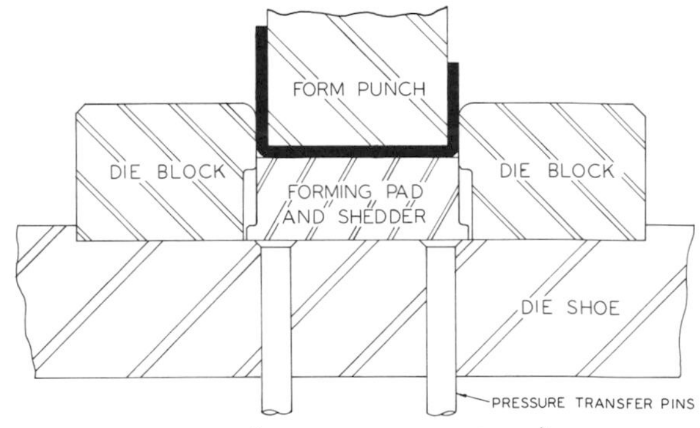

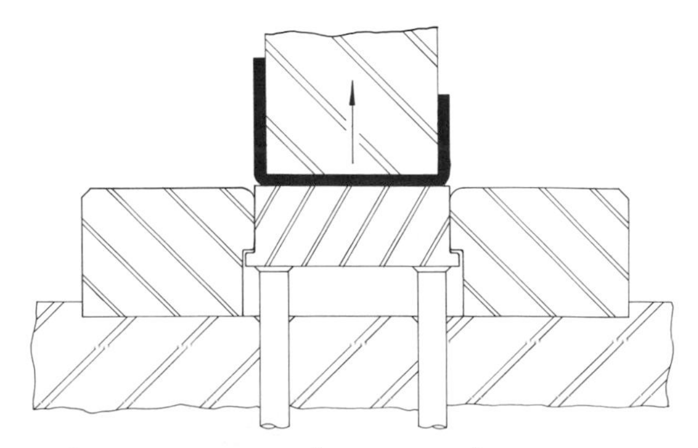

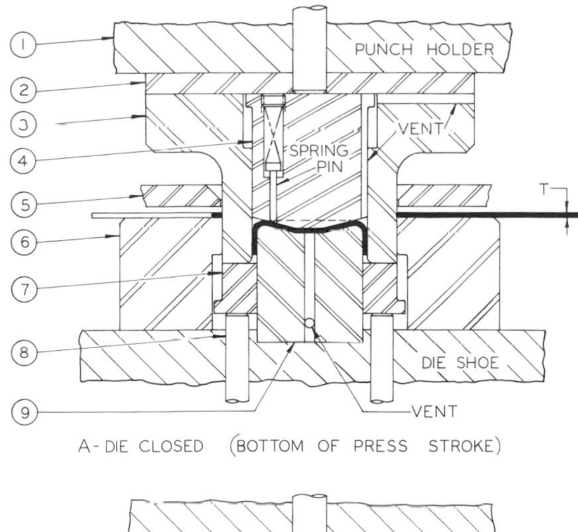

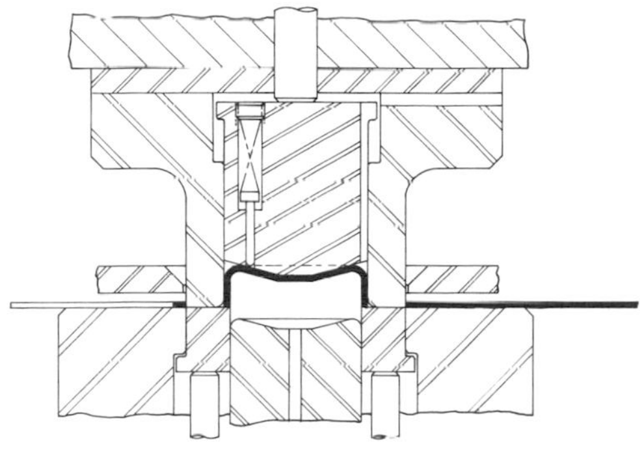

Figure 11·2 Stripping and shedding sequence in combination die.

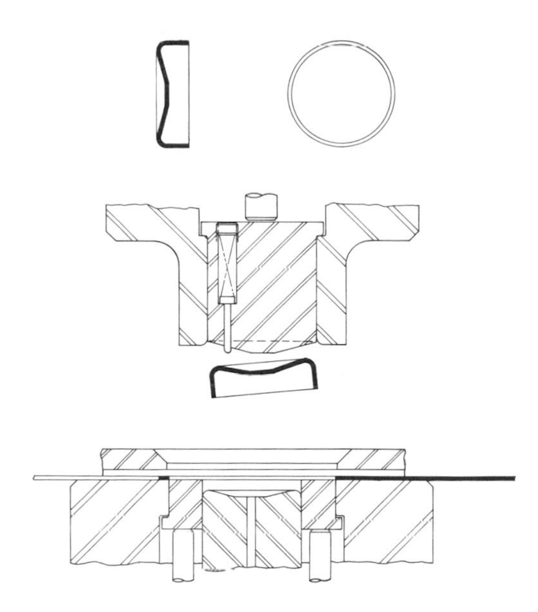

Figure 11·3 Stripping and shedding sequence (continued from Fig. 11·2).

for inverted-type dies. This shedder is a flanged shedder. The flanges, an integral part of the shedder, act as keepers, retaining the shedder within the die cavity.

In Fig. 11·4, the assembly is shown as it would appear with the punch holder removed from the die shoe. This is the condition when the assembly is being checked for proper relationship between the die block, shedder, and knockout. The shedder flanges are resting on the die-opening ledge at X. The shoulder on the knockout rod is seated against the bottom of its counterbore at C. In this condition, the shedder should protrude beyond the die face a distance A. The knockout rod should not be in contact with the shedder; a clearance gap B must be present. Minimum for these dimensions should be

$$A = 3/64 \text{ in., min.}$$
$$B = 0.005 \text{ in., min.}$$

Avoid working to the minimums whenever possible. Normal optimum allowances are

$$A = 1/16 \text{ in.}$$
$$B = 1/64 \text{ in}$$

These allowances may be increased for relatively crude work, provided, however, that gap D is adequate for the specific application.

Figure 11·5 depicts the shedder relationship at the bottom of the press stroke (die closed). The punch entry distance is shown at E. The stock material thickness is T. The shedder is forced upward into the die opening a distance $A = E + T$. With conditions as shown, gap D should normally be a minimum of $2\frac{1}{2}T$ or ⅛ in., whichever is greater. Gap D is an important safety factor: In

Figure 11·4 Relationship of shedder to die block and knockout.

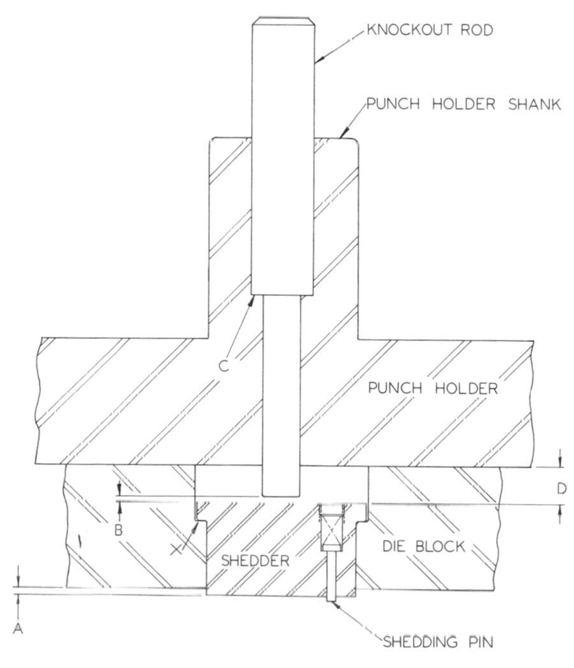

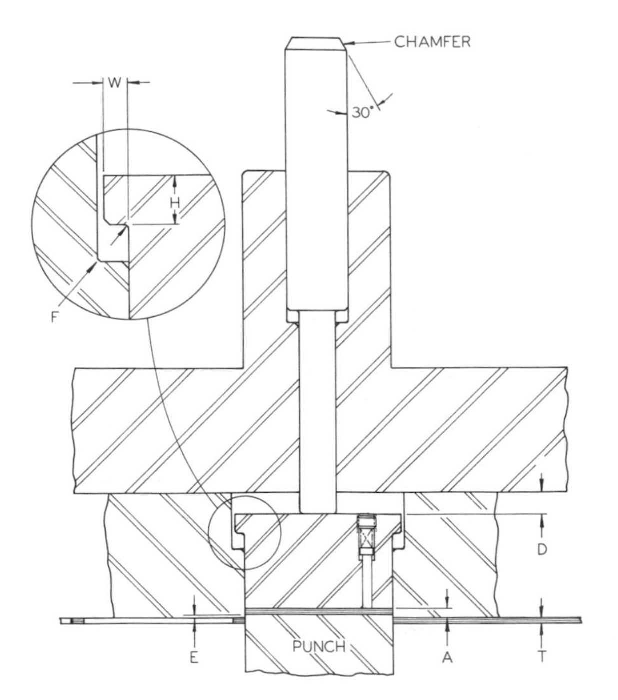

Figure 11·5 Shedder relationship, die closed (bottom of press stroke).

the event of either shedding-pin or knockout malfunction, gap D should permit the accumulation of at least two extra piece parts within the die cavity. This gives the operator an opportunity to notice the malfunction and stop the press. It is obvious that an accumulation of doubles can have serious consequences, including damage to the die and/or punch press as well as possible injuries to persons in the vicinity.

For certain forming operations (see Figs. 11·1 and 11·2), it is necessary to eliminate gap D in order to hit home with the shedder. In such applications, bottoming is an essential function of the shedder. These dies are unavoidably susceptible to the hazards associated with bottoming and must be treated accordingly. However, the diemaker should avoid bottoming whenever it is possible to do so.

Flange proportions should be

$W = \frac{1}{16}$ in.	min.
$H = 1\frac{1}{2}\,W$	min.
$H > 1\frac{1}{2}\,W$	whenever practical

Always provide a slight fillet radius where shown at F. Do not make these corners sharp; a fillet radius as small as 0.010 in. is far superior to a sharp corner.

The top end of the knockout rod is chamfered. A 30° chamfer, as shown, is generally optimum. It will eliminate peening and spalling while providing maximum end area for contact against the punch press knockout bar.

Figure 11·6 indicates the necessary relationship at the top of the press stroke. At this stage

$$A = B \quad \text{approx}$$
$$C = \tfrac{1}{2}B \quad \text{approx}$$

Using the recommended optimum dimensions (Fig. 11·5), the relationship at the top of the press stroke would be

$$A = \tfrac{1}{32} \text{ in.}$$
$$B = \tfrac{1}{32} \text{ in.}$$
$$C = \tfrac{1}{64} \text{ in.}$$

Immediately following the condition shown here, the knockout rod and the shedder drop free to the condition shown in Fig. 11·4.

Do not underestimate the significance of the relationships shown; they are vitally important to successful shedder functions. Note also the function of the shedding pin, and remember that the shedding pin in this and similar applications is mandatory. The importance of an adequately functional shedding pin cannot be overemphasized; an inadequate one is all too often the primary cause of a wrecked die.

The assembly in Fig. 11·7 is somewhat more sophisticated. This is necessary because of the nature of the shedder contour. Since it is essential for knockout force to be evenly distributed with respect to shedder contour, the knockout-

Figure 11·6 Shedder relationship, die open (top of press stroke).

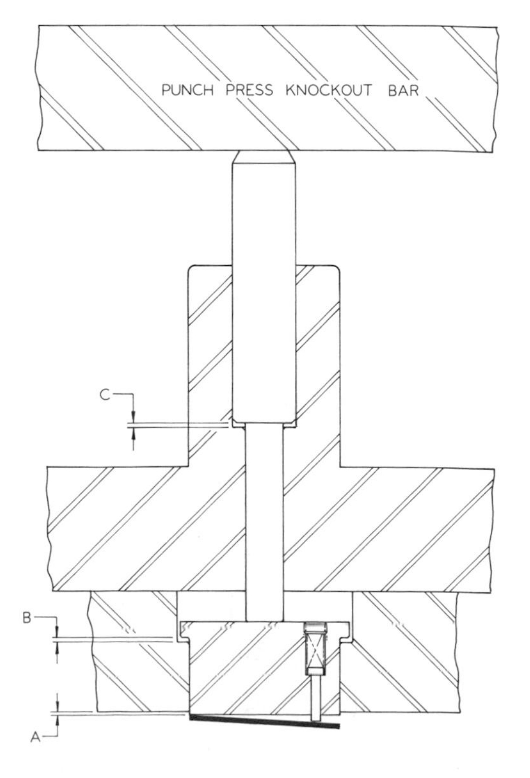

shedder assembly must be designed specifically for each application. This, of course, leads to considerable variation in construction. It does not, however, change the basic principles associated with positive-shedder assemblies. In principle, the relationships described for Figs. 11·4 to 11·6 are essential for all positive-shedder applications. The requirements for dimensions A and B in Fig. 11·7 are the same as for A and B in Fig. 11·4. Gap D can be derived, by association, from Fig. 11·5. In Fig. 11·7,

$$D = E + T + A + 2\tfrac{1}{2}T \qquad \text{min.}$$

where E = punch entry distance
T = stock material thickness

Space G must not be less than D. To facilitate diemaking, G should be made 1/32 in. or more larger than D. However, when determining the required pocket depth J, always check and be certain that dimension K will be adequate. Do this before machining the pocket.

Shedder contours determine the location of shedder flanges. Locate the flanges to facilitate grinding and/or machining. Flange locations should be balanced out around the periphery of the shedder. The number of transfer pins required, as well as their location, must be appropriate for the specific shedder. The transfer pins must distribute the knockout force evenly throughout the shedder. Avoid unbalanced conditions. Be es-

Figure 11·7 Shedder and knockout assembly using loose transfer pins.

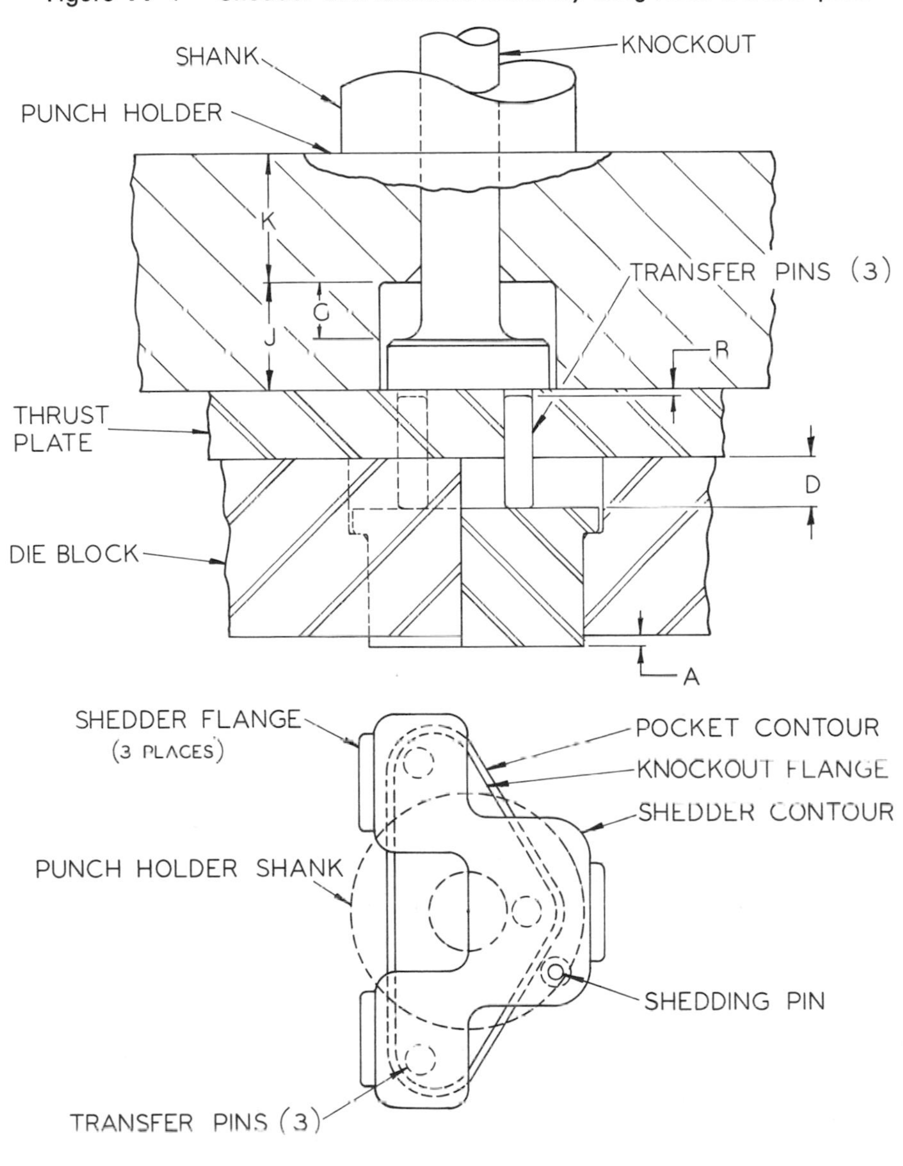

pecially careful in cases where shedders have areas of cross section which are comparatively delicate.

Among the many possible knockout variations are those in Fig. 11·8. Here the transfer pins are assembled by peening them into a pin plate. In view *A*, the pin plate and knockout rod are also fitted together and secured by peening. This method is practical only for light duty, where the knockout force is evenly distributed and very well balanced in relation to the shedder contour. The complete upper die assembly is shown here in order to emphasize the existence of the normal shedder-knockout relationships, even though the dies differ in construction. The knockout assembly shown in view *B* is much stronger. Here, the knockout rod is welded to the pin plate. This knockout would be capable of withstanding more severe service.

A generalized restriction to transfer-pin location is indicated in Fig. 11·9. Do not permit more than one-half of the pin diameter *A* to bear against the flanged portion of the shedder. Whenever practical, have the entire pin diameter located behind the body of the shedder.

On occasion, laminated construction may be used to advantage. The shedder in Fig. 11·10 is for a compound pierce and blank die. The flange is a separate ring, secured to the shedder body by screws. The size and contour of this particular shedder make laminated construction feasible. Its size permits installation of screws which are adequate in size, sufficient in number, and strategically disposed. Its contour is such that the straight-through machining permitted as a result of laminating is a definite asset. However, in spite of its advantages, this construction should not be used indiscriminately. It can at times be tempting enough to encourage gambling upon its adequacy, but keep in mind that positive shedders are subject to severe impact conditions, and employ laminated construction only where it is practicable.

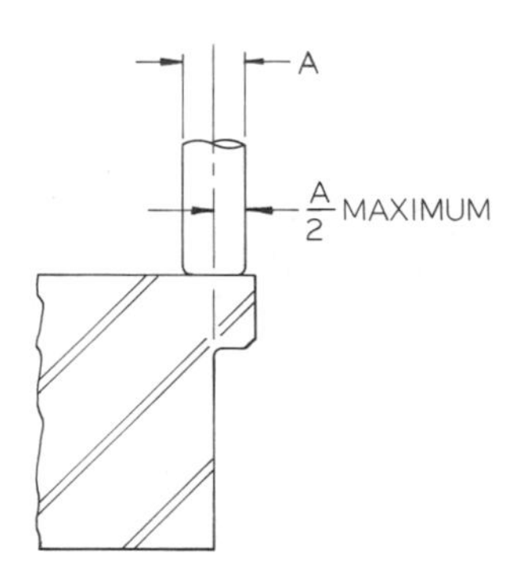

Figure 11·9 Transfer-pin location in relation to shedder flanges (limiting condition).

Figure 11·8 Knockout constructions.

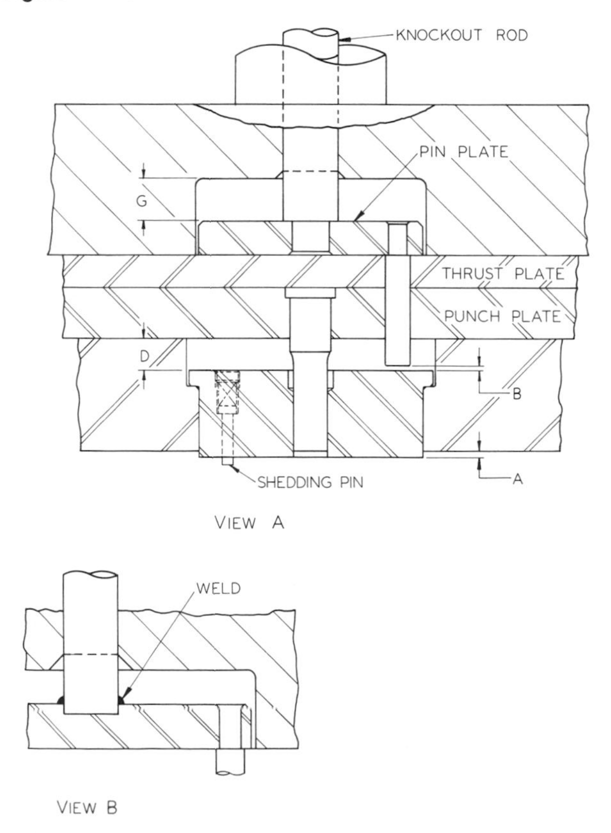

Figure 11·10 Laminated shedder.

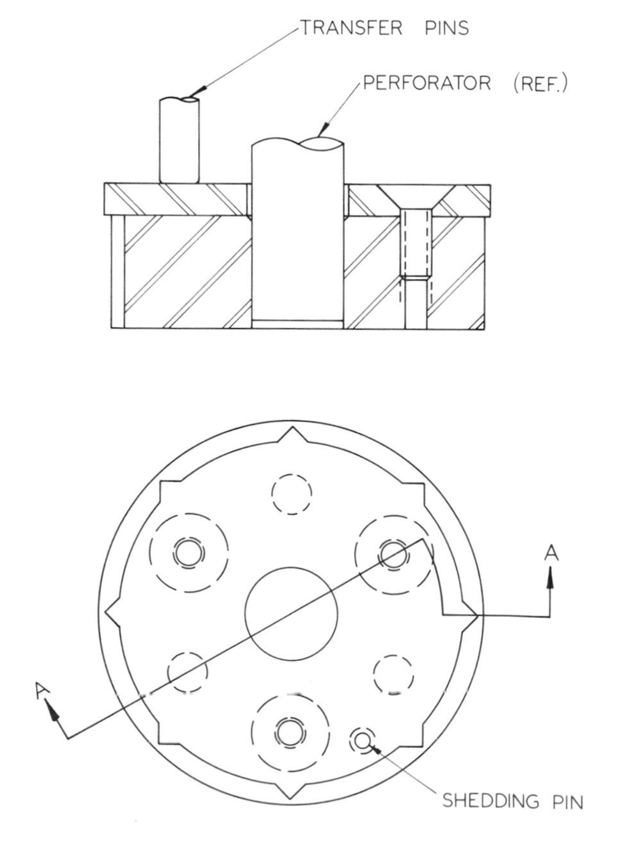

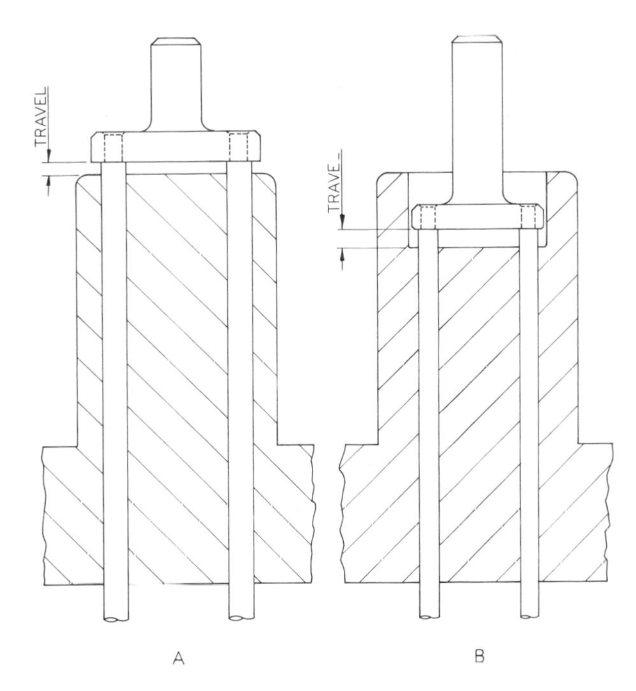

Figure **11 · 11** Top bridged knockouts.

The knockout constructions in Fig. 11 · 11 are variations of transfer-pin knockouts. They are not heavy-duty knockouts but are limited to applications where the span of the transfer pins can be contained within the punch-holder shank. They are desirable from the die cost standpoint and are practical for many applications. The knockout in view *A* is the more economical to make. In addition to the cost advantage, it permits a wider span for the transfer pins. However, in some punch presses, the cavity in the ram may not be deep enough to accept the top bridge. This condition worsens in proportion to the knockout travel. Such an objection, of course, does not apply to the knockout in view *B*. For applications which require a long knockout travel, the long, slender transfer pins may be too susceptible to bending, especially if considerable knockout force is required.

Keeper lugs (flanges) are not required on the shedder which appears in Fig. 11 · 12. The knockout rod 1 is peened into assembly with the shedder 5. A collar ring 2 is assembled to the knockout rod by means of a cross pin 3. The collar acts as a stop, limiting the protrusion distance *A* of the shedder. The collar impacts against the top of the shank and subjects the cross pin to considerable force, which is suddenly applied. Therefore, the cross pin should be made of drill rod and left soft. Do not attempt to use a standard dowel pin for this or any similar application.

In the case illustrated, the shedding-pin installation differs from those previously shown. This type of installation is not feasible in round shedders unless some means of preventing shedder rotation is provided. A much stronger shedder assembly is pictured in Fig. 11 · 13. Here the knockout shaft is engaged (threaded) in the shedder and secured by a lock nut. Two lock nuts are jam-tightened on the shaft, stopping the shedder travel as required. Wrench flats should be provided on the knockout shaft to aid in positioning and tightening the lock nuts, especially the lower lock nut. The shedding pin 1 shown here is actuated by a flat spring 2.

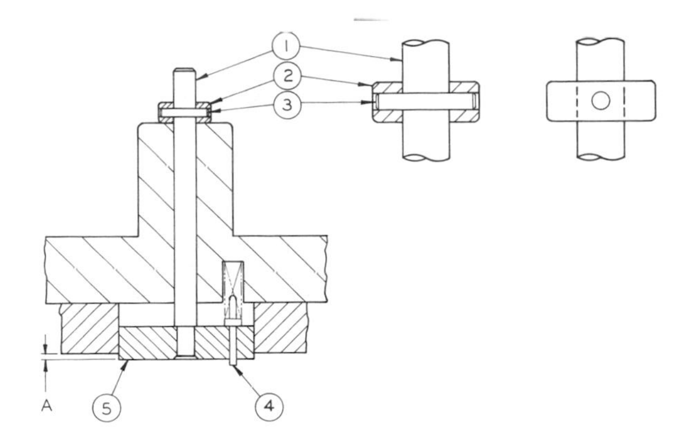

Figure **11 · 12** Flangeless shedder.

By definition, the arrangement in Fig. 11 · 14 is a stripper. Its function is to strip either the workpiece or the stock material scrap (whichever the application requires) from the punch. It is included here because its construction and action exactly parallel those of a positive shedder. Thus the allowances and relationships given earlier for positive shedders apply, as equivalents, to positive knock-off strippers.

A spider-type bridge plate 2 is shown. The spider operates within a suitably contoured recess which is milled in the punch holder. In this case, the transfer pins are threaded studs. The studs

Figure **11 · 13** Flangeless shedder, locknut retained.

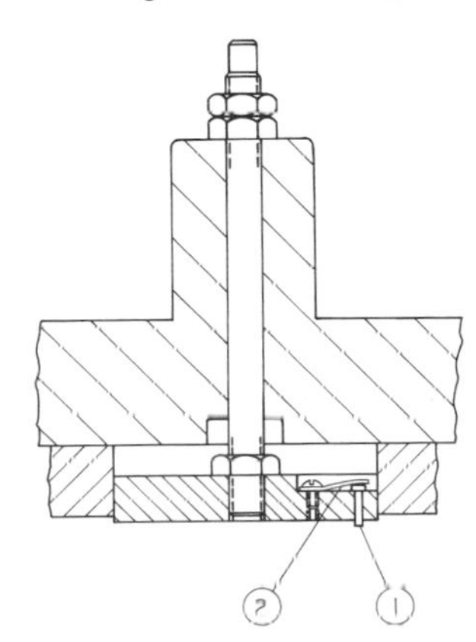

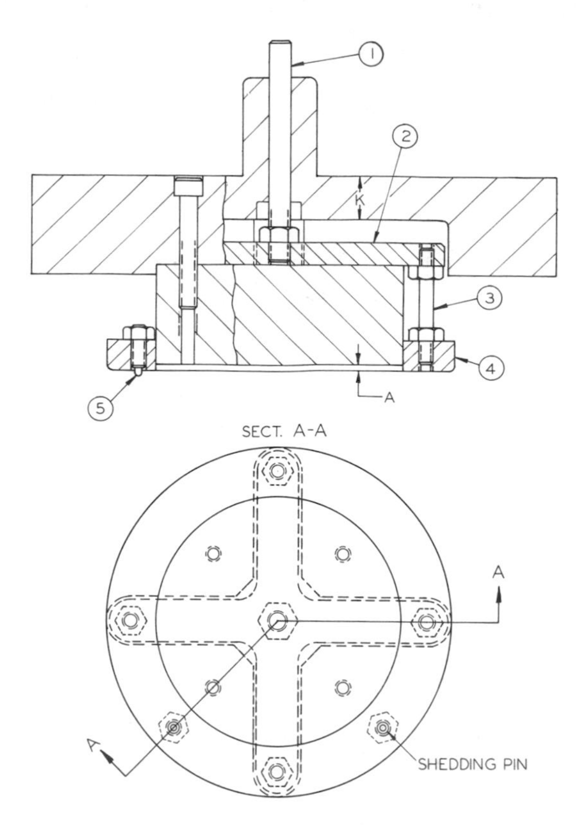

Figure **11 · 14** Positive knock-off stripper driven by spider-type knockout.

are secured with lock nuts to maintain the required spacing between the stripper plate 4 and the spider.

Spiders, which are often used in conjunction with large shedders and strippers, permit a long span between the transfer-pin centers. In addition, the contoured recesses permit punches (or other components) to be mounted directly to the punch holders. Before milling the recess, check and be certain that the security of the shank will not be endangered (see *K*).

One or more shedding pins are required for this kind of stripper. In the case illustrated, standard commercially available spring pins 5 are employed. They are of the type which is installed in tapped holes and secured with lock nuts, as shown.

A knockout arrangement for large work is represented in Fig. 11 · 15, where it is applied to a positive knock-off stripper. It would apply to large shedders, as well. The mechanical relationships are the same as for the center-poised knockout-shedder arrangements previously described. Center distance *M* between the knockout rods must be made to suit the specific punch press in which the die will be operated.

Shankless punch holders are often used for large dies. In other instances, the shanks are used only to position the die centrally in the press.

Figure **11 · 15** Outboard knockout rods.

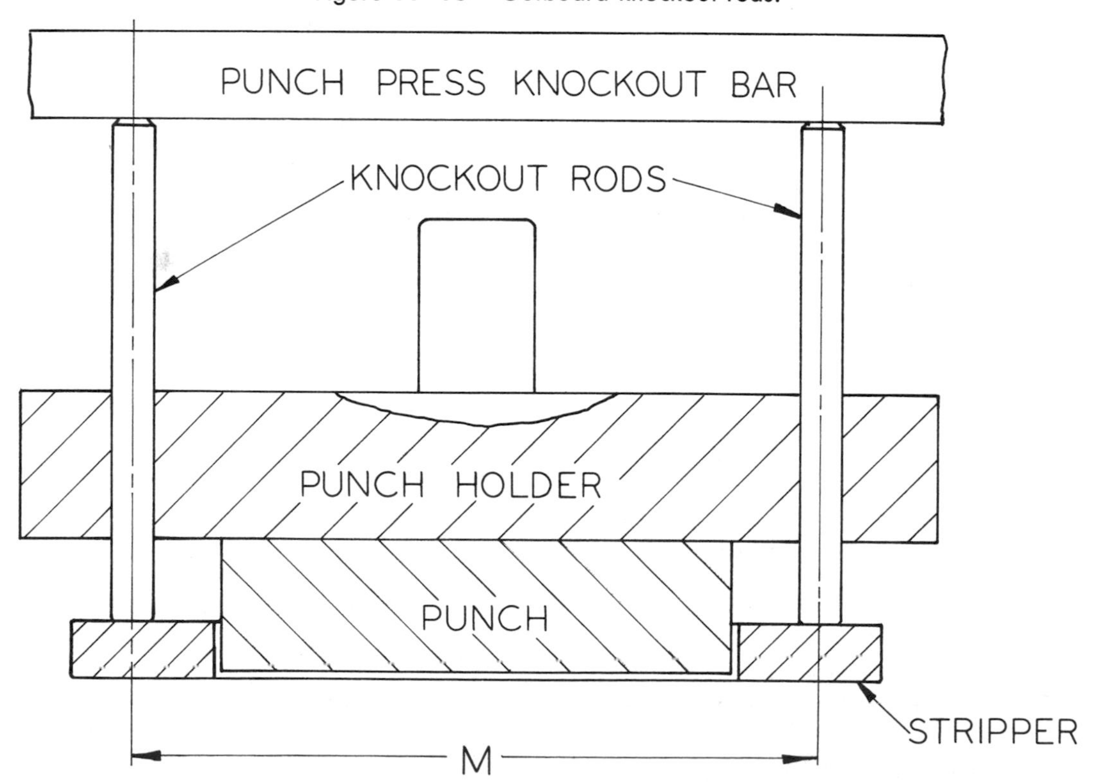

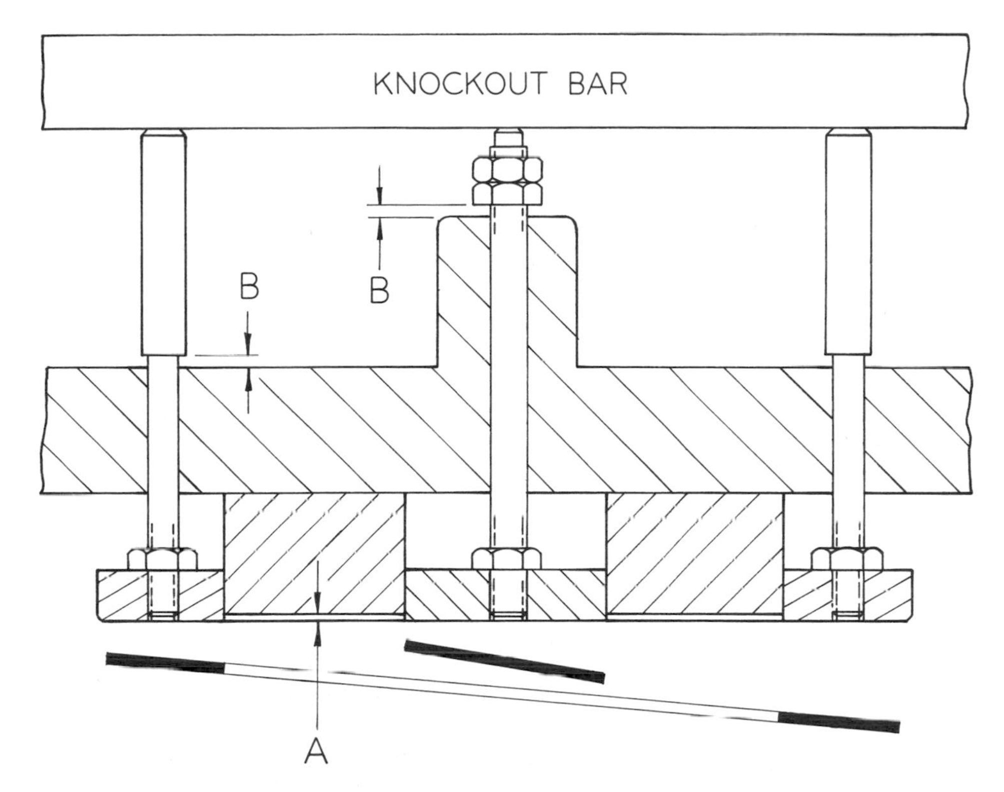

Figure 11 · 16 Combination knockout arrangement.

In these situations the punch holders are normally screw-attached to the ram. Also some presses may be equipped with special adapters. All of these factors must be taken into consideration when determining specific knockout constructions.

As with many other aspects of diemaking, variations of shedder-stripper-knockout constructions can be virtually endless. However, the basic relationship principles may, by association, be generally applied to all positive constructions of this type. The combination arrangement shown in Fig. 11 · 16 serves to emphasize the fact. The illustrated arrangement cannot be considered a common application, and yet in spite of this, the *A* and *B* relationships shown here are the same as those shown earlier in Fig. 11 · 6.

Two common shedding-pin installation errors are pictured in Fig. 11 · 17, view *A:*

1. The shedding pin is located at the center of the shedder. Because of this, there is no leverage advantage for the shedding pin. This error can result in failure of the shedding pin to displace the blank from the face of the shedder. The final result could be just as disastrous as if the shedding pin had been omitted entirely.
2. Length *L* is too short. It does not provide grinding life on the shedder. This would not apply, of course, to situations where grinding life is not required.

In view *B*, the shedding pin is correctly installed. It is located off-center, which permits it to lever the blank away from the shedder face as shown. Also, length *L* will permit the shedder face to be ground away, as required, in the course of die maintenance.

SPRING SHEDDERS

A basic spring-actuated shedder appears in Fig. 11 · 18. It is a flanged shedder. Flange proportions are indicated at *H* and *W*. These should be

$$H = 2W \qquad \text{min.}$$

Figure 11 · 17 Shedding-pin installation.

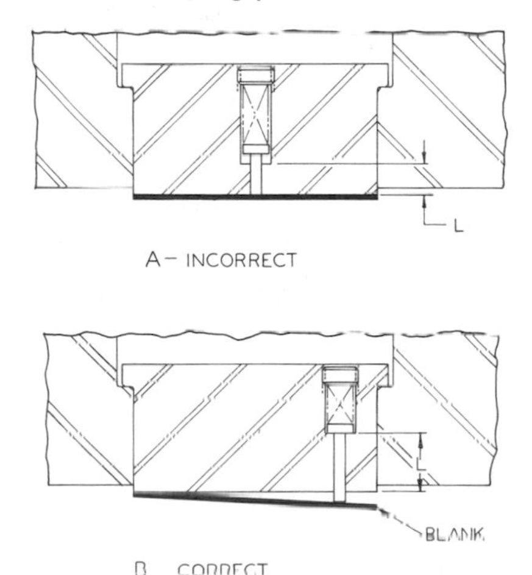

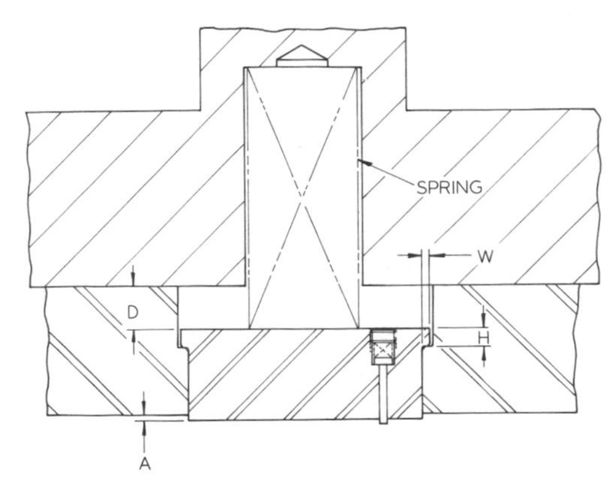

Figure **11 · 18** Basic spring shedder.

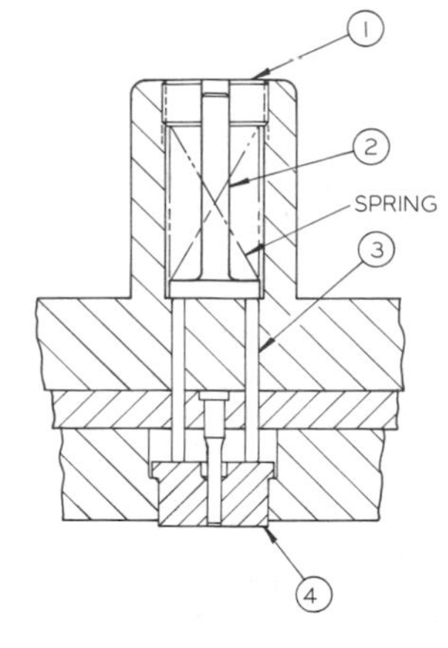

Figure **11 · 19** Shedder actuating spring located in shank of punch holder.

Space *D* should be treated the same as for a positive shedder.

Protrusion distance *A* will vary according to the specific application. Range of *A* will be from zero (flush) to approximately 1/32 in., as dictated by the requirements of the specific application. If the shedder is intended to push the blank back into the stock strip, protrusion *A* should range from zero for light-gage stock material to 0.005 in. for heavier stock. For light-gage material, two or more shedding pins should be installed in the shedder. They should be symmetrically located. The springs which actuate the shedding pins must not be too strong. If they are, or if the shedding-pin locations are not balanced, the blank may be pushed through the stock strip. For heavier material, shedding pins may not be necessary. However, holes should be provided in the shedder anyhow, to permit the installation of shedding pins if this should prove necessary.

For applications where the shedder is not intended to be a push-back shedder, protrusion *A* may be 1/32 in., if desired, to facilitate diemaking. In the case of heavy, crude work, this may be increased, if desired, to 1/16 in., which should be ample to facilitate diemaking even on crude work. For shedders which are not push-back shedders, only one shedding pin is generally required. Installation is generally the same as for a positive shedder.

In Fig. 11 · 19, the shedder actuating spring is contained within the punch-holder shank. The spring applies pressure to the plunger 2, and transfer pins 3 transmit the spring pressure to the shedder 4. The plunger flange is a slip fit in the spring pocket. The tail, or shank, of the plunger is a slip fit in the hole in the screw plug 1. This stabilizes the spring action, eliminating cocking or tilting forces on the shedder and transfer pins. The shedding pins were omitted from this drawing for purposes of simplification.

Stripper bolts are a practical means of suspending larger shedders in the manner typified in Fig. 11 · 20. Construction practices for such shedders are identical with those described for spring strippers. Remember to associate the spring and stripper-bolt locations to provide balance, in order to prevent deflection of the shedder plate. Keep in mind that the most common errors in this type of construction are:

1. Poor stripper-bolt-to-spring association.
2. Stripper bolts which are too small for the application.

Figure **11 · 20** Large spring shedder suspended with stripper bolts.

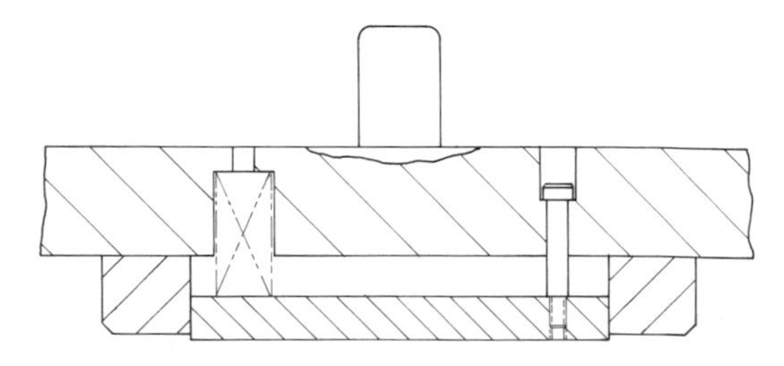

CHAPTER 12

NEST GAGES

Nest gages are commonly used on dies which perform secondary operations, such as piercing, forming, etc. Nest gaging is also required for dies in which unit stock is run, since this too requires the feeding of individual pieces into the die.

The purpose of a nest gage is to locate and position the workpiece properly in the die. The workpiece may be unit stock, or it may be an unfinished piece part.

In principle, a nest gage surrounds the workpiece, confining it within the nest. Because of this, the work performed by the die is usually done in relation to the outer periphery of the workpiece. However, approximation nests are an exception to this.

NEST-GAGING REQUIREMENTS

There are four important conditions which must be satisfied in order to provide suitable nest gaging: accuracy, loading ease, unloading ease, and last but not least, foolproofing.

Nest-gage accuracy factors. Nest gages must register the workpiece safely within the dimensional limits imposed by the piece-part specifications. The nest must perform this function consistently. For gaging purposes, it is not necessary for a nest to fit the entire contour of the workpiece. All that is required is for the nest to provide a sufficient number of contact spots (gaging points) strategically located in relation to the outer periphery of the workpiece. The spots must, in effect, surround the workpiece to prevent it from shifting.

The number of contact spots required for any given nest gage varies according to size and contour of the workpiece. For round work and triangular work, a minimum of three contact spots will be required. For other contours, the minimum is four. Keep in mind that these figures are minimums and are based on an exact fit between the workpiece and the contact spots. In order to ensure accurate nesting, it is often necessary to provide more than the so-called minimum number of gaging contact spots.

At times, a workpiece must be processed through more than one die where nest gaging is necessary. When this situation prevails, the location of the gaging contact spots on all of the nests concerned should coincide, if possible. If the workpiece is gaged at the same points in each successive operation, it will be easier to maintain correct dimensional association between the operations performed in the different dies.

In many instances, for the sake of accuracy it will be necessary to have the gaging spots fit the workpiece exactly. However, in cases where the nature of the work permits, it is both practical and desirable to allow a slight clearance between the gaging spots and the workpiece. The amount of clearance will, of course, vary in accordance with the accuracy requirements of each job.

Loading ease. The facility with which an operator can feed or load work into a die is an important consideration in the design and construction of nest gages. If a job is relatively difficult to feed, it will take longer to run and thus will involve higher production costs. When feeding is difficult there is also more likelihood that workpieces will be incorrectly loaded into the nest. Improper loading, besides slowing down the job, can cause the production of unsatisfactory piece parts. Moreover, there is always the possibility that the die will be damaged if a workpiece is incorrectly loaded. In addition to these factors, there is a direct relationship between operator safety and loading facility. A die which is easy to load will be much safer to operate than one which is difficult.

Some factors which influence ease of loading are: height (thickness) of the nest arrangement, shape of the workpiece, visual identification of the nest with the workpiece, and adequate lead. Suitable nest height depends on the particular job. Generally, larger work requires deeper nests. The shape of a workpiece may be such that it is easier to load one way than another. When this condition exists, it should be turned to account, since it will certainly expedite production.

Obvious visual association between the nest and the workpiece simplifies loading. If the press operator can readily see which way the workpiece fits into the nest, it is easier for him to place the piece correctly.

Adequate lead on the inner periphery of a nest facilitates placing the workpiece in the nest. Since the lead angle causes the nest opening to be larger at the top surface, it creates a larger area in which the operator can more easily place the workpiece. The lead angle then directs the course of the piece into its final gaged position. The principle is that the lead supplies a funneling effect which contributes to ease of loading.

Unloading ease. The preceding section has stressed the importance of making it easy for the operator to place a workpiece in the nest. It is also important that the work can be removed from the nest with as little difficulty as possible. When unloading is awkward or difficult, the lowered production rate brings the inevitable increase in cost of production.

In some dies, stripping (or ejecting) is accomplished by means of a positive knockout, which is actuated at the top of the press stroke. With dies of this type, the workpiece is not left in the nest when the die opens. Occasionally, the knockout will eject a workpiece back into the nest, or perhaps athwart the nest. This problem is normally eliminated by employing an air jet to blow the piece clear of the die.

If a die is equipped with a spring-actuated stripper or shedder, the workpiece will remain in the nest when the die opens. The piece must then be removed from the nest before another can be loaded. Unloading can be accomplished in various ways, such as mechanical ejection, air-jet ejection, manual removal, or combinations of each. Choice of methods depends on factors appropriate to the given job, some of which are: high or low production, type of die, size and shape of the workpiece, relative delicacy of the workpiece, type of press in which the job is run.

The workpiece should be evacuated from the nest as freely as possible. Sticking and binding in the nest is a source of distorted and spoiled work. The shape of the workpiece should be considered in this connection. The nest should be oriented in such a manner that the piece will clear the die without interference. Check for openings or projections which would interfere with freedom of unloading or ejection.

Foolproof nesting. If a nest is constructed in such a manner that the workpiece cannot be incorrectly loaded, the nest is said to be foolproof. It behooves the diemaker to be conscious of the importance of foolproofing. He should be capable of determining, by examination, the gaging and foolproofing possibilities of any given workpiece. To do this efficiently and reliably requires a systematic approach.

One method is to consider the workpiece in relation to its axes. Any piece part (workpiece) can be assigned three axes similar to those illustrated in Fig. 12·1. The X axis extends centrally through the piece from left to right. The Y axis extends from front to back, bisecting the X axis at a 90° angle. The vertical axis Z passes through the intersection of X and Y. Axis Z is perpendicular to both X and Y. Any one axis is, of course, perpendicular to the other two. If the workpiece is now considered in relation to these axes, gaging and foolproofing possibilities can readily be determined by examination.

Illustrations of various typical contour-to-axis relationships are shown in Fig. 12·2.

View A. The workpiece illustrated in this view is symmetrical about all three axes. It may be rotated 180° about any of the three and it will still fit in the nest. There is no truly positive method of preventing a symmetrically contoured workpiece from nesting even though it has been rotated about one or more of its axes. Therefore, a piece of symmetrical configuration cannot be positively foolproofed insofar as restricting it to one nesting position is concerned. Fortunately, this does not matter in many instances, since the work performed by the die will have the same relationship to the outer contour of the workpiece no matter which way the piece is loaded into the nest. Caution! Keep in mind that there is one exception, and that exception is burr side. Generally, from the standpoint of nest construction, it is not considered practical to attempt to achieve absolute foolproofing of symmetrical work in relation to burr side.

To repeat, a die operation performed on the

Figure 12·1 Major axes of piece part or workpiece.

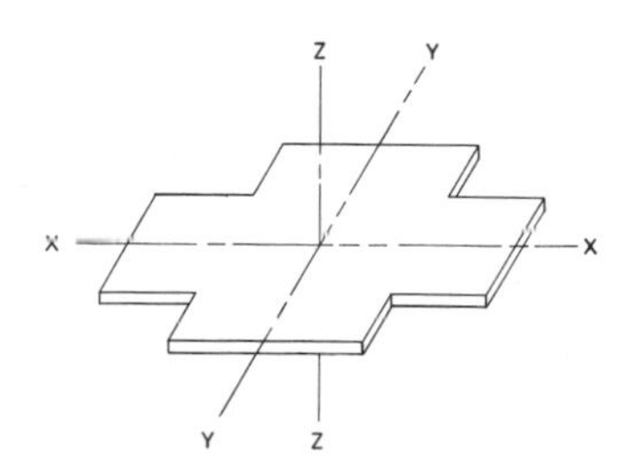

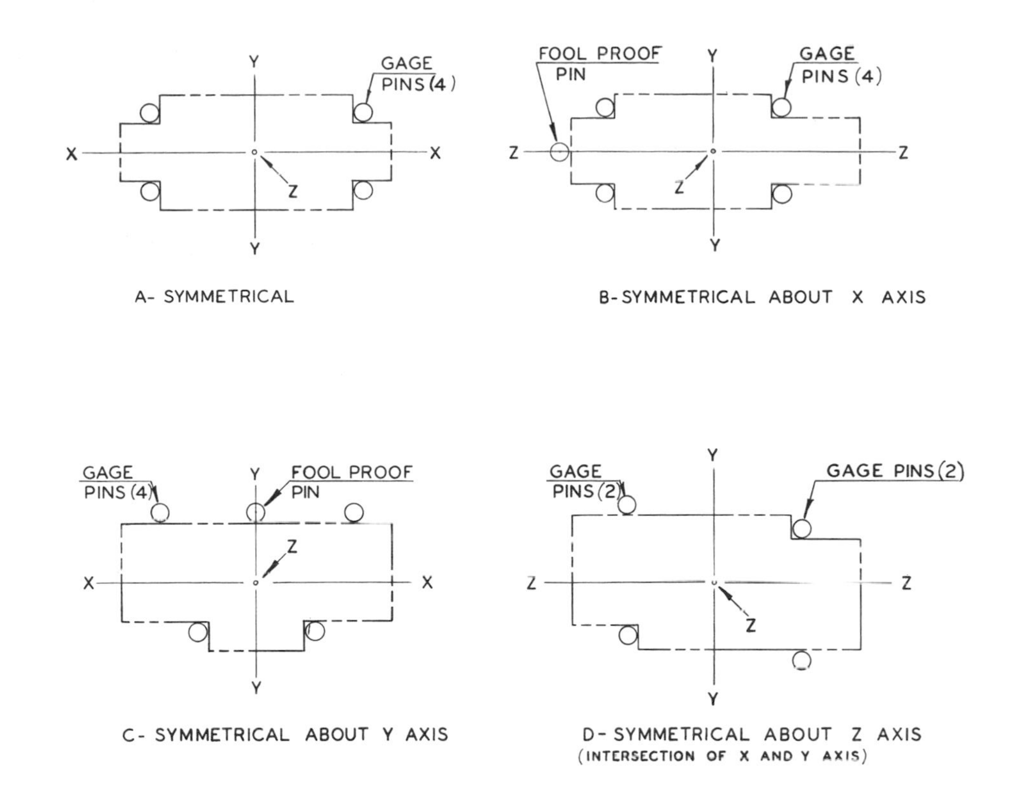

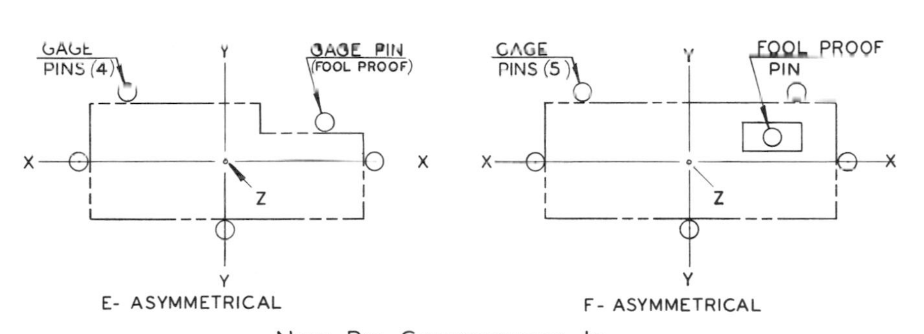

NEST PIN CONFIGURATION IN
RELATION TO AXIAL SYMMETRY

Figure **12·2** Nest gaging related to axial symmetry.

workpiece illustrated will be in proper relation to the outer contour of the piece no matter which way the piece is loaded into the nest. The nest itself, however, does not foolproof the piece as far as burr side is concerned.

View B. This workpiece is symmetrical about the *X* axis only. It may be rotated 180° about this axis and placed in the nest. The rotation will not affect the relationship of the die operation to the contour of the piece. However, it can affect the relationship of the burr side to the die operation, since the workpiece can be loaded with its burr side either up or down.

Since the workpiece is not symmetrical about its *Y* axis, it cannot, owing to the presence of the foolproof pin, be fitted between the gage pins if it is rotated about the *Y* axis. Since the workpiece is also asymmetrical in relation to axis *Z,* the foolproof pin also prevents loading if the piece is incorrectly rotated about its *Z* axis.

Thus the workpiece illustrated in view *B* is adequately gaged and foolproofed in respect to its periphery. It is not foolproofed in relation to burr side.

View C. The workpiece is symmetrical about its *Y* axis only. It can be rotated about the *Y* axis and nested with either face up. The rotation will not alter the relationship between the workpiece contour and the work performed in the die. However, a constant burr-side relationship is not assured, because the piece can be nested burr side up or burr side down. The foolproof pin shown will prevent loading if the piece is incorrectly rotated about either the *X* or the *Z* axes. Thus this workpiece, too, is foolproofed relative to its outer contour but not relative to burr side.

View D. In this case, the contour of the workpiece is symmetrical about the *Z* axis only. It can be rotated in 180° increments about the *Z* axis, and it will nest in either position. This rotation is acceptable, because the relationship of the workpiece contour to the die operation is the same in either position. The piece will not nest between the gage pins if it is inverted (rotated) about either the *X* or the *Y* axes.

Revolution about the *Z* axis does not produce inversion of the workpiece. Because of this, and because the piece cannot be loaded if it is inverted about the *X* or *Y* axes, the nesting can be considered foolproof in relation to both peripheral contour and burr side. The nest arrangement shown in the illustration does not require extra pins for foolproofing, since the two gage pins which are located in the notch corners serve as foolproofers.

View E. The workpiece illustrated in view *A* is symmetrical. In views *B*, *C*, and *D*, the workpieces are partly symmetrical. In view *E*, the contour of the piece is asymmetrical. Since it is not symmetrical in relation to any of the three axes, absolute foolproofing is possible. Nesting for asymmetrical workpieces may be arranged in such a manner that the piece will gage in one position only. This is true for inversion of the piece, as well as for horizontal rotation. Therefore, a properly constructed nest for an asymmetrical workpiece should be foolproof for burr side as well as for contour.

The amount of asymmetry must, of course, be great enough to be effective; the foolproofing device must cause obvious interference with nesting when the piece is not correctly oriented in the nest. For example, in view *E* the notch in the piece is large enough to be practical even for a pin-type nesting arrangement, as shown.

View F. View *E* demonstrates that absolute foolproofing is possible if the contour of a workpiece is asymmetrical. In view *F*, the contour of the outer periphery is fully symmetrical. However, the opening within the piece is asymmetrically located in relation to the axes of the piece. If the diemaker takes advantage of this condition for foolproofing purposes, then absolute foolproofing is possible for work of this type also.

Foolproofing for burr side. Figure 12·2 showed that asymmetrical workpieces and workpieces which are symmetrical only in relation to the *Z* axis can be positively foolproofed for burr side as well as for contour. In regard to workpieces which are symmetrical about the *X* and/or *Y* axes, however, a different condition was revealed. It was shown that this type of symmetry allows the work to be placed in the nest with the burr side facing either way, up or down. When such a piece has a definite burr-side requirement, and when it must be manually loaded into a secondary-operation die, it is necessary to provide some means of assuring correct burr-side relationship.

One method of controlling burr-side relationship in secondary-operation dies is to rely upon the operator to load the workpieces correctly into the nest. When this method is chosen, an instruction plate should be installed and prominently displayed on the die. It should also be easy to read and understand. Bear in mind that this procedure, although it is practiced on occasion, is not foolproof. It is effective only for heavier work, where the operator can easily identify the burr side of the workpiece.

Another procedure which the diemaker may occasionally encounter is that of stamping an identifying mark on the piece in the previous operation. The operator can then tell, by visual reference to the mark, which way the work should be loaded into its nest. This method, too, requires that an instruction plate be installed on the die.

Neither of the above methods are foolproof, but are rather a substitute for foolproofing, to be used only when warranted by production and die construction circumstances. Whenever possible, die operations should be processed in such a manner that it is not necessary to resort to substitute foolproofing.

Foolproof nesting summarized. The following points are summarized because they are worthy of emphasis:

1. Foolproofing is important. The diemaker must be conscious of this fact and must know how to proceed accordingly.
2. Nest arrangements for symmetrical work can be, practically speaking, foolproofed in relation to contour but not in relation to burr side. This holds true for workpieces which are symmetrical about only their *Y* axis or about their *X* and *Y* axes as well.
3. Nesting for asymmetrical workpieces can be foolproofed for integrity of burr side as well as for contour relationship. From the practical standpoint, this is also true for work which is symmetrical about the *Z* axis only. In order to foolproof asymmetrical work, the asymmetry

of the contour must be pronounced enough to assure the practicality of using the nest contour for foolproofing.

PIN-TYPE NEST GAGES

Nests of this type are simply an arrangement of cylindrical pins mounted in the appropriate die member. In effect, the pins surround the workpiece, providing a framework within which the piece can be confined. The gaging pins are located tangent to the required outline position of the piece, restricting it laterally within the limits imposed by the piece-part specifications. Foolproof pins, if their function is solely that of foolproofing, are not required to be tangent to the peripheral location of the piece. (See Fig. 12·2 for an example of gage-pin and foolproof-pin locations in relation to various types of piece-part contours.)

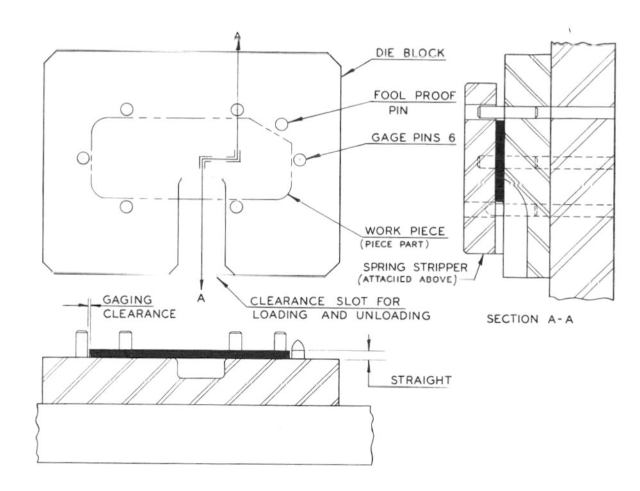

Figure **12·3** Typical fixed-pin nest gage.

Fixed-pin nest gages. Figure 12·3 shows what is probably the simplest, or at least most basic, style of pin-type nest gage. Straight cylindrical pins are pressed or driven into the appropriate die component—in this case, the die block. Incidentally, standard dowel pins are convenient to use and for many applications will serve very nicely for this purpose.

Gage pins are located either tangent to the periphery of the workpiece or at the tangent point plus desired gaging clearance. Clearance between workpiece and gage pins is desirable to facilitate loading and unloading. The amount of gaging clearance which can be provided is, of course, limited by the dimensional specifications of the piece part. However, it should be kept in mind that, for smaller work, a clearance of only 0.0005 in. is enough to assure freedom of the workpiece within the nest. Generally, greater clearances are both feasible and desirable for large work.

Providing gaging clearance yields another advantage of direct benefit to the diemaker. When a nest is provided with gaging clearance, gage-pin installation becomes a less exacting task. The accuracy with which the gage pins must be located can usually be considered proportional to the amount of gaging clearance.

When gage pins are being installed for very accurate nesting, they should be deliberately located in such a manner that the resulting nest will be slightly too tight for the workpiece. Then the gaging portion only of the pins should be polished (stoned or lapped) to the desired fit. This is the safest and most logical procedure.

The nest illustrated has an extra pin installed for the sole purpose of foolproofing. The need for it could have been eliminated by arranging the pin layout in such a manner that one of the gage pins would have been located along the angled edge of the workpiece. That gage pin would then have served as a foolproofer, in addition to functioning as a gage pin. However, since this would have required calculating the tangential location of the pin to the angled edge, it was more practical to arrange the pins as shown, even though an extra pin was required. Note the extra space between the foolproofer and the edge of the workpiece. This extra clearance facilitates installation of the foolproof pin.

An additional advantage which justifies the installation of an extra pin in this nest is identity. The extra pin enhances visual association between the nest and the workpiece, and this association tends to expedite feeding.

Various tools, such as tweezers for small work and pliers and tongs for larger work, are often employed to aid in the feeding of secondary-operation dies. On occasion, special tools are devised for this purpose. Then again, there are jobs where the operators simply use their hands. It is assumed, of course, that proper safety devices and precautions are used in all cases. In any event, the diemaker should ascertain that loading and unloading conditions are adequate. It behooves the diemaker to be operator-conscious—to consider each die from the standpoint of the press operator. That is, he should anticipate possible operational difficulties and either eliminate or alleviate them.

In the case of the die in Fig. 12·3, the press operator was supplied with a tool similar to a pair of pliers. This tool was used to place the workpiece in the nest and also to retrieve it from the nest after completion of the press operation. As shown, a clearance slot was milled in the die block to provide room for the nose of the feeding tool. Incidentally, this procedure was used for the initial production run only. For subsequent runs, where production requirements were higher, the operator was not required to retrieve the piece part manually from the nest. Instead, it was air-ejected. To do this required no structural changes or modifications of the die. The air jet was simply directed into the existing clearance slot, which allowed the air stream to get under the piece part and flip it out of the nest.

In cases where an opposing die member covers the area occupied by the nest pins, clearance must be provided in that member to prevent it from striking the nest pins. In the die illustrated, it was necessary to provide holes drilled through the stripper for this purpose.

Lead is provided by rounding, or bullet-nosing, the gage pins. A truncated conical tip is also satisfactory. The straight (gaging) portion of these pins should extend above the die surface only the minimum distance necessary for effective gaging. For most applications, this straight distance will be equal to the thickness of the workpiece or 1/32 in. whichever is greater.

The nest in the illustration has lead provided on the front and right-hand-side gage pins only. The back gage pins and the one on the left side do not have lead. Also, they protrude farther above the die face. This expedites feeding by providing a back stop and side stop for the workpiece. The foolproof pin, too, has been made higher, in order to cause immediate interference if the piece is not correctly oriented.

Ordinarily, in small-to medium-size dies, the diameters of the gage-pin installation holes should provide a light drive fit for the gage pins. For large work, a medium drive fit should generally be provided. Whenever it is practical, clearance holes for the gage pins are drilled through the die shoe. The presence of these holes allows the height of the gage pins to be more easily adjusted, since it is not necessary to remove the die block from the shoe for this purpose.

In some dies, the gage-pin clearance holes in the die shoe are a safety factor. If a workpiece is inadvertently loaded on top of one or more gage pins, it is possible for the opposing die member to drive the pin (or pins) down into the clearance holes. Such an action can reduce the hazard of possible die damage when a misfeed occurs. This is not to imply that the clearance holes are a panacea which will prevent die damage. However, there are many instances where clearance holes will either prevent or minimize die damage. If one or more gage pins have been driven down because of a misfeed, the fact is immediately apparent to the operator. He knows that a misfeed has occurred, and the job can then be checked for possible damage before continuing with production.

Disappearing pins for nest gaging. Turn to Fig. 12·3 and note that it was necessary to provide the stripper with clearance holes in order to avoid striking the gage pins. However, situations may occasionally be encountered where it is not practical to provide gage-pin clearance in the opposing die member. An example of such a condition is illustrated in Fig. 12·4. The die shown is an inverted trimming die, which trims a narrow band of material around the entire periphery of the workpiece. Because of the narrow trim scrap, the gage pins must be located quite close to the cutting edges. The proximity of the gage pins to the cutting edges does not permit the provision of gage-pin clearance holes in the die block. Therefore spring pins were used to create a nest for the workpiece, since they preclude the need for clearance in the opposing die component. In the illustration, the die block is descending and is about to contact the nest pins. After contact, the nest pins will retract until the tops of the pins are flush with the top surface of the stock material.

To facilitate loading, all nests must be provided with adequate lead-in. For nest applications, lead is provided on spring pins in the same manner as on fixed pins. When a spring pin is depressed in the normal course of its function, gaging contact is lost, as illustrated in Fig. 12·5. The use of spring pins for nesting must be restricted to applications where this condition can be tolerated.

In the course of loading, the press operator may inadvertently place the workpiece on top of one or more of the nest pins instead of within the nest. He may then trip the press without noticing the misfeed. In this event the work will be spoiled, and in some cases the die may be damaged. If a fixed pin had been driven down because of such a misfeed, it would have remained down, acting

Figure 12·4 Disappearing gage pin.

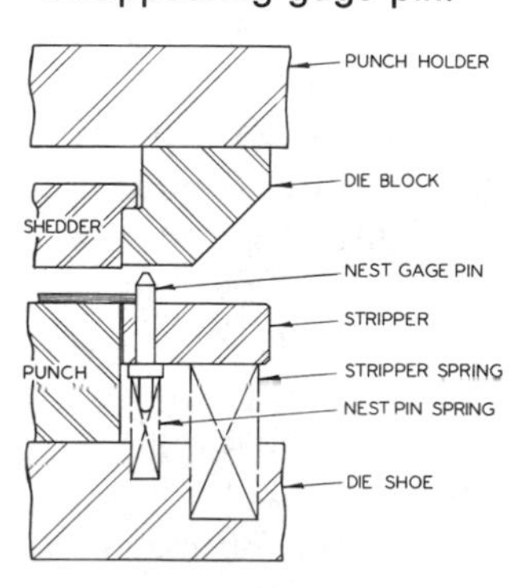

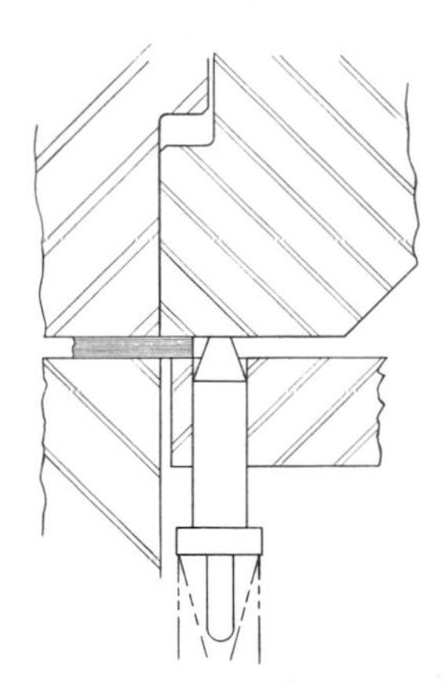

Figure **12·5** Disappearing pin in depressed position, showing loss of gaging contact.

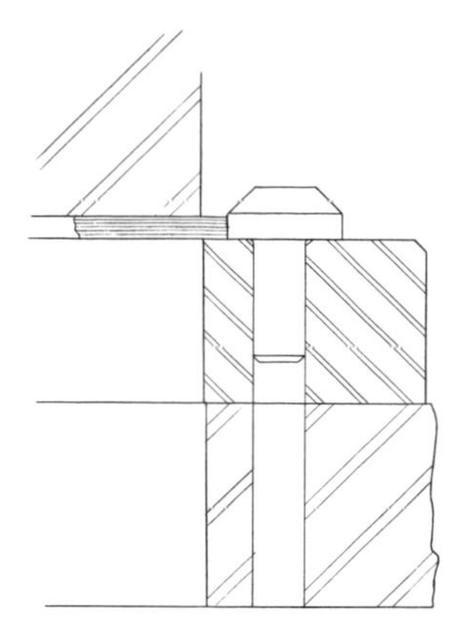

Figure **12·6** Headed, or shoulder-type, nest pin.

as a telltale to advise the operator that a misfeed had occurred. Spring pins do not give this advantage since, unless they become jammed, they will return to their original position.

Nests composed of spring pins generally require more gaging clearance than other types of nests. Keep in mind that when used for nest-gage purposes, spring pins require relatively strong spring pressure and, in addition, must have adequate travel. In the making of spring-pin nests, the most common construction error is insufficient spring pressure and room for travel. The type of spring pin shown in Figs. 12·4 and 12·5 is the most efficient for nest gages, because it permits maximum pressures and deflections.

To summarize: As far as nest-gage applications are concerned, spring pins have one inherent advantage in that they do not require clearance in the opposing die components. However, this is their only advantage. In all other respects, fixed pins may be considered superior to spring pins for nest-gage purposes. Spring pins should not be used indiscriminately for nest-gage applications just to avoid the necessity of providing clearance. This does not imply that disappearing nest pins should never be used; it means simply that careful consideration must be given to the suitability of their application.

Shoulder-type nest pins. In Fig. 12·6 is another condition where the gaging point (or spot) must be located close to a die edge. If a pin-type nest is incorporated in a die where this condition exists, then shoulder pins are employed to make up the nest. The use of shoulder pins allows the gage-pin mounting holes to be located a safe distance from the die edge.

Nest-gage pins must be accurately located in order to gage the workpiece properly in the die. However, it is often desirable or even necessary to build a secondary-operation die completely before the exact dimensions of the workpiece are known. If a pin-type nest is used in such a case, then shouldered nest-gage pins offer an advantage over straight cylindrical pins, because mounting holes for shoulder pins do not need to be tangent to the gaging edges. Therefore they can be located safely outside the gaging area. Figure 12·7 can be considered an example of this situation. Here it can readily be seen that, except for gage-pin diameter *D*, the die can be completely finished in every respect even though the exact gaging dimensions are not known. Trial blanks (workpieces) can then be formed in the die in order to ascertain the exact required dimensions of the flat blank (workpiece). After these dimensions are developed, the shoulder diameter *D* is finished to suit. Gaging clearance is provided by making the shoulder diameter smaller by an amount equal to twice the desired gaging clearance.

A typical shoulder-type gage pin is shown in Fig. 12·8. In the majority of cases drill rod, either oil-hardening or water-hardening, is the most suitable material for making these pins. Drill rod gives excellent service and is readily available in a large selection of standard sizes. It is also easy to machine and to heat-treat.

Except, of course, for dies of a temporary na-

Figure **12·7** Headed nest pins used to expedite making of form die.

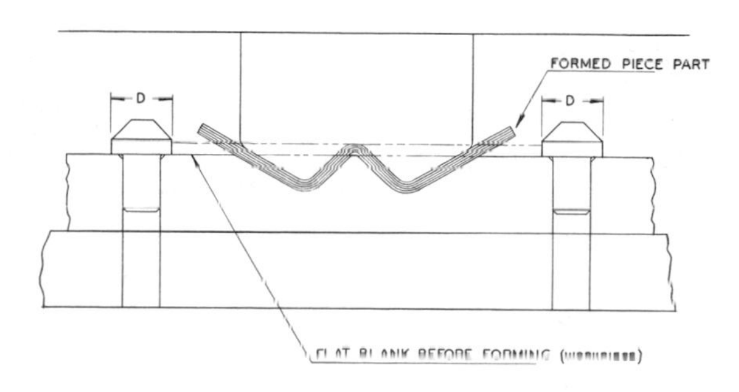

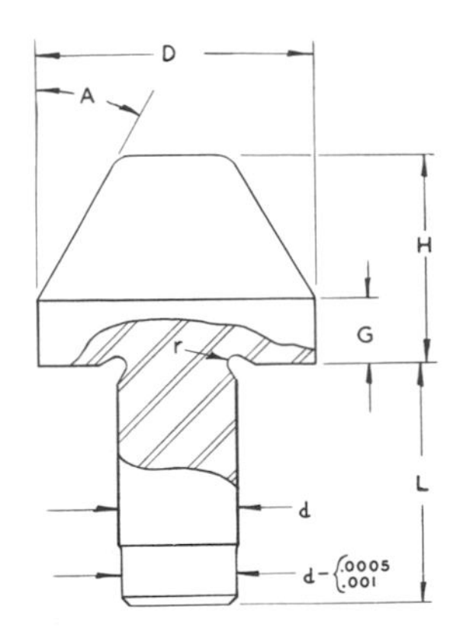

Figure **12·8** Detail of undercut shoulder pin (cone-type head shown here).

ture, nest pins are heat-treated. The degree of hardness depends upon two factors: first, the proportions of the pin, and second, the production requirements of the die. A harder pin is more wear-resistant; it is also more brittle. A well-proportioned pin may be heat-treated to a hardness of 60 to 62 on the Rockwell C scale. A pin whose proportions are not so favorable should be held to 56 to 58. If the proportions of a pin are really poor, the hardness should be correspondingly reduced to a range of 46 to 50. After all, it is better to have a pin that is somewhat less wear-resistant than it is to have one that is dangerously fragile. Barring exceptional cases, however, a hardness of Rockwell C 56 to 60 will satisfy most conditions.

Shoulder diameter *D* (see Fig. 12·8) must be made to suit each particular gaging situation. However, the mounting holes should be located at a distance from the gaging edges that will allow diameter *D* to be made from a standard rod size with a minimum of machining. For many applications, if the mounting holes are properly located, it will not be necessary to machine diameter *D* at all. Do not locate the mounting holes too far out from the gaging edges or it will be necessary to make the shoulder diameter disproportionally large.

Height *H* should be in proportion to the size of the job. Large work and thick stock material require higher gage pins. Lead angle *A* may vary between 15° and 45°, depending on the nature of the work. For average conditions 30° seems to be a good compromise lead angle. The lead angle is often omitted on the pins which are located at the back and at one side (usually the left side), and these pins are then made higher than the others. This procedure provides a back and side stop for the workpiece and thereby expedites loading.

Effective gaging height *G* should be equal to or slightly higher than the stock material thickness. For stock material less than 1/16 in. thick, use a minimum height of 1/16 in. for the *G* dimension. For very close-fitting nests, where there is little or no gaging clearance, it is often beneficial to give a slight taper to the gaging band *G*. This angle should, of course, be in the same direction as the lead angle. A taper of ½ to 1° will be satisfactory, on the average. Tapered gaging bands are especially desirable for thin stock material. When the gaging band is tapered, the height of the band *G* may be made greater than otherwise, if desired.

Stem or shank diameter *d* should be a light tap fit in the mounting hole. Mounting holes should be a standard reamer size. Wherever practical, stem length *L* should equal twice the stem diameter. It is a good idea to provide a drive lead at the entering end of the stem. The drive-lead diameter should be 0.0005 to 0.001 in. less than stem diameter *d*. Length of drive lead should be approximately one-fourth *L*.

A small but important construction feature is the radius *r* at the junction of the shoulder and the stem. The gage pin illustrated has an undercut at the junction. Undercutting tends to weaken the pin. Because of this, the depth of undercut should be held to a minimum. In addition, the junction must never be sharp, but should be blended with a radius as shown.

Instead of undercutting, the diemaker may fashion shoulder pins with a small fillet radius as shown at *F* in Fig. 12·9. Smaller pins and pins which have a relatively large disparity between diameters are often made this way in order to enhance their relative strength. In many instances the fillet radius must, of necessity, be quite small.

Figure **12·9** Detail of filleted shoulder pin (bullet-nose head shown here).

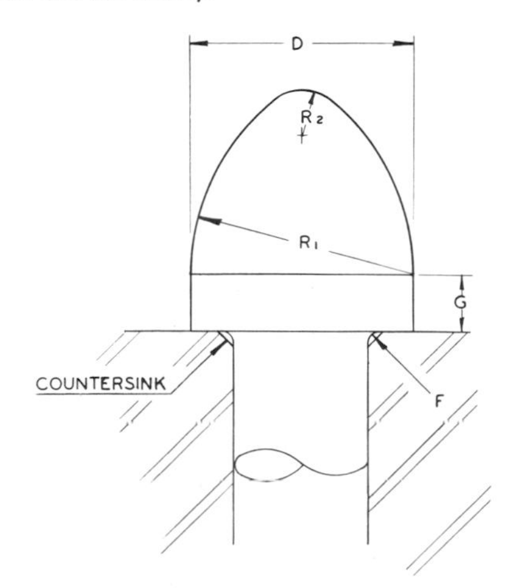

However, a fillet of 1/64 in. radius gives considerable strength to smaller pins. The mounting hole must have an adequate countersink to clear the fillet radius. On dies where the top surface is sharpened away in the course of maintenance, the countersink will have to be renewed as required for clearance. Hardened components require either hand grinding or the use of a carbide burr for countersink renewal.

Figure 12·9 also illustrates the proportions of a typical bullet-nose gage pin. Radius R_1 originating at the top of the gaging band is approximately equal to D. A smaller radius R_2 is used to blend away the sharp point at the top which results from the intersecting R_1 radii.

PLATE-TYPE NESTS

In concept, a nest of this type is a plate with an opening worked in it. The opening is such that it will, like any other type of nest, suitably receive and gage a given workpiece. Gaging is accomplished by confining the piece in the proper location and position.

It is not necessary or even desirable for the nest to fit the entire periphery of the workpiece. Gaging fit should exist only at the required minimum number of contact spots (gaging points). This is advantageous to both the construction and the function of the nest. The number of contact spots required, as well as their location, depends upon the nature of the workpiece, as was explained under "Nest-gage accuracy factors" above. The principles are the same for all types of nests. The contact spots in a plate-type nest are the equivalent of the gage pins in a pin-type nest.

A plate-type nest may be of one-piece construction or it may be made up of a number of sections. Nest plates are generally made of steel. If the sole function of a nest is gaging, and production requirements are low, the plates may be made of cold-drawn steel and left soft. However, most nests require heat treatment (hardening). The type of steel and how to heat-treat it are, of course, interdependent and are determined by the particular requirements of each job.

As far as efficiency of gaging is concerned, only the gaging contact spots need to be hard. The balance of a nest plate can be left soft without affecting the gaging function. Because of this fact, localized hardening techniques (spot hardening) can be employed to harden the required contact areas only, leaving the balance of the plate soft. This will expedite mounting of the nest, since the dowels can be installed through the soft portion of the plates. Important! The dowel holes must be a safe distance away from the hardened area. This is true for any and all machine operations which may be performed on a spot-hardened component.

Soft plugging is another method which can be employed on appropriate occasions to provide soft areas for doweling. With this procedure, the nest section is hardened in its entirety, after which soft plugs (usually cold-drawn steel) are peened into the holes which have been provided for them.

The above procedures are, most generally, employed in situations where the nest plates are mounted on hardened components. This is especially true in cases where soft plugs are used for doweling. If a nest plate is mounted on a die member which is not hardened, then the nest plate can be hardened in its entirety and the dowel holes transferred into the soft component.

Where the size of the work permits, the best method for doweling two or more hardened components together is to jig-grind the dowel holes. This is true for nest plates as well as for other components. In most cases, jig grinding can be the most economical method, as well as the best.

If nest plates are relatively large in size, hardened inserts can be installed at the required gaging spots. The plates can then be made of mild steel and left soft.

If plating facilities are available, nest plates can be chromium-plated. It is necessary to plate only the gaging contact spots. In fact, since chromium is a very hard material, it is necessary to be certain that the plating will not interfere with any machine operations which may be required after plating. The plater should be advised of the areas where plating would be objectionable. He will then simply mask these areas to prevent the chromium from being deposited. Excess chromium, or chromium deposited on surfaces where it is objectionable, can be removed by grinding. Hard chromium, deposited 0.0005 to 0.005 in. deep, will provide an excellent gaging surface. The op-

Figure 12·10 Ring-type nest plate.

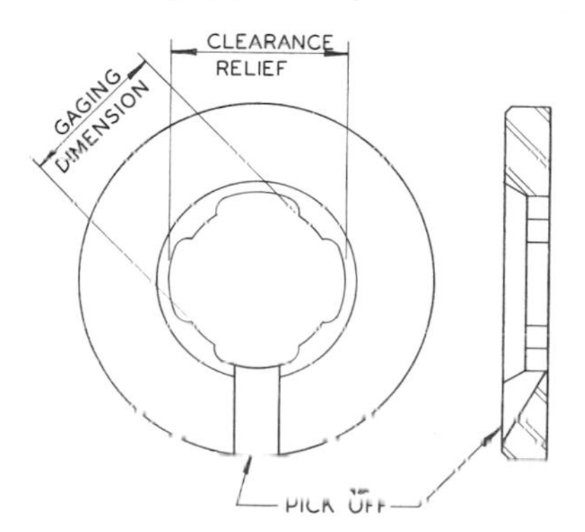

timum plating depth for nest-gage applications is 0.001 to 0.002 in. Don't forget to allow for the plating thickness, when and where required. For example, the gaging dimensions in Fig. 12·10 would have to be machined oversize by an amount equal to twice the plating depth. Plating would then bring the gaging dimension to size. A nest made of mild steel, chromium-plated at the gaging area, provides a soft mounting surface together with wear resistance where required. In contrast to heat treating, plating does not distort the component. On this basis, chromium plating is worthy of consideration in the design and construction of nest gages.

A typical plate-type nest is illustrated in Fig. 12·11. The nest shown is a two-piece affair. The back gage is higher than the front gage. Angular lead is provided at the front and on the right-hand side. A shallow pick-off slot is provided in the die block. This slot may be renewed by grinding, when required. Note the relieved portions in the back gage. The relief shown at the sharp corners of the workpiece is especially important. Whenever sharp corners of a piece are included within the confines of a nest plate, the plate must have adequate clearance at these points. In the case of the plate illustrated, corner clearance was provided by drilling a hole through the plate at the site of each corner. When using this method, remember to drill the clearance holes before attempting to machine the balance of the nest opening. The nest illustrated is inherently foolproof, since the workpiece is asymmetrical and cannot be successfully loaded in any position other than the position shown.

Figure 12·11 Two-piece flat-plate nest, tailored to contoured workpiece.

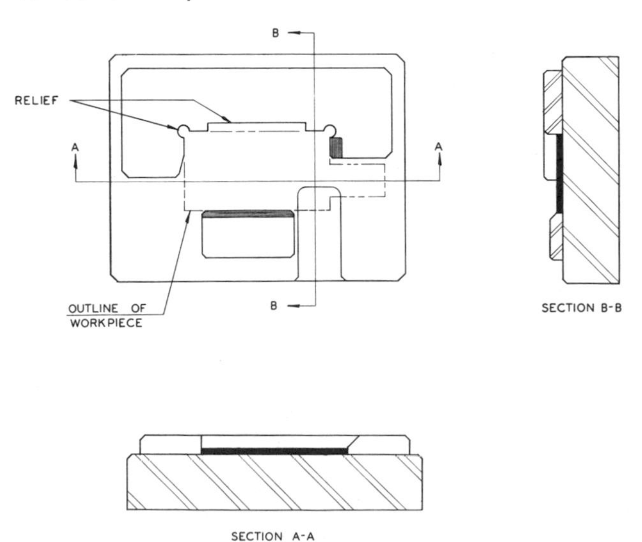

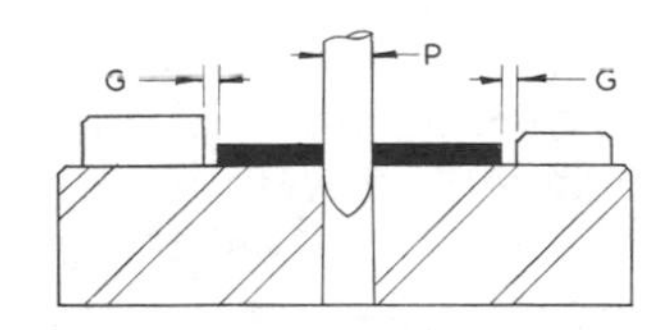

Figure 12·12 Clearance gaps for approximation nesting (traveling pilot shown).

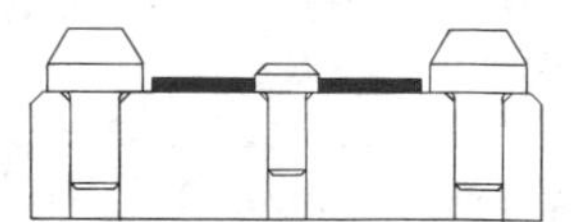

Figure 12·13 Approximating nest in conjunction with fixed pilots.

APPROXIMATION NESTS

These nests are appropriately named in accordance with their purpose. Unlike other nests, they are not intended to confine the workpiece accurately in gaging position. Their function is only to confine it within limits that will permit the actual gaging to be performed by other die components, such as pilots.

The pertinent features of a typical approximation nest are shown in Fig. 12·12. It is essential, of course, that the pilot dimensions *P* are the required gage fit for the openings in the workpiece. The nest, though, is not closely fitted to the piece; the nest opening is made deliberately larger. When the piece is positioned (gaged) by the pilots, a gap *G* should exist equally at all nest spots.

Maximum gap is an advantage as far as running the die in production is concerned, because the extra clearance facilitates feeding. However, excessive gap can be the source of damage to the workpiece and/or the die itself. How much gap to provide depends to a great extent on pilot size. Larger pilots permit the use of wider gap. As a rule, for average conditions a gap range of 7 to 10 per cent of pilot diameter will be satisfactory for pilots up to ½ in. diam. For large pilots, gap should generally be restricted to a maximum of 1/16 in. An exception is large, heavy work where wider gaps may be desirable, provided that piloting is adequate. The gap is also related to workpiece strength and hardness. If the stock material is light in gage and/or relatively soft, the gap must be minimized accordingly.

The only feature which distinguishes approximation nests from other nests is the gap. In all other respects they are the same. Approximation nests are sometimes referred to as rough nests or rough gages, because they do not gage the workpiece exactly. The principles of approximation nesting remain the same, even though the nests are applied to dies which differ in various con-

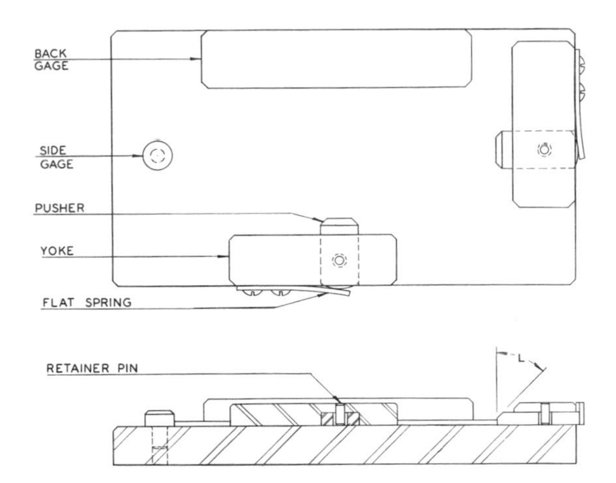

Figure 12·14 Nest gage incorporating pushers

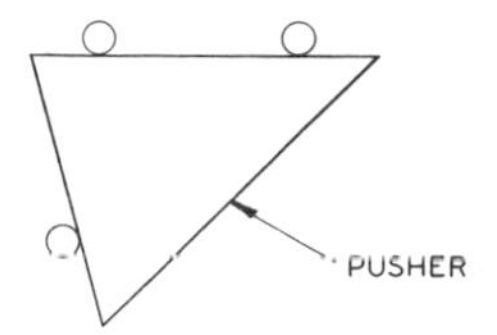

Figure 12·15 Single pusher nest gage for triangular workpiece.

struction features. For example, in the gaging configuration shown in Fig. 12·13, both the pilots and the nest pins are located in the lower die block.

PUSHER-TYPE NESTS

Pusher-type nests are so called because pushers are employed in conjunction with the other gage members. The function of the pushers is to hold the workpiece against the gaging members as shown in Fig. 12·14. In principle, pusher-type nests differ from other types in that the workpiece is gaged from two edges only. At first thought, this might not seem very desirable, but further analysis will reveal that a properly constructed pusher-type nest can gage just as accurately as any other type, if not more so.

Pusher-type nests are, generally, more expensive to make than pin-type or plate-type nests. Also, in many cases, pusher-type nests require more operator time for loading and unloading when the die is running in production. For these reasons, they are usually restricted to that class of work which makes their use more or less mandatory.

Pusher-type nests are particularly well suited to work where variations or discrepancies in the overall dimensions of the workpiece may be encountered. This is a common condition where the piece is unit stock, sheared to size. Pusher-type nests will gage such work with consistent accuracy in relation to two of the edges. Discrepancies in the overall workpiece size will then affect the opposite edges only. There is a large variety of work where this condition exists and where it is acceptable.

The nest arrangement illustrated is made up of a back gage, side gage, and two pusher assemblies. The back gage is a rectangular plate, or bar. Side gage is a shoulder pin. Each pusher assembly consists of pusher, yoke, flat spring, and stop pin.

The stop pin is a drive fit in the yoke. The pusher has a clearance hole for this pin. Size and location of the clearance hole is fairly critical: It must be large enough to permit the necessary pusher travel, but not so large as to allow the pusher to snap too far forward, where it would interfere with loading. As a bonus benefit, the stop pin acts as a retainer, keeping pusher in assembly when the die is being handled and transported, or stored. Each pusher must slide freely in its yoke. It is not necessary to heat-treat (harden) the yokes.

Note the lead angle L on the nose of the pushers. This lead is absolutely essential. Without it, the workpieces cannot be loaded into the nest. Angle L should not be more than 45°. The lead angle, pusher travel, and spring pressure should all be in proportion to the strength of the workpiece. For light-gage stock material, the lead angle should be smaller, the pusher travel should be minimized, and spring pressure should be lighter. For more detailed pusher information, see Chap. 13, *Pushers.*

The workpiece in Fig. 12·14 is rectangular in shape. Gaging is accomplished with reference to the back edge and the left side. The gaging will be parallel to the back edge but not necessarily to the sides or front edge. These relationships will depend on the squareness with which the blank is sheared.

The number of pushers required varies with the size and shape of the workpiece. Large work will generally require additional pushers. The triangular piece in Fig. 12·15 requires only one pusher, provided, of course, that the piece is not too large.

Important! Before proceeding with construction of the die, check and verify the following:

1. Is the workpiece being gaged from the edges it should be gaged from?
2. Are the pushers properly located and aimed? Pushers should not oppose each other; they should always act to force the proper edges of the piece against the gages.

CHAPTER 13

PUSHERS

Pushers are installed for the purpose of holding the required edge or edges of the work securely in contact with the appropriate gaging member or members.

There are two basic pusher applications: (*a*) to assist in the gaging of individual workpieces, and (*b*) to help locate and align strip or coil stock while it is being fed through a die. Pushers are most often required on progressive dies and on primary-operation dies where the workpiece is unit stock. They are required on other dies too, but not as frequently.

When pushers are installed in conjunction with nest gages, they are considered part of the nest assembly. Such combinations are referred to as pusher-type nests.

FLAT-SPRING PUSHERS FOR TOP FEEDING

The pusher assembly illustrated in Fig. 13·1 is part of a nest arrangement which is intended to gage individual workpieces. The pusher assembly is composed of pusher, yoke, and spring.

In order to load or feed the work into a die equipped with spring pushers, it is necessary to provide a lead on the pushers. The manner in which the lead should be provided is determined by the basic pusher application. These applications are defined at the beginning of this chapter.

On dies where the workpiece is loaded from above, pusher lead is provided as indicated by lead angle *L*. In view *A* the pusher is extended in its full forward position. This is the normal position when the nest is empty. To load the nest, a workpiece is placed in the position shown. Then it is forced down onto the die face until it snaps into place. The lead angle acts as a cam surface, causing the pusher to slide outward against the resistance of the spring. When the loading sequence is completed, the workpiece is held against the gage by spring pressure, which is transmitted via the pusher (view *B*). Angle *L*, therefore, functions as a cam surface in addition to providing lead. This condition is somewhat paradoxical, since larger angles are desirable for lead, while smaller angles provide a more efficient cam action.

For applications of this type, the size of the lead angle will usually range between 30 and 45°, depending upon the individual requirements of the job. Lead angles should be held to the smaller end of the range for light-gage (delicate) work.

In view *A* the pusher is shown in its closed position; in view *B* the pusher is open. The distance between these positions is the travel. Optimum travel distance varies in accordance with individual job requirements. For average work, a travel distance of 1/64 to 1/32 in. will usually be satisfactory. More travel (sometimes considerably more) may be required for large, heavy work. For delicate work, the travel should be minimized.

Excessive travel indicates that the pusher is intruding too far into the nest. This condition is a source of undue loading interference and is generally detrimental to die operation.

Figure 13·1 Pusher actuated by flat spring.

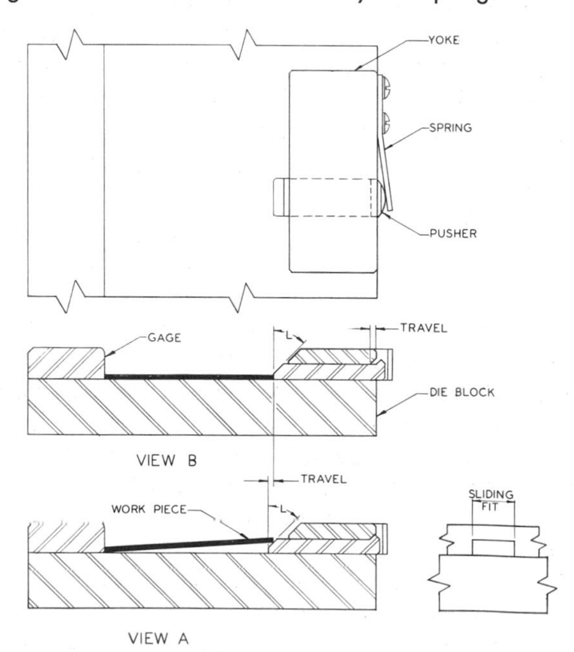

Spring pressure must be strong enough to assure gaging, yet not so strong as to have an adverse effect upon the loading sequence. These conditions make it necessary to more or less tailor the spring pressure to the particular application. Heavy work requires strong spring pressure; moderate pressures are used for average work; delicate work requires light pressures.

Spring-actuated pushers should normally be preloaded. In view *A* it is discernible that the spring is exerting pressure against the back end of the pusher when the pusher is in its closed position. The applied pressure in this position is the preload. It is obligatory that the preload be in proportion to the spring pressures described above.

The pusher in Fig. 13 · 1 is actuated by a flat spring. In order to facilitate the action of this type of spring, the back end of the pusher is made in the shape of an arc. It has a peened head which acts as a stop, preventing the pusher from intruding too far into the nest.

The spring, of course, is spring steel, spring-tempered. In general, pushers should be made of tool steel. Oil-hardening ground stock is ideally suited to most applications. The nose (front end) of this type of pusher should be hardened, on the average, to Rockwell 60 to 62 on the C scale. The tail (back end) should be drawn back to a spring temper (Rockwell C 45 to 48) or lower. Another procedure is to harden the nose and leave the tail soft.

Because it acts as a cam surface, the face of the lead angle should be smooth, preferably polished. The pusher must slide freely in the yoke slot. It is not necessary to harden the yoke, and it is certainly much easier to mount the assembly if the yoke is left soft. In view of this, the yoke may be made of low-carbon steel. Cold-drawn steel is most often used, because it requires a minimum of machine work.

NOSE CONTOURS FOR SPRING PUSHERS

It has been noted above that a large angle is desirable for lead purposes and that a small angle gives a better cam effect for snapping the workpiece into position. This paradox is normally resolved by using a compromise angle. However, the diemaker may occasionally encounter situations where a double angle (Fig. 13 · 2, view *A*) will prove to be an advantage. Here, the greater angle *L* provides a large lead to aid in guiding the workpiece into the die. The lesser angle *K* forms a more effective cam surface for completing the loading sequence.

If necessary, the ability of a pusher to hold a workpiece in place can at times be augmented by making the nose of the pusher as shown in view *B*. Here, angle *M* tends to hold the piece against the die face. Angle *M* should be small; angles of 1 to 3° will usually suffice. A similar angle can be provided on the opposing gage. Such an angle would usually be held to 1°. Slightly larger angles are sometimes used, but only if proved necessary. It is possible that a pusher may, on occasion, benefit by using a combination of all three angles (*K, L,* and *M*).

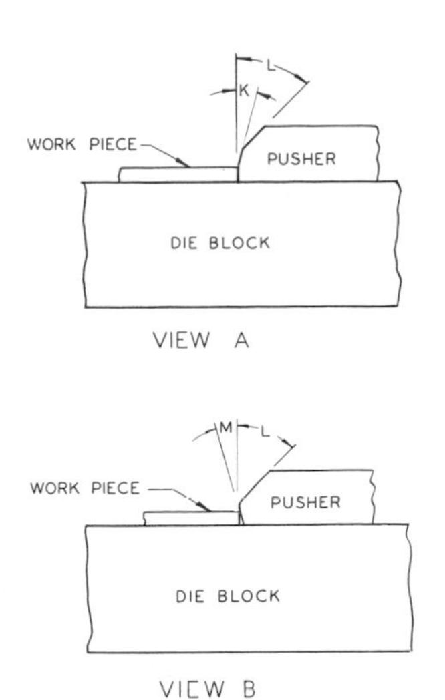

Figure 13 · 2 Pusher nose contours (for top feeding).

FLAT-SPRING PUSHERS FOR SIDE FEEDING

The pushers in Fig. 13 · 3 and 13 · 1 are identical except for the position and size of the lead angle *L*. The difference results from the fact that they have different basic applications, as described at the beginning of this chapter.

The pusher in Fig. 13 · 3 is intended to function as an aid to feeding strip stock through the die. To accomplish this, the lead angle *L* must be

Figure 13 · 3 Flat spring-actuated pusher for side feeding.

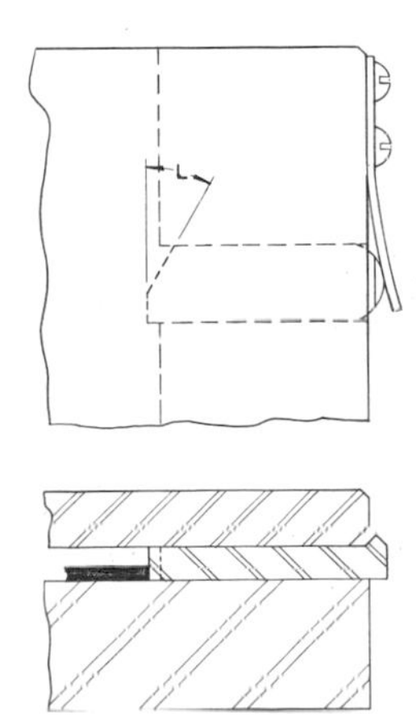

positioned as shown. Smaller angular increments are generally more common in this type of application, since large leads are not required for loading. Also, smaller lead angles provide a better cam effect, which at times can facilitate starting the stock strips through the die. Lead angles greater than 30° will seldom be necessary. Keep in mind, nevertheless, that the lead angle must be large enough to accept the widest stock strip for which the die is intended.

PUSHER CONSTRUCTIONS

Pushers can be, and are, contrived in numerous different ways. Some typical examples are shown in Fig. 13·4. The drawings are for the most part self-explanatory.

View A. Here, the pusher is actuated by a spring which is formed from flat spring stock and fastened to the die shoe. As shown in the plan view, the back end of the pusher is shouldered. The shoulders act as a stop, limiting the depth of entry. As shown in the section view, the back end is rounded to suit the spring action. The use of this type of spring is mostly restricted to dies where the height H remains constant throughout the entire life of the die.

View B. This pusher is not shouldered, nor is the back end peened. Instead, a stop pin is pressed

Figure **13·4** Various pusher constructions.

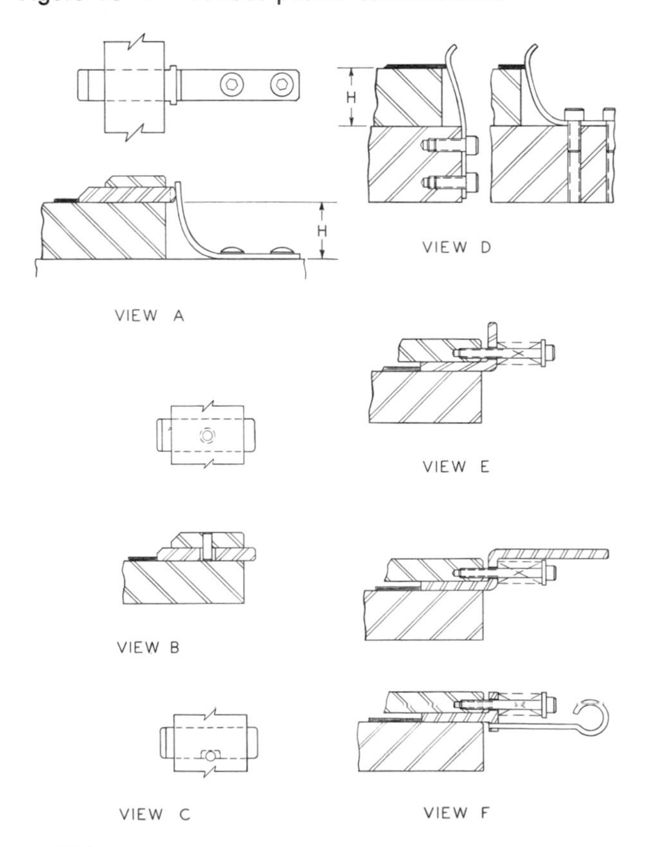

into the yoke. A clearance hole for this pin is provided in the pusher slide. The diameter and location of the hole is such that it controls the pusher travel. The back wall of the hole contacts the pin, preventing the pusher from entering too far. In some cases, it may be necessary to provide an elongated hole instead of the round hole shown.

View C. This method is particularly well suited to narrow pushers. Here a slot is provided in one side of the pusher slide. The function of the slot is the same as the function of the hole shown in view *B*. Incidentally, it is not mandatory that this method be restricted to narrow pushers; it will serve quite well with wider pushers also. The pusher must, of course, be long enough to accommodate the slot while still providing adequate bearing.

View D. In this view, the springs themselves are the pushers. It is an economical method wherever the die construction and operating conditions make it practical. Before installing these pushers, be certain that they will be satisfactory from the standpoint of die life.

View E. Here the pusher is actuated by a coil spring. The use of this type of pusher, or various modifications of it, is widespread, especially on progressive dies. One reason for its popularity is the versatility of the coil spring. For pusher applications, coil springs will generally give better performance than will springs made from flat spring stock.

These pushers can be machined from rectangular bar stock, or they can be made from flat stock bent to shape. They can be made of tool steel or of mild steel, pack-hardened. It seems that, for the majority of tool and die shops, the most practical procedure is to make pushers of oil-hardening ground stock, machined to the desired shape and finished by grinding after hardening. Pushers made of tool steel are superior to pushers made of mild steel.

View F. The pushers illustrated in this view are evolved from the pusher shown in *E.* These pushers provide a distinct safety advantage over all the preceding types. Here, the operator can safely grasp the pusher and pull it back when starting the stock strip through the die. It is often impossible, or at least impractical, to attempt to force the leading end of a stock strip by a pusher. This is especially true for progressive dies, and the condition is worsened when there is more than one pusher in the die.

Obviously, these pushers are somewhat more expensive to make. Nevertheless, they are used quite extensively because of their high safety factor and because they tend to make the die easier to operate.

Note that on these pushers and the one shown in view *E*, the spring-retaining screws are all tapped into the strippers. The pushers are actually part of the stripper assembly. This allows the die to be sharpened as required without disturbing the function of the pusher. The pusher must not be made inverted!

HEAVY-DUTY PUSHERS

The pushers previously described are not heavy-duty pushers. It is obvious that strong spring pressures would cause objectionable resistance to feeding or would even make feeding impossible. For heavy-duty pushers, it is mandatory that the pusher force be applied in a different manner that will eliminate resistance to feeding.

Figure 13·5 typifies a heavy-duty pusher assembly. The pusher is driven by a spring plunger which is mounted to the punch holder and thus travels with it. When the die opens, the spring plunger is withdrawn from contact with the pusher, removing all pressure from the pusher. This leaves the stock strip entirely free for feeding.

With this type of design, the spring pressure is applied only when it is needed. Consequently it is possible to apply very strong spring pressures to the pusher. In fact, with this type of pusher design, it is possible to apply pressures far in excess of the amounts that are ever likely to be required.

Because the plunger acts as a cam against the back end of the pusher, the back end of the pusher must be hardened, as well as the front end. The plunger should also be hardened.

A specific advantage of this pusher design is that there is no pusher resistance to piloting. Important! For dies equipped with pilots, the stroke *S* of the plunger must be restricted to suit the pilots. It is absolutely essential that the pilots position the stock strip (or workpiece) before the plunger contacts the pusher.

Figure 13·5 Heavy-duty pusher arrangement.

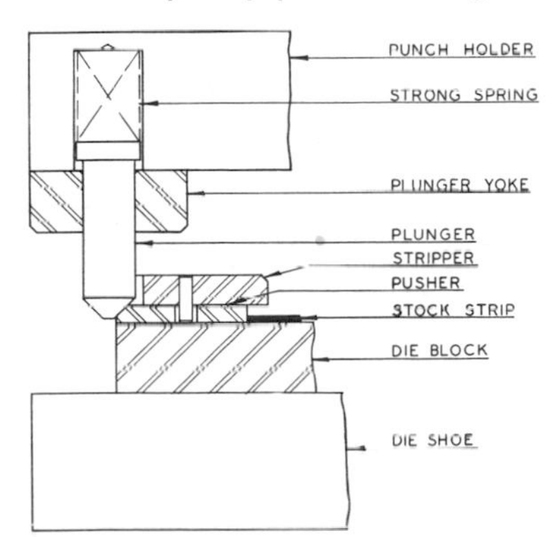

AIR-CYLINDER-DRIVEN PUSHERS

Instead of springs, small air cylinders are sometimes used to actuate pushers. Although definitely more expensive than springs, there are cylinders which are not prohibitive in cost.

In order to apply air cylinders to dies, it is essential that compressed air be available at the press in which the die is to be run. In addition, the press must have suitable automatic controls for timing the operation of the air cylinders.

One version of an air-cylinder-actuated pusher is illustrated in Fig. 13·6. In this case, the cylinder is mounted on an angle bracket which in turn is screwed to the front of the die shoe. The screw clearance holes in the bracket are elongated to permit adjustment and to allow for die life. The air cylinder is linked to the pusher by a clevis and clevis pin. The clevis pin should be safetied with a cotter pin.

This pusher is retracted with the back stroke of the air cylinder, which thus removes all pusher pressure from the stock strip. Therefore, pusher drag is nonexistent and cannot affect either feeding or piloting. Spring pushers of the types which exert constant pressure on the stock strip tend to cause rapid wear on the nose of the pusher and on the back gage. This condition is eliminated with pushers actuated by air cylinders. Because the pusher is retracted when the stock strip is being started (and during the advance), it is not necessary to provide lead on the pusher.

Timing the pusher is an important aspect of setting up a die equipped with this type of pusher. It is mandatory that the pilots properly position the stock strip before the pusher exerts pressure on it.

WIPERS

Some die operations do not lend themselves to the installation of pushers on the lower die assembly. Where a pusher effect is required for such a situation, installing a wiper (Fig. 13·7) may well be a satisfactory solution for the problem. Wipers are, actually, just another type of pusher. They are different in that they are mounted with the upper punch-holder assembly and travel with it.

Figure 13·6 Pusher actuated by air cylinder.

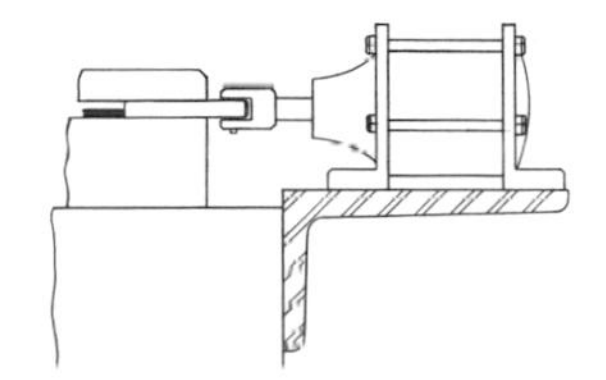

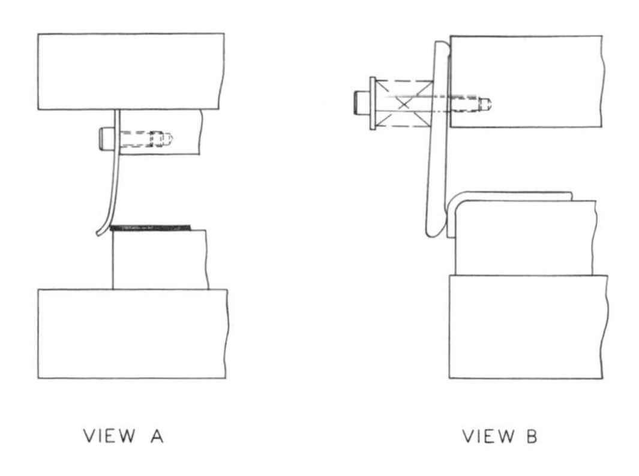

Figure **13·7** Wipers.

For light duty, a wiper may be simply a flat spring with a lead formed on it at the working end (see view *A*). The wiper shown in view *B* is intended for heavier applications. The illustrations are self-explanatory. However, keep this in mind: Adequate lead is important to the function of wipers. Be certain that the lead is in proportion to the requirements of the job.

HEAT-TREAT REQUIREMENTS FOR PUSHERS

It is conceivable that, for very limited production, a pusher can be left soft. This situation is somewhat of a rarity, though, since pushers are seldom required for such applications.

For low-production dies, pushers are sometimes made of mild steel and cyanide-hardened. Where pushers are required for either low-production or temporary dies, the use of inexpensive commercially available pushers should be considered. They will often be both better and cheaper.

If a pusher is made of mild steel, the nose of the pusher should be pack-hardened to a case depth of 1/32 in. or more. To avoid the possibility of grinding through the case, the pusher should be finished to size before hardening. If it should happen that correction grinding is necessary on the nose of the pusher, such grinding should be limited to a depth of 0.005 to 0.010 in.

When pushers are made of tool steel, the front end (nose) of the pusher may be hardened as required. For the majority of pusher types, the back end should be drawn (tempered) to Rockwell 45 to 48 on the C scale. At the front end of the pusher, a hardness of Rockwell C 60 to 62 will be satisfactory for most applications.

PUSHERS SUMMARIZED

There are, certainly, many more pusher design variations than the group presented in this section. However, if a diemaker understands these functions and constructions, he should be able to handle any pusher application in a competent manner.

Tool steel is superior to mild steel for pusher slides. Generally, oil-hardening ground stock is most convenient to use. Pusher slides should be hardened, as described. Where applicable, coil springs are superior to so-called flat springs.

A number of different pusher types are commercially available. These range from very inexpensive types for low-production dies to roller-type pushers for high-production, high-quality dies. Their use can result in definite cost advantages.

CHAPTER 14

DIE STOPS

PURPOSE OF STOPS

In the majority of applications, stops are installed on dies for the purpose of arresting the feeding movement of the stock strip. Stops are used for the same purpose in relation to the feeding of workpieces, but less often. However, the underlying principles of stopping are the same in either application.

BASIC STOP PRINCIPLES

It is essential that two basic definitions be associated with the fundamental principles of stops:

1. Stop Position. This is the location of the actual stopping point or surface against which the stock strip is halted.

2. Registry Position. This is the exact location in which the stock strip must be established in order that the work will be dimensionally correct. The registry position may or may not be the same as the stop position.

In a sense, stops can be considered gages as well as stops. Some stops perform a true gaging function. Others act as approximation gages, permitting different components (usually pilots) to perform the final, accurate registry of the stock strip. Thus, there exists a fundamental difference: In one case, the work is registered by the stop; in the other, the work is registered by pilots or by components which act as pilots.

The relationship between stop position and registry position depends upon the function of the stop. If a stop acts as a true gage, the stop position and the registry position are one and the same. If a stop functions as an approximation gage, the stop position does not coincide with the registry position. It can be said generally that if a stock strip is piloted, it is necessary for the stop to act only as an approximation gage, allowing the strip to be overfed. If a stock strip is not piloted, the stop then functions as a true gage.

Figure 14·1 is a comparison drawing illustrating this difference in the function of stops. In each case the stop depicted is a final stop, solid type. However, the same principles apply to all stop categories.

The die shown in view *A* does not have pilots. The stop position must coincide with the registry position, because the stop is the sole means of positioning the stock strip lengthwise. In operation, the leading end of the strip is advanced to the stop and remains against the stop during the press cycle. This stop acts as a true gage, registering the stock strip.

Incidentally, examination of the illustration reveals a weakness of this type of die. If the operator does not feed the leading end of the strip against the stop and hold it there, the resulting piece part will not be to dimension; it will be too short. It is necessary to depend entirely upon the

Figure 14·1 Fundamental principles of stops.

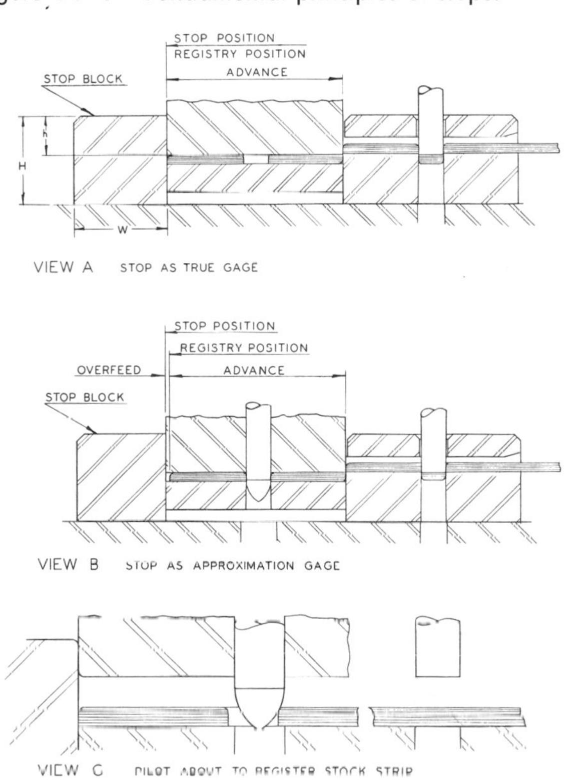

press operator to guarantee dimensional integrity of the piece part.

The die represented in view *B* is equipped with a pilot. Here too the operator advances the lead end of the stock strip against the stop between each press stroke. However, in this type of application the stop face is positioned slightly beyond the required registry position. This extra distance is called the overfeed. Thus, the operator feeds the stock strip a distance equal to the advance plus the overfeed. On the downstroke, the pilot enters the previously pierced opening in the strip. In so doing, the pilot displaces the strip longitudinally back to the registry position. This stop acts as an approximation gage, permitting the pilot to register the stock strip. When a strip is piloted, final stops and secondary stops must permit the stock strip to be overfed. Do not attempt to make the stop position coincide with the registry position.

View *C* illustrates the pilot function. The pilot is shown as it is about to displace the stock strip from stop position to registry position. Note that the pilot leads the punches. This is absolutely necessary, as the pilot must register the strip before the punches make contact with the strip.

To repeat, even though the stops depicted in this illustration are identical in construction, they differ in principle. The difference is manifested in the location of the stop (stop position).

Optimum overfeed. Overfeed is the difference between the stop position and the registry position. It has been established in the previous section why and where overfeeding is necessary. Overfeeding, like so many other facets of diemaking, is contingent upon conditions which vary according to the individual application. In order to make and install stops properly, the diemaker must be capable of forming an adequate decision about the optimum amount of overfeed.

Determination of optimum overfeed is governed by two major interrelated factors. These are:

1. Pilot size
2. Thickness and hardness of the stock material

Relatively small pilots require that overfeeding be held to a minimum. This is true also for light-gage stock material. Conversely, it is desirable to increase the amount of overfeed for large pilots in heavy stock material. The table in Fig. 14·2 is suggested as a guide to determining overfeed allowances. The amounts given in the table have proved generally satisfactory.

PERMISSIBLE OVERFEED AMOUNTS (INCHES)						
	STOCK MATL. THICKNESS	.005	.015	.030	.060	.125
PILOT DIAMETER	1/8	.002	.003	.005		
	3/16	.003	.005	.008	.010	
	1/4	.004	.008	.010	.015	
	5/16	.004	.008	.010	.015	
	3/8	.004	.008	.012	.020	.030
	1/2	.005	.010	.015	.030	.030
	3/4	.005	.010	.015	.030	.040

Figure **14·2** Permissible overfeed amounts.

Stops and feeds. With the exception of primary stops, automatic press feeds as a rule eliminate the need for die stops. In many instances, however, the die must be capable of operating either as a hand-fed die or in conjunction with a feed, as required. When this situation is encountered, the stops should be installed to accommodate hand feeding.

STOP CATEGORIES

All stops may be classified as belonging to one of the following categories:

1. Primary
2. Secondary
3. Final

The categories are definitive: A primary stop is the first stop, and a final stop is the last stop in a die. The stops in between are, of course, secondary stops. Primary stops act as true gages, registering the stock strip. When installing them, locate the stopping position to coincide with the registry position. Secondary stops serve, normally, as approximation gages. Therefore, allow for overfeeding when installing them. As described under "Basic Stop Principles" above, final stops may or may not register the stock strip. When mounting them, locate the stopping position as required.

Stop types. Quite a number of different stop types are in common use—for example, solid stops, pin stops, finger stops, pivoted auto stops, etc. In addition, there are many variations of each type. This diversity exists because of the wide variety of applications to which stops must be adapted.

SOLID STOPS

The stop which was used as an example in Fig. 14·1 is a final stop, solid type. It is simply a hardened steel block securely mounted at the required location. This block, however, performs other functions in addition to acting as a stop. It retains the lower spring pad, and it also serves as a thrust block, supporting the chopoff punch against lateral thrust engendered by the cutting action.

Although the stop block serves three purposes, its prime function is to act as a stop. Therefore, the block must be correctly mounted at the required stopping position. The chopoff punch and the spring pad must be made to suit.

Combination blocks of this sort are inherently well suited to chopoff dies and parting dies. Figure 14·3 is an example showing a solid-stop and thrust-block combination as it could be applied on a parting and forming die.

Either tool steel or mild steel may be used to make the stop block. The block must be hardened for the following reasons:

1. In the process of feeding, the leading end of the stock strip repeatedly strikes against the stopping surface, producing an eroding effect which tends to cavitate the stop face and destroy the accuracy of the stop.
2. The stop is subject to friction induced by the motion of the spring pad.
3. Due to its additional function as a thrust block, the stop is also subject to very high frictional loads imposed by the chopoff punch.

If the stop block is made of tool steel, it should be hardened to Rockwell 60 to 62 on the C scale. If made of mild steel, it should be pack-hardened to a depth of 1/32 in., min. Cold-drawn steel, pack-hardened, is used quite extensively, because it minimizes the machining time required to make stops of this kind.

Figure 14·3 Solid-stop and thrust-block combination (shown as applied to a parting and forming die).

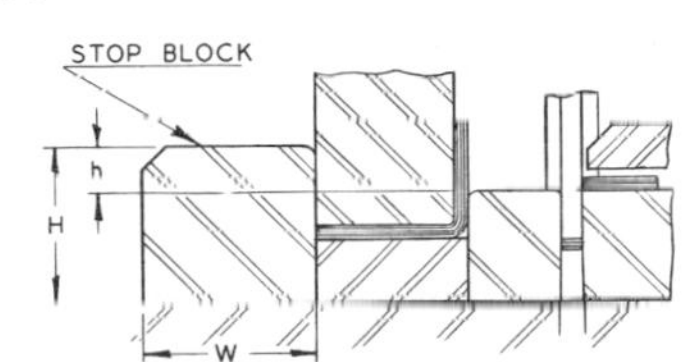

Caution! Height h is important. The h dimension must be high enough to preclude any possibility of feeding the stock strip over the stop. Because the block serves as a thrust block as well as a stop, the width w should be at least equal to, and preferably greater than, the overall height H.

Solid stop mounted on hardened die block. In concept, the stop indicated in Fig. 14·4 is a very simple stop. However, it should be kept in mind that applications of this kind require the doweling together of two hardened components. Because of this, it is recommended that such a stop block be made of a nondeforming tool steel. The stop can then be mounted (including doweling) before it is hardened.

On occasion, it may be desirable to leave 0.005- to 0.010-in. grinding stock on the stopping face. The hardened stop can then be assembled to the die block and checked to ascertain the exact amount of grinding required to make the stop face coincide with the stopping position. Another method is, of course, to jig-grind the dowel holes in assembly.

The dowel holes in the stop block should fit the dowels tightly. The mating dowel holes in the die block should provide a slip fit for the dowel pins. This procedure simplifies assembly and disassembly of the stop to the die block, which in turn facilitates building the die. Also, the procedure is a distinct advantage from the standpoint of die maintenance.

Do not attempt to make such a stop block of mild steel, pack-hardened. Generally, the combination shown (hardened stop, hardened die block) is avoided as much as possible. However, it can be done as described above, and conditions

Figure 14·4 Solid stop mounted on hardened die block.

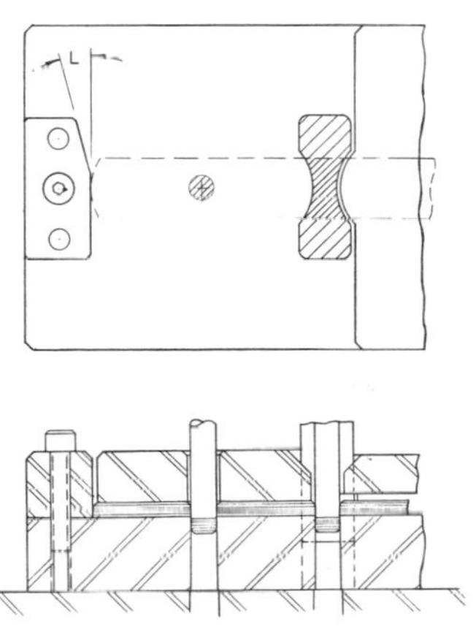

can exist where this will actually be the most practical method.

Note angle L, which was provided to facilitate air ejection of the piece part to the rear.

Solid stop, end-mounted. Where applicable, this kind of solid stop is highly regarded. As Fig. 14·5 shows, it is simple in concept and in construction. Setting these stops to the proper stopping position is easy, since the position of either the stopping face or the offset mounting face can be readily changed by grinding, as required. This facilitates the original installation of the stop. In addition, the stopping face can be renewed and/or the stop position readjusted later during the life of the die.

If the end of the die block happens to coincide with the required stopping position, the stop will not need to be offset. However, for most applications it will be either necessary or more convenient to offset the stop. Stops can be offset as shown in views A and B or in views C and D, whichever method is required.

In view A, the bottom of the form block will be ground for maintenance purposes. This avoids the necessity of renewing the forming edge at each sharpening. Since maintenance grinding will not be performed on the top surface of the form block, the relationship between this surface and the stop will remain unchanged throughout the anticipated life of the die. The stop may therefore be mounted as shown. If, however, the intention is to grind the top surface of the block, it is necessary to provide the stop with an elongated clearance hole for the mounting screw, as indicated in view B.

The stop in view C has its stopping face extended below the top surface of the die block. This provides die life by allowing the top of the die block to be lowered by maintenance sharpening without affecting the stopping position. If, on the other hand, it is intended to maintenance-grind the bottom of the die block, the stop is made

Figure 14·5 Solid stops, offset and end-mounted.

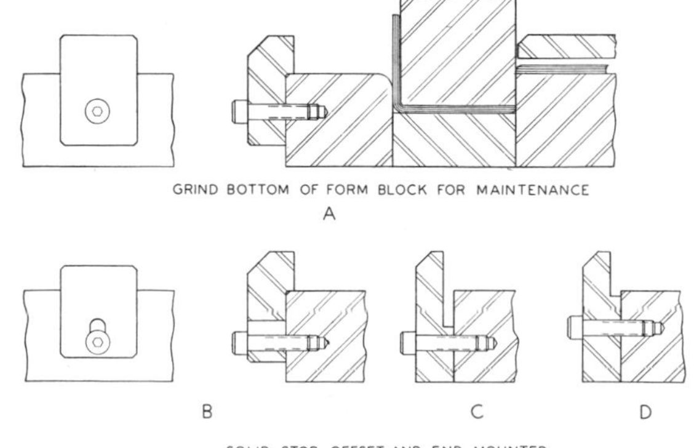

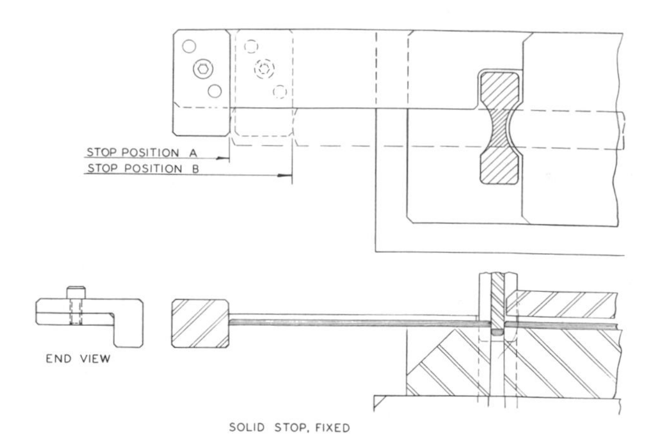

Figure 14·6 Solid stop, fixed.

and assembled as shown in D. Note that the mounting-screw center line is higher in views A and D in order to be consistent with the required grinding procedure.

Solid stop, fixed position. This stop (Fig. 14·6) is a solid stop as typically applied to a parting die. If it were applied to a chopoff die, the construction would be practically identical.

With this construction, the back gage is made long enough to provide a mounting place for the stop block. The stop block is fixed in position on the back gage by dowel pins. One or more screws complete the attachment of the stop to the back gage. The stop block is offset as shown at the end view, in order to form an obstruction against which the leading end of the stock strip will strike.

Occasionally, these dies are intended to produce piece parts of two or more different lengths. This is accomplished by providing mounting holes at the desired locations along the back gage. The stop block can then be fixed at the different positions, as required. The stop in the illustration can be fixed in two positions: stop position A and stop position B.

When using this stop, be certain that the back gage is strongly constructed and mounted. Since the back gage extends outside the die-set area, it is almost inevitable that it will eventually be used as a handle to lift or move the die.

Although such a stop is adjustable in a sense, the adjustment is limited. Therefore, it is not called an adjustable stop but rather a single-position fixed stop, a two-position fixed stop, etc. A truly adjustable solid stop is shown in Fig. 14·7.

Solid stop, adjustable. The parting die illustrated in Fig. 14·7 is equipped with an adjustable solid stop. This feature was necessary to enable the die to produce piece parts of various lengths, as re-

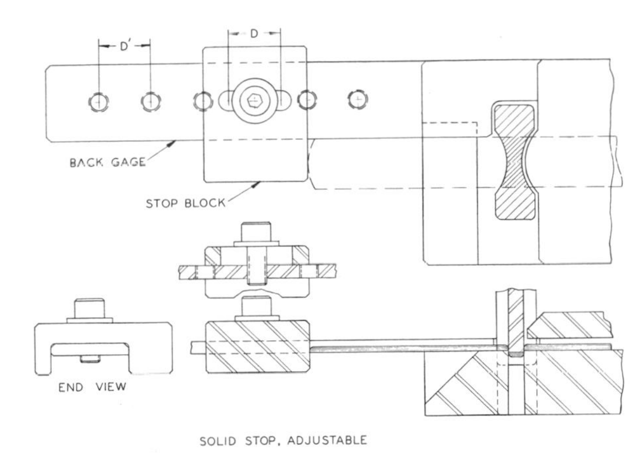

Figure **14·7** Solid stop, adjustable.

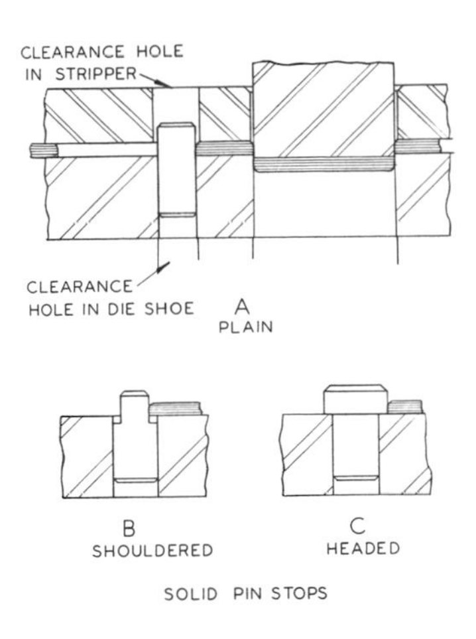

Figure **14·8** Solid pin stops.

quired. Range of adjustment is provided by extending the back gage to the length necessary to accommodate the longest anticipated piece part. Individual adjustment is controlled by the position at which the stop block is screwed to the back gage. An elongated clearance hole for the mounting screw is provided in the stop block to make possible an adjustment of distance *D*. A series of tapped holes, covering the overall adjustment range, is provided in the back gage. Center distance *D'* between the tapped holes is equal to center distance *D* in the stop block. This combination affords a continuous series of settings throughout the entire adjustment range.

The channel in the stop block (end view) positions the stop at right angles to the back gage. The stop block should slide easily along the back gage to facilitate adjustment. The size of the mounting screw must be adequately in proportion to the requirements of the job. Be certain that the back gage is strong enough, and that it is securely mounted.

PLAIN PIN STOP

One definition of the word plain is "not complicated; simple," and according to this definition, the plain pin stop certainly merits the name. As shown in Fig. 14·8, it is without a doubt the simplest stop in both conception and application.

The stop is a plain cylindrical pin. A mounting hole is provided at the desired location in the required die component. The stop pin shown in view *A* is mounted in a die block. The pin is a light drive fit in the mounting hole. The mounting hole is, generally, made to suit the stop pin so that a standard pin size can be used. A clearance hole for the pin should be provided in the die shoe for three reasons:

1. To permit adjusting the height of the stop pin without removing the die block from the die shoe
2. To allow the stop pin to be removed in order to sharpen the die with the die block fastened to the die shoe
3. To allow the pin to be driven down in the event of a misfeed, thus reducing the likelihood of damage to the die

It is always necessary to provide clearance in any opposing members which cover the site of the stop. In view *A*, the spring stripper has a hole drilled through it to afford clearance for the stop pin.

The stop pin can be altered as shown in view *B* to effect a change in the stopping position, if necessary. If a standard dowel pin is employed as a stop pin, do not alter it this way; make a new pin of drill rod. Altering the dowel pin would expose the soft core of the dowel, rendering it unfit for stopping purposes. Changing the stopping position in the other direction may be accomplished by substituting a headed stop pin as shown in view *C*.

Headed pin stop. It frequently occurs that a stop must be located close to a die opening. In such cases, the use of a plain stop pin is prohibited, because the proximity of the mounting hole to the die opening would be structurally unsound. For these applications a headed pin stop (Fig. 14·9, view *A*) may be employed. The mounting hole can then be located at a safe distance from the die opening, while the proximity of the stopping position to the die opening is maintained. Heads of stop pins may be made to suit the particular requirements of the application. In view *B*, for example, the head of the stop pin has been adapted to a relatively small opening in the scrap skeleton.

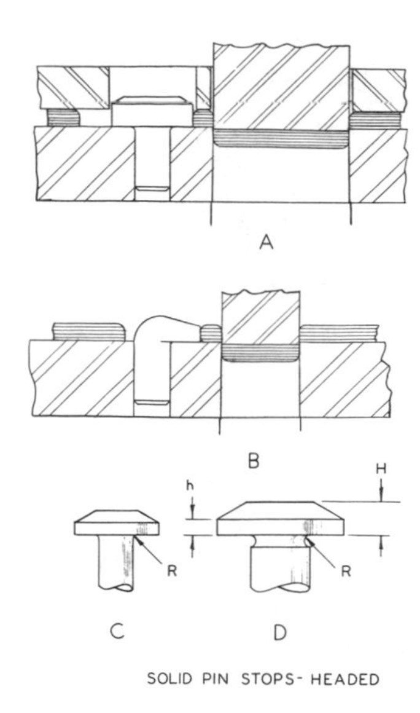

Figure **14·9** Headed solid pin stops.

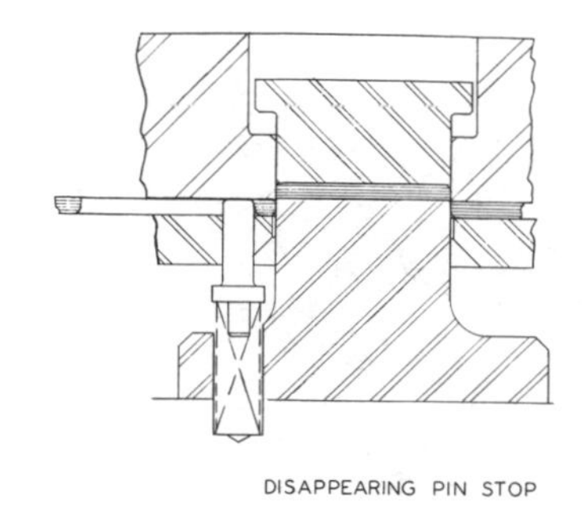

Figure **14·10** Disappearing pin stop.

For all but the most temporary applications, stop pins are heat-treated (hardened). Height of the head *H* is usually restricted to the lowest practical height consistent with structural soundness. This is done to facilitate feeding of the stock strip, which must be lifted over the stop at each advance. As a rule, effective stopping height *h* may be 1 to 1½ times the thickness of the stock material, with a minimum of 1⁄16 in.

Small pins should be made with a fillet radius *R*, as shown in view *C*. The fillet should be smooth. Size of the fillet depends upon the differential between the head size and the shank diameter. These fillets generally range from 1⁄64 to 1⁄32 in. radius, proportionate to the difference between head size and shank diameter. The mounting hole must be countersunk to clear the fillet. With some types of dies, such as the die represented in this illustration, the countersink will have to be renewed when the die is sharpened. Because the die block is hardened, the renewal will have to be by grinding or by the use of a carbide burr.

If the shank diameter is adequate, the pin may be undercut as shown in view *D*. Undercutting must be done carefully, with the depth of undercut held to the minimum and smoothly finished to a slight radius *R*. Radius *R* is equal to, or less than, the depth of undercut. Undercut depths commonly range from 0.010 to 1⁄32 in., depending upon the shank diameter. A filleted pin, as shown in *C*, will always be stronger than an equivalent pin which is undercut.

Disappearing pin stop. A typical disappearing pin stop is seen in Fig. 14·10. It is simply a spring pin located at the required stopping position. Disappearing stops offer one important advantage over other pin stops in that they do not require clearance in opposing die members. In many cases, as in the illustration, it is not feasible to provide this clearance. For such applications a disappearing stop is a necessity. In other cases, although not truly necessary, a spring pin may often be employed as a stop simply for the sake of convenience.

As far as stopping is concerned, spring pins perform the same as solid pins. The only difference is the so-called disappearing feature. Disappearing pin stops may be made in various ways. The type shown here is in very common use, probably because it is relatively simple and trouble-free.

FINGER STOPS

Some typical finger-stop constructions are shown in Fig. 14·11. Stops *A* and *B* are held in assembly by screws. The retaining screws limit the outward travel of the stops. Tapped holes for the screws are provided in the strippers. The holes are tapped to provide a jam fit for the retaining screws. *Do not* attempt to invert the stop and provide the tapped hole in the die block instead of the stripper!

OPERATION

The stop is actuated manually. It is pushed inward until the stop shoulder contacts the front edge of the stripper. This distance is the travel, as indicated in the illustration. When the stop is in closed position, the nose of the stop extends into the stock channel, obstructing the stock strip. The stop is held in closed position and the leading end of the stock strip is fed against the stop. Then the operator trips the press and releases the stop. The spring returns the stop to its open position, where it remains until a new stock strip is fed into the die.

When making a stop as shown in view *A*, be certain that dimension *H* and/or dimension *W* are adequate. This is necessary to enable the op-

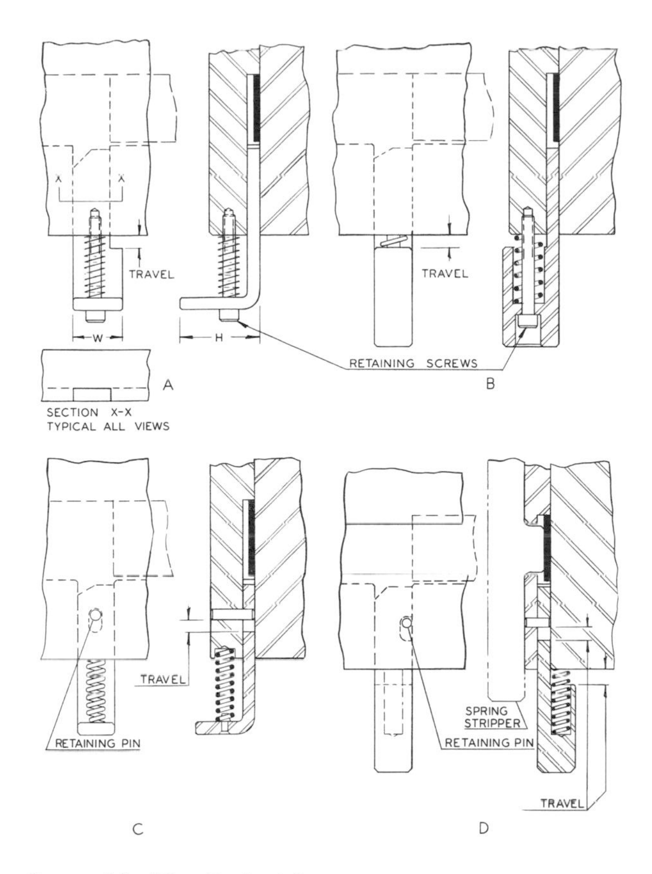

Figure **14·11** Typical finger stops.

erator to manipulate the stop without undue interference from the head of the retaining screw.

Stops *C* and *D* differ in that they are held in assembly by retaining pins instead of screws. Stop *C* has an elongated hole in it. The retaining pin is a drive fit in the stripper and intrudes well into the elongated hole in the stop. Thus, the elongated hole controls the extent of travel.

In stop *D*, the retaining pin enters a notch in one side of the stop. Inward travel can be controlled by either the notch or the stop shoulder. This die has a spring stripper, which influences the construction of the guide rails and the operation of the stop. Because of these conditions, stop *D* is inverted. Stops which are made and assembled as shown in this view may be inverted provided that the die block is thick enough to allow adequate die life between the stop and the die shoe. Keep in mind, however, that inversion is not practical for stops which are made and assembled in the manner shown in views *A*, *B*, and *C*.

Note that the stops are assembled into either the stripper or the guide rail. Thus, when the die block is sharpened, there is no change in the structural relationship between the die block and the stops.

Finger stops are sometimes called starting stops because it is their function to aid in starting the stock strip properly through the die. However, other stop types can (and do) serve the same purpose, and therefore finger stops are best referred to as finger stops.

Materials for finger stops. Tool steel and mild steel are both used to make finger stops. If finger stops are made of tool steel, the material should be in the form of ground stock in order to save machining time. For the same reason, if the material is mild steel, it should be cold-drawn bar stock.

As a rule, only the front end (nose) of the stop needs to be hard. The rest of the stop should be either drawn to a lower degree of hardness or left soft, depending upon the type of material and the type of heat treatment.

If made of mild steel, the stop may be carburized and hardened (pack-hardened), or it may be cyanide-hardened. If carburized and hardened, the case depth should be 1/64 in., min., to 1/32 in. The chief objection to the pack-hardening method is brittleness. If the nose of the stop is relatively fragile or delicate, there is a possibility of breakage. Superficial hardening of the stop nose by cyanide treatment is a convenient method used for low-production applications.

Finger stops made of tool steel should generally be heat-treated to a hardness of Rockwell C 58 to 62 at the front end. The exact degree of hardness to specify varies according to the relative delicacy of the nose contour. The rest of the stop should be drawn to Rockwell C 45 to 48. This is particularly true for stops made as shown in views *B* and *D*. It can be generally assumed that the best material for making finger stops is oil-hardening ground stock. This is true considering both quality and convenience.

Installing finger stops. Finger stops are installed in slots (Fig. 14·12) which are machined into the front guide rail. The front guide rail may be an integral part of the stripper, or it may be a separate rail. The stops must slide freely in their slots.

Slot width *W* (view *A*) should be of a dimension that will permit the use of standard stock sizes for making the stop. Where feasible, the predominant practice is to make the thickness of the stop at *H* equal to the gap height between the ceiling of the stripper channel and the face of the die block. Construction may be simplified somewhat, however, by offsetting the stop slot a slight distance from the ceiling of the stripper channel, as indicated by *G* in view *B*. A step of as little as

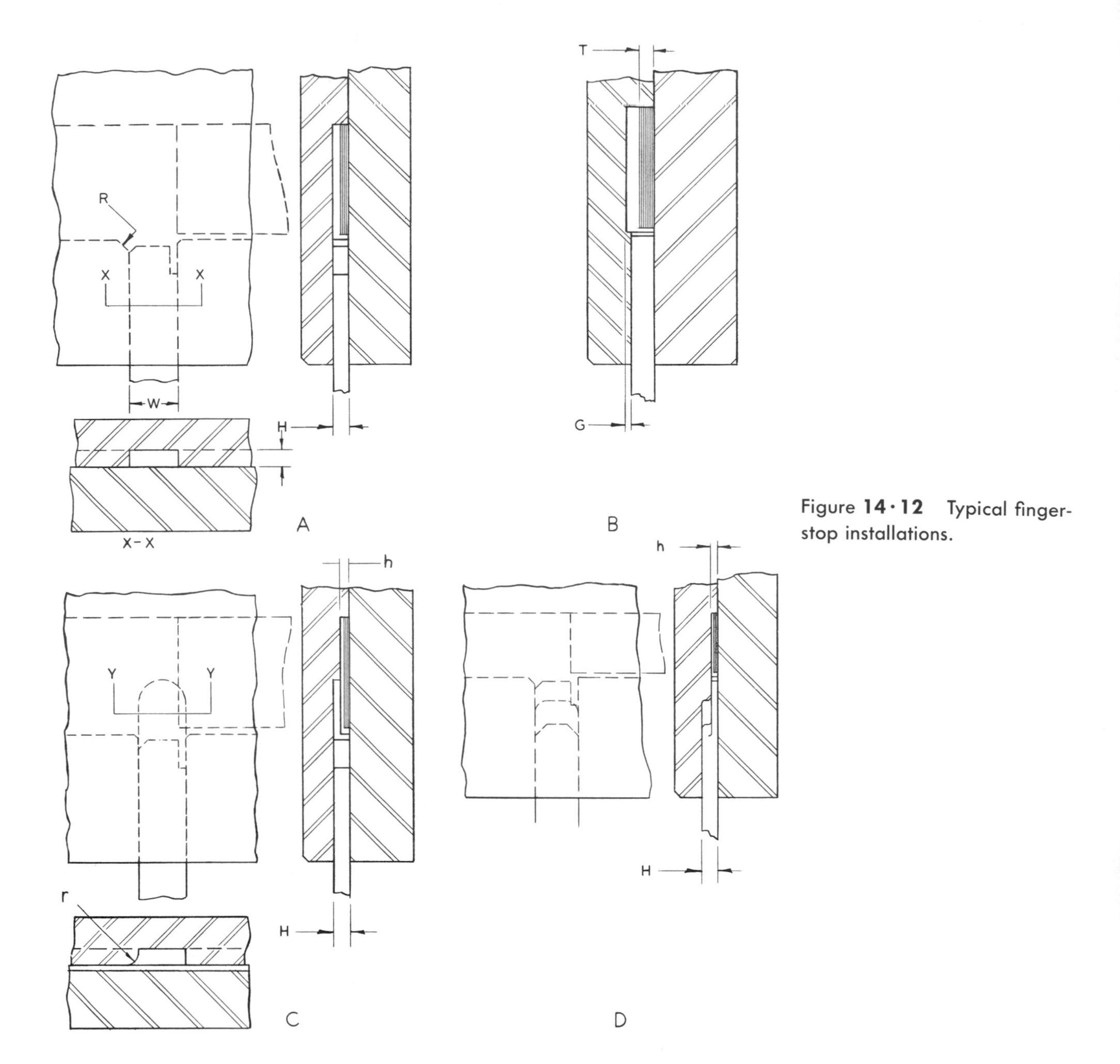

Figure **14·12** Typical finger-stop installations.

0.001 to 0.002 in. will facilitate machining the slot.

Although a slight step is desirable, it must be restricted or the resulting gap *G* will be too large. A possibility that the stock strip can ride up over the stop must never exist. When stop slots are machined in this manner, the step dimension *G* should be restricted to a maximum of one-fourth *T*, where *T* represents the thickness of the stock material. For stock material less than 0.005 in. thick, the top surface of the stop nose should be flush with the ceiling of the stripper channel, or even let into the ceiling in a manner similar to view *C*.

The front edge of the guide rail must be relieved at the slot intersection as indicated by *R* in view *A*. The relief may be in the form of a lead angle as shown, or it may be a radius, if preferred. Such relief is necessary to eliminate the possibility that the leading end of the stock strip may strike against this corner and hang up.

Instances may be encountered where the gap height *h* of the stripper channel in views *C* and *D* will be quite small. In such cases, the practice of making the thickness of the stop at *H* equal to or slightly less than the gap height *h* may result in objectionably flimsy stop proportions. For applications of this kind, finger stops can be made and installed as shown in views *C* and *D*.

In view *C*, the thickness of the stop at *H* is made to a dimension judged proportionately compatible with the stop construction. The stop slot is milled to suit by extending the slot into the stripper channel. When this method is employed, the edge of the slot should be relieved as indicated at *r* to eliminate possible interference when starting a stock strip. Another method used in conjunction with shallow stripper channels is illustrated in view *D*. Here, the nose of the finger stop is stepped down to match the stripper gap. The stop slot is milled accordingly.

When installing finger stops, the diemaker will find the procedure indicated in Fig. 14·13 helpful. In this illustration the stop is shown pushed inward, obstructing the stock strip at stopping position.

Since finger stops operate in slots which are provided in the front guide rail, the slot location determines the location of the stop. Instead of machining the edge of the slot precisely at the stopping position, the diemaker deliberately locates the slot edge out of position a slight distance ahead of the required stopping point. The nose of the stop can then be offset at *S* to coincide with the required stopping position. An offset dimension *S* of 1/64 to 1/32 in. will be generally satisfactory. Offsets larger than 1/32 in. may be employed, provided that the resultant stop nose profile is not too fragile. This procedure should, in most instances, facilitate the installation of finger stops.

Pusher stops. These stops are a special type of finger stop. As implied by their name, they serve a dual purpose as both stops and pushers. A pusher stop is typified in Fig. 14·14. Essentially, the construction is that of a pusher located to act as a stop. The spring forces it inward, where it obstructs the stock-strip channel. In operation, the leading end of the stock strip is fed against the pusher stop. After the press cycle, the stop is manually pulled outward, permitting the strip to advance to the next stop. When released, the stop in effect becomes a pusher.

Pusher stops, like conventional finger stops, are employed either for primary stopping or as secondary stops. When making and installing pusher

Figure **14·13** Finger-stop installation.

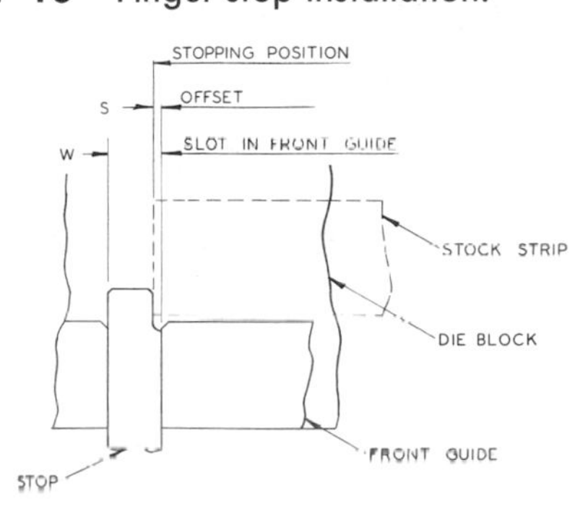

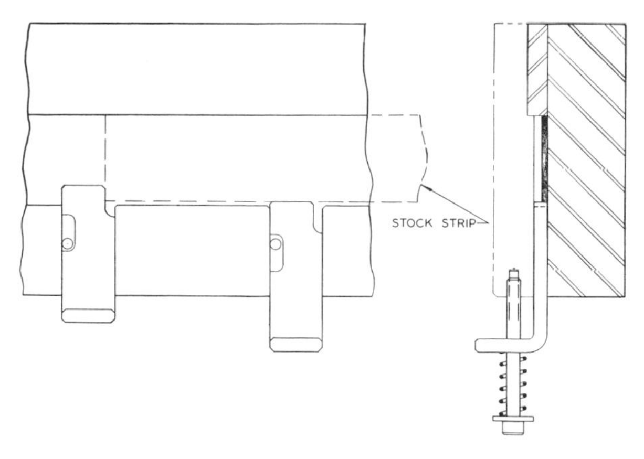

Figure **14·14** Pusher stop.

stops, keep in mind that, of the dual functions served by these stops, the prime function is to provide a stop.

PIVOTED AUTO STOPS

In the past, these stops have been designated by a variety of different names, such as trigger gage, tripper stop, swinging latch, pivoted latch, and automatic stop. If a die stop is specified by any of these terms, the chances are that the stop intended is a pivoted stop, automatic in operation.

For the type of stop illustrated in Fig. 14·15, the term pivoted auto stop is accurately descriptive and not susceptible to being confused with other components. A pivoted auto stop assembly consists of the stop 1, return spring 2, pivot pin 3, and actuator 4. The actuator is commonly referred to as the tripper.

OPERATION

In the illustration, the press ram is descending and the tripper 4 is about to contact the stop. As the ram continues its descent, the tripper depresses the tail of the stop 1. Depressing the tail of the stop causes it to rotate on the pivot pin 3, elevating the nose of the stop. As soon as the nose is lifted clear of the stock strip, the return spring causes the stop to pivot laterally, swinging the nose to the right through distance *D*. The stop nose is then positioned above the scrap bridge, as depicted in view *B*. At this point, the press ram has reached the bottom of its stroke. Then, as the punch holder ascends, the return spring lifts the stop tail, rotating the nose of the stop downward onto the scrap bridge. When the operator advances the stock strip, the stop nose snaps down onto the die face within the area of the blanked-out opening in the strip. The oncoming edge of the scrap opening then strikes against the stop nose, swinging the stop laterally to the stopping

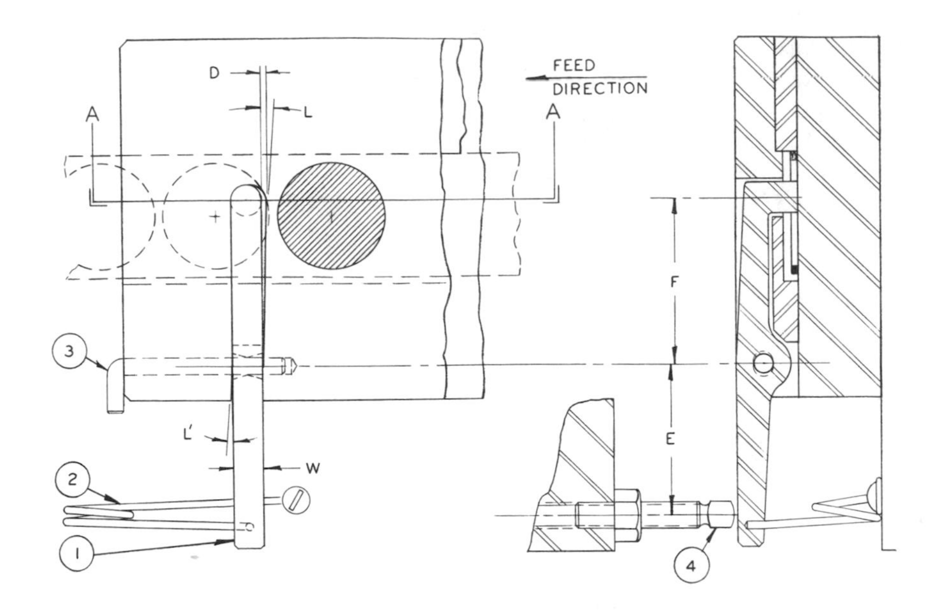

Figure 14·15 Typical pivoted auto stop.

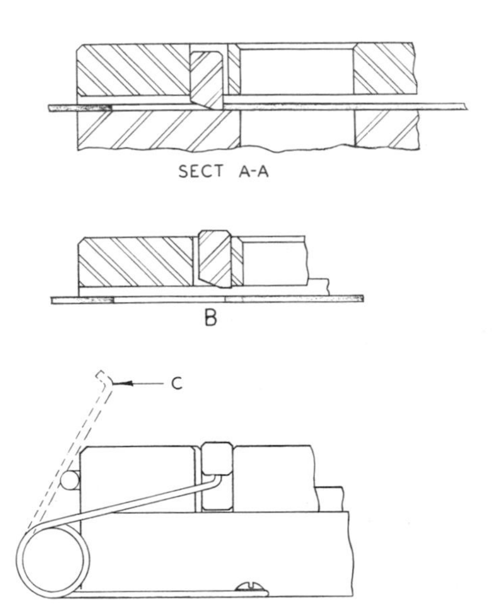

position, as shown in the plan view and at section *A-A*.

The return spring is music wire, formed as shown at *C*. The dotted outline indicates the spring in free position before its tip is inserted into the tail of the stop. Resultant action of this spring contour is a vector from the compressed position to the *C* position. A return-spring action of this kind is absolutely necessary to proper stop performance. For a stop installed as shown, the return spring must act to elevate the strip tail and move it to the left simultaneously.

Return springs should be resilient, limber in action, and yet positive. They should not be stiff or harsh. For light through average conditions, music wire 1/32 to 3/64 in. in diameter will be generally satisfactory, assuming that spring proportions approximate those illustrated.

Spring contours different from the form shown here may be employed, provided that they satisfy the requirements of resiliency combined with adequate action in the proper direction. The safety-pin-type spring in the illustration is a well-tried and proved spring design for this purpose.

The pivot pin is simply a piece of 1/8-in.-diam cold-drawn steel, bent as shown. Bending the pin facilitates assembly and disassembly. The pivot pin should slide easily (not sloppily) in its mounting hole. A pin of this type will slip easily into the mounting hole until it reaches a point approximately 1/8 in. or 3/16 in. from the bend. At this point, resistance will be encountered due to the distortion created when bending the pin. The pin is then driven home. The drive fit should, of course, be a light drive fit. A pivot pin of this type, properly made and assembled, will not shake loose in service, yet may be easily disassembled if required.

For average applications, a pivot-pin diameter of ⅛ in. is both adequate and convenient. It is very seldom necessary to use a smaller pin, and smaller pins are not easily installed. For really large stops and heavy service conditions, the diameter of the pivot pin should be increased in proportion to the requirements of the particular situation.

The tripper in this case is a standard square head set screw of suitable length. These screws are very convenient for this purpose. Securing the tripper screw at the desired setting is mandatory. A locknut, as shown, is the most reliable method of locking such screws at the proper setting.

The stop lever may be made of mild steel or of tool steel. If mild steel is used, it should be cold-drawn stock. If the stop is made of tool steel, oil-hardening ground stock is most desirable. For low-production requirements, mild steel, cyanide-hardened, may be employed. For high production, a mild-steel stop should be carburized and hardened to a case depth of 1/32 in. Tool-steel stops, generally, should be heat-treated to a hardness of Rockwell C 60 to 62 at the nose, decreasing progressively to Rockwell C 45 to 48 at the tail.

The stop lever is, in principle, a typical first-class lever. The pivot (fulcrum) is located between the force (tail end) and the work (nose end). The proportional relationship of dimension *E* to dimension *F* determines whether the lever will be fast-acting or slow-acting. A slow action where *E* is longer than *F* is not desirable for a stop lever. For optimum die operation, the tripper should not contact the tail of the stop lever until the pilots have registered the stock strip. If the die is not piloted, the stop lever should not be actuated until the punch or punches begin to cut into the stock strip. In order to assure satisfactory stop performance, dimension *E* should not be greater than dimension *F*. A procedure which will give good average results is to make dimension *E* 10 to 25 per cent less than dimension *F*. A comparatively fast-acting stop is easier to set and is less susceptible to operating difficulties.

The discussion above concerns the vertical pivot action of the stop lever. As described under the heading "Operation," the stop lever must also pivot laterally so that the stop nose will descend on the scrap bridge and permit the stock strip to be advanced. Lateral movement of the stop nose is indicated by *D* in the plan view of the illustration.

For plain blanking dies (no pilots), *D* can range from 1/64 in. for light-gage work to 1/16 in. for large, heavy work. More than 1/16 in. is not ordinarily required. For pilot-equipped progressive dies, it is necessary to increase distance *D* by an amount equal to the overfeed.

To allow the stop lever to swing sideways, the stripper slot is angled as shown at *L* and *L'*. It is essential that this be done carefully, to avoid making the slot too wide at the pivot line. The stop lever should fit the slot closely at the pivot line. It is suggested that the difference between the width of the stop lever *W* and the slot width at the pivot line be held to a maximum of 0.005 in. (A clearance of 0.002 in. is even better.) Stop levers will function nicely with a clearance of only 0.001 in. at the pivot line.

Angle *L'* is shown exaggerated for clarity. In actuality *L'* is the same as, and parallel to, *L*. Angle *L* is tangent to the stop nose opening in the stripper and extends to the pivot line. Angle *L'* extends from the pivot line to the front edge of the stripper.

Construction and installation. Figure 14·16 depicts construction features of a typical pivoted auto stop lever. Width *W* is governed by the requirements of the individual job. Width should not be too narrow in proportion to the overall length. Speaking generally, widths ranging from 3/16 to ⅜ in. will satisfy the majority of applications. Height *H* should also be adequately in proportion. Again, heights of less than ⅜ in. are in general uncommon. In the majority of cases, stop levers will be somewhere between ⅜ and ¾ in. in height.

There will, of course, be exceptions to these general height and width dimensions. When stops of exceptional size are required, the need will be obvious. In principle they will be the same as stops of average size.

Whenever practical, stop levers installed in box-type strippers are made approximately flush with the stripper. If the top surface of the stop lever rises above the top plane of the stripper, check and ascertain that there is sufficient open space above the stop to satisfy die life requirements. It may be necessary to slope the forward end of the lever, as indicated at *K*. Slope angle *K* permits the front end of the lever to pivot upward without rising above height *H*.

Purpose of the bulge, or boss, indicated by radius *R* is to provide adequate structural mass around the pivot-pin hole. In some cases, the structural proportions of the lever are such that the bulged profile is not required for strength. The bulge contour *R* can then be eliminated in order to simplify making the lever.

For box-stripper installations, an *h* dimension high enough to avoid breaking through the ceiling of the stripper channel is desirable (see Fig. 14·15). However, the stop lever should not be

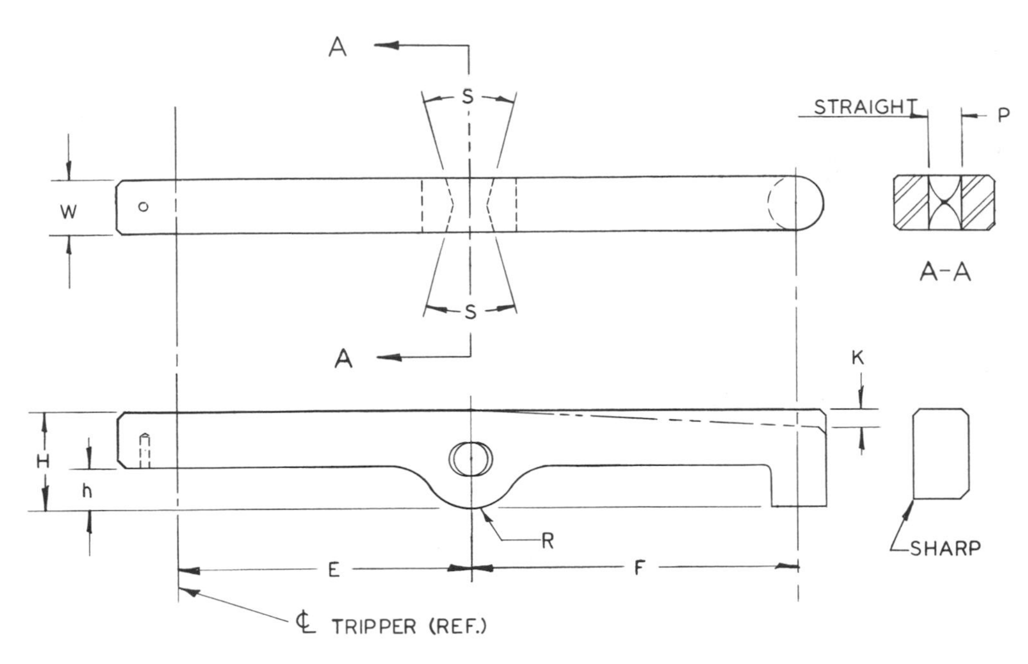

Figure **14 · 16** Typical stop lever.

made unnecessarily weak just to accomplish this. For installations other than box strippers, h should be a minimum of twice the thickness of the stock strip. Incidentally, or perhaps coincidentally, an h dimension of ½ H seems to work out very nicely for most box-stripper applications.

As with any other die component, sharp edges and corners should be rounded off or beveled, except where required. The bottom corner of the stop nose against which the stock strip is fed is required to be sharp, as noted (Fig. 14 · 16). Here, the contour of the nose is cylindrical. However, for many installations, the nose contour will have to be tailored to suit the requirements of the particular application.

The return-spring hole at the tail of the stop lever must be deep enough to assure retention of the spring. If the tail of the lever is less than 5⁄32 in. thick, drill through. However, if this is done, do not allow the spring end to protrude above the top surface when the stop is assembled.

Pivot-pin hole P, section A-A, should be reamed to rotate easily (not loosely) on the pivot pin. The hole should then be relieved angularly, as indicated at S. This angular relief allows the lever to pivot laterally. Note that the hole is not countersunk. It is elongated at an angle from each side. Direction of elongation is from front to back only. The resulting hole is round at the center. When viewed in cross section the hole is straight, as shown in section A-A. In the illustration, angle S is exaggerated for clarity. In actuality only a small angle is required. Included angle S should be slightly more than twice angle L in Fig. 14 · 15.

When making a stop lever, check for balance. The lever when finished should be slightly nose-heavy.

Very thin stock strips may have a tendency to slide under the stop nose. This condition can be corrected by grinding a shallow notch in the die face similar to that shown in Fig. 14 · 17, section A-A. The slight depression formed into the die face allows the nose of the stop to descend below the stock line and eliminates the possibility that the stop strip may slip under the stop nose. Such notches should be shallow: The notch shown in the sectional view is exaggerated for illustrative purposes.

Flat-spring trippers. In Fig. 14 · 15, the stop lever was actuated by a square head set screw. Another type of tripper which is quite popular is the flat-spring tripper (Fig. 14 · 18). These springs are usually made of phosphor bronze. For most applications, a phosphor-bronze spring 1⁄32 in. thick

Figure **14 · 17** Undercut in die face for stopping thin stock.

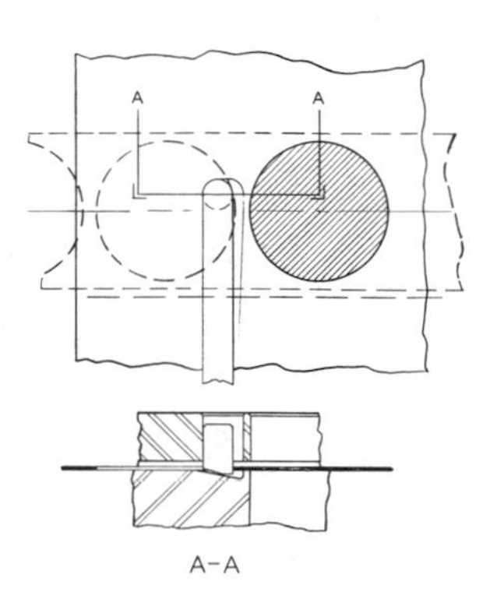

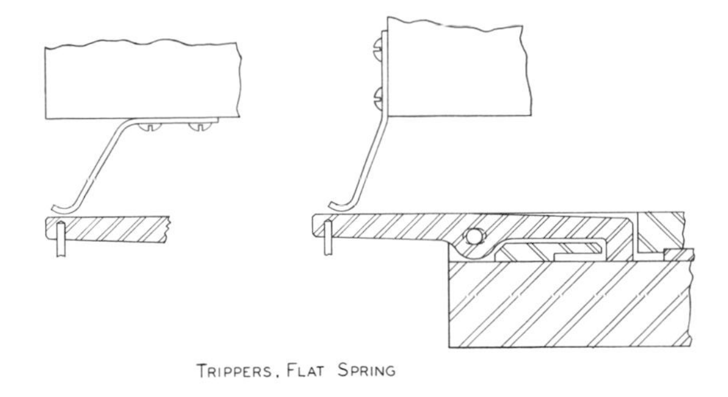

Figure 14·18 Flat-spring-type trippers.

by ½ in. wide will be satisfactory. Springs of this type are commercially available at very nominal cost. They can be readily altered to suit individual applications, as indicated in the illustration.

If the assembled die is mishandled and the punch holder allowed to descend too far, a flat-spring tripper will not damage the stop lever. Flat-spring trippers are also desirable from the standpoint of safety for the press operator.

Coil-type return springs. Coil-type tension springs are often employed as stop-lever return springs. They are well suited to this function. They are especially desirable for larger stops and heavier work.

Figure 14·19 exemplifies a coil type return spring. The spring anchor is a cold-drawn steel pin pressed into the die shoe. The height and location of the pin are such that the required motion is imparted to the stop lever by the spring in tension.

INVERTED PIVOTED AUTO STOPS

Certain types of dies (compound dies, etc.) are inverted in construction. These dies are predominantly equipped with pin stops, especially disappearing pin stops. However, pivoted auto stops are at times employed on such dies.

Figure 14·19 Coil-type return spring.

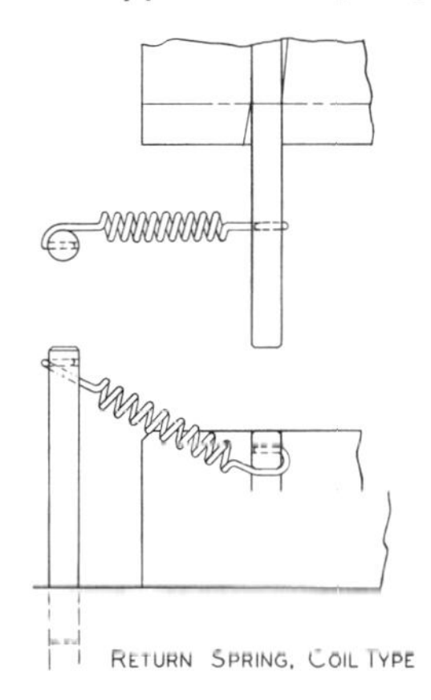

A pivoted auto stop mounted for inverted action is illustrated in Fig. 14·20. The stop lever is conventional. The pivot pin has a threaded shoulder. The pivot-pin hole in the stripper is tapped (jam fit) to receive the pivot pin. The return spring is fastened to the die shoe so as to exert a pulling action downward on the tail of the lever. Inverting the stop does not affect the relative feed direction. Therefore the lateral direction of the return spring action is the same as that of the conventional stops previously described. Inversion does, however, reverse the vertical pivot action of the lever. For this reason, the return spring acts to pull downward on the tail of the lever. The tripper is mounted to the die shoe immediately under the tail of the lever in order to actuate the lever vertically in the required direction.

Action of inverted auto stop. In view *A*, the press ram is descending. The face of the upper die block has just contacted the top surface of the stock strip. As the press ram continues to descend, the die block forces the stripper downward. Since the stop lever is mounted in the stripper, it too travels downward. The relative position of the stop lever in the stripper remains unchanged until the tail of the lever contacts the tripper. Then, as

Figure 14·20 Inverted pivoted auto stop.

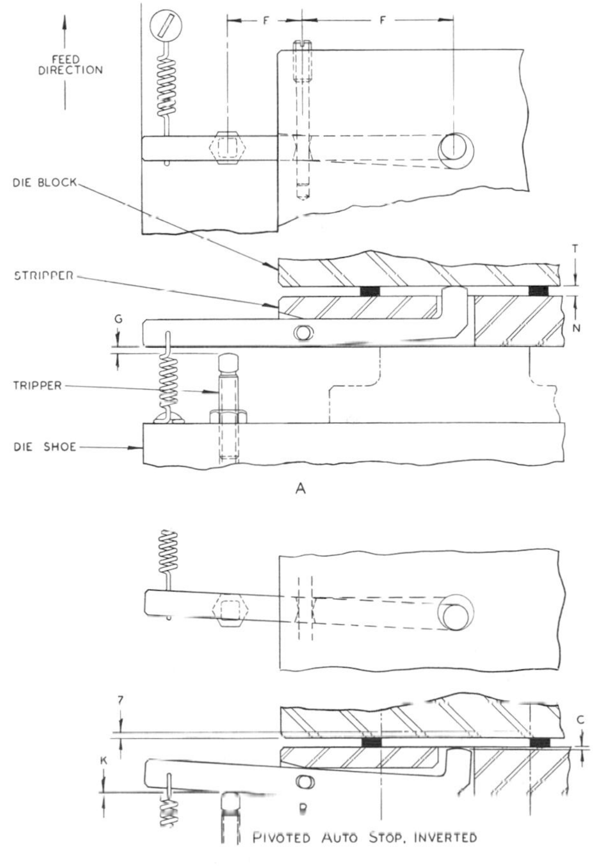

the die continues to close, the tripper, in conjunction with the return spring, actuates the stop. When fully actuated, the nose of the stop swings to the required position beneath the scrap bridge. This condition, shown in view *B*, coincides with the bottom of the press stroke.

The effective stopping height *N* is the distance the nose protrudes above the top plane of the stripper when the die is open. In the case of the die here illustrated, $N = T$, where *T* is the stock material thickness. For proper stop action, the nose of the lever must pivot downward a distance equal to *N* plus a very small clearance *C* (view *B*). For discussion purposes, clearance *C* can be disregarded, since the punch entry distance will assure this clearance.

In order to pivot the nose downward, the tail of the lever must be actuated upward, as indicated at *K*. The actuating distance *K*, which is necessary to achieve the required reaction *N*, is directly related to the leverage proportions *E* and *F*. This relationship may be expressed $K/N = E/F$ or $(E/F)N = K$.

The illustrated stop lever has an *E* dimension equal to one-half its *F* dimension; therefore, $E/F = ½$. Applying the formula $(E/F)N = K$ yields $½N = K$. Therefore, in this case, the required actuating distance *K* is $½N$.

Disregarding the punch entry, the total available actuating distance is equal to *T*. Thus, $T - K = G$, where *G* is the gap at which the tripper is set as shown in view *A*.

The above *E*/*F* proportion was ½. Suppose, for a similar stop, $E = F$. Then $E/F = 1$. Again applying the formula $(E/F)N = K$, we now have $N = K$. As stated in the previous example, $T = N$. Therefore, the required actuating distance *K* is equal to the available actuating distance *T*. Gap $G = T - K$; thus in this case, gap *G* is zero. The tripper can be set so that it just touches the lever when the die is open, or the punch entry distance *Z* can now be taken into consideration. If gap *G* is set at $½Z$, then since $E = F$, clearance *C* will also be $½Z$.

For stops of this type, where *N* is equal to or less than *T*, *E*/*F* proportions ranging from ½ to ⁱ⁄₁ will be generally satisfactory. Stops made to these proportions will have good action and service characteristics. They will also be simple to make and install. Stop proportions where *E* is less than $½F$ should be used only when necessary. Stop proportions where *E* is longer than *F* should be avoided.

When setting the tripper height, check and ascertain that the nose end of the lever does not

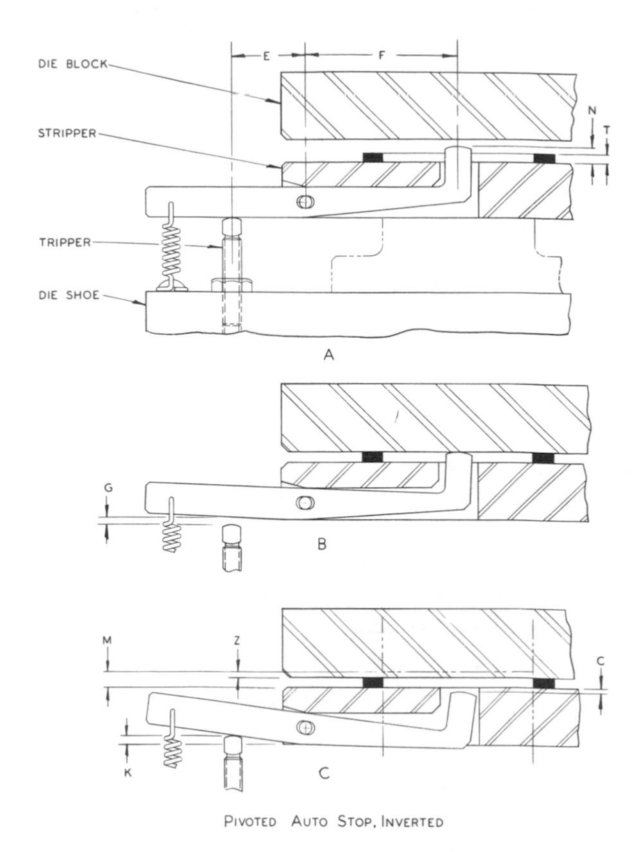

E—Distance between centers, pivot and tripper.
F—Distance between centers, pivot and nose.
N—Stopping height: distance nose protrudes above stripper.
T—Stock material thickness.
G—Induced gap.
K—Distance stop tail is actuated by tripper.
M—Total distance required for complete actuation.
Z—Entry distance: punch into die opening.
C—Clearance between nose and bottom of stock strip.

Figure **14·21** Action sequence for inverted auto stop.

swing too far downward when the stop is fully actuated.

Because *N* is the effective stopping height, it is not always practical to make *N* equal to or less than *T*. For dies in which light-gage material is processed, it is necessary to make *N* greater than *T* in order to provide adequate stopping height. This is the condition illustrated in Fig. 14·21.

When the difference between *N* and *T* is relatively large, the tripper can be set tangent to the lever tail, as shown in view *A*. This will minimize the total required actuating distance.

In view *A*, the die is still open. The die block, which is descending, is about to strike the stop nose. As the die block continues to descend, it depresses the stop nose until the nose is flush with the top surface of the stock strip (view *B*). Depressing the nose causes the tail of the lever to react upward, creating a gap G between the lever and the tripper. Then as the die continues to close, the stop lever is carried downward along with the stripper. In the course of its descent, the stop tail strikes the tripper, which, in conjunction with the return spring, actuates the stop (view C).

Using the identifications in the legend of Fig. 14·21, the following proportions may be considered valid within the normal range of stop action.

$$\frac{E}{F}(N - T) = G$$

$$\frac{E}{F}T = K$$

$$G + K = M$$

$$M - T = Z$$

An understanding of these relationships is an asset to the diemaker, since it can eliminate guesswork and cut-and-try procedures when he is making and installing stops.

A simple relationship can be applied to determine whether or not the proportions of a given stop are practical:

$$\frac{E}{F}N = M$$

PROCEDURE:

Substitute values and find M. Subtract T from M to ascertain punch entry distance Z. If Z is excessive, stop proportions should be altered. To reduce Z, either decrease the N dimension or increase the ratio between E and F, or both. For applications of this kind, E/F ratios will generally range between 1/2 and 1/4.

Factors which may affect stop proportions can be many and varied. Because of this, it is not practical to establish and adhere to rigid rules for many aspects of stop construction. Rather, it is essential that the diemaker attain a thorough understanding of relative stop proportions and be able to associate them with different situations. This is necessary because many stops must be either custom-made or tailored to the particular requirements of individual situations. The following generalities are offered, not as positive rules, but as a guide to stop construction and installation.

N Dimension. This dimension is the effective stopping height. It must be high enough to provide adequate stopping. However, if N is too high, punch entry Z will be excessive, or else the ratio between E and F may be increased to an undesirable figure. Generally, for N a height ranging between T and $2T$ with a minimum of 1/16 in. is satisfactory. Occasionally for heavy work, N may be less than T.

Z Dimension. This is the distance the punch intrudes into the die opening at the bottom of the stroke. Punch entries of 0.005 to 1/32 in. are considered average. Shallower entries are sometimes desired. Deeper entries are at times necessary, but should be avoided if possible.

C Clearance. This clearance permits the nose of the stop to swing under the scrap bridge. A clearance C of 0.002 in. will permit the stop to function. However, to facilitate setting and timing the stop, clearances of 1/64 to 1/32 in. are commonly applied.

Nose contoured for actuation. For inverted installations, pivoted auto stops are sometimes made as shown in Fig. 14·22. Like the previous stops, this stop has a return spring. However, it differs in that it does not require a tripper at the tail of the lever.

With this method, the stop nose is contoured in a manner similar to that shown. The high point serves as an actuator. The face of the opposing die block contacts this peak to force the nose of the stop lever downward. Return-spring action at the stop tail then pivots the nose of the lever sideways to the position shown under the scrap bridge. A stop lever of this sort must be wide enough, in proportion to the stock material thickness, to provide favorable ramp effect for feeding.

Inverted auto stop with integral actuator. Here, a separate tripper is not required, since the actuator is built into the stop lever (Fig. 14·23). At

Figure 14·22 Inverted auto stop, nose contoured for actuation.

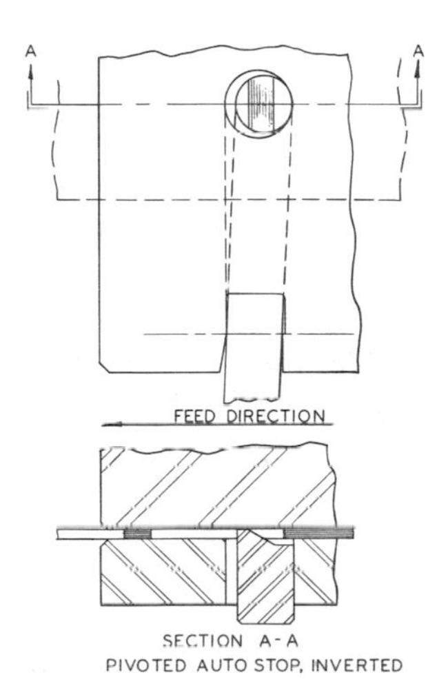

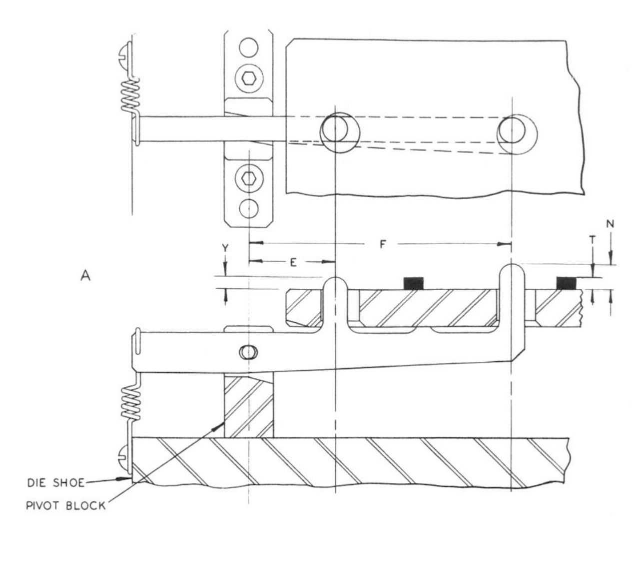

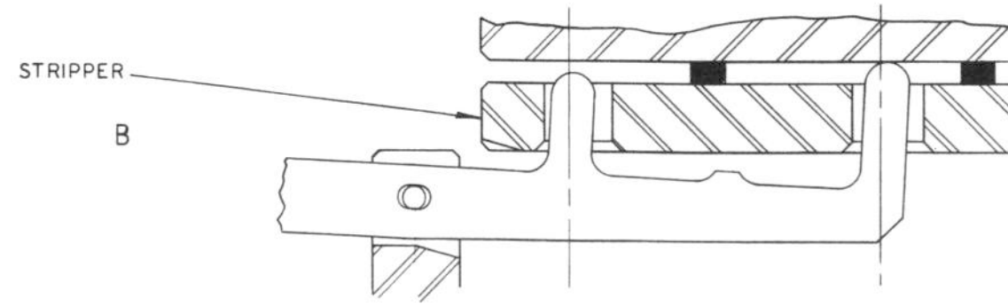

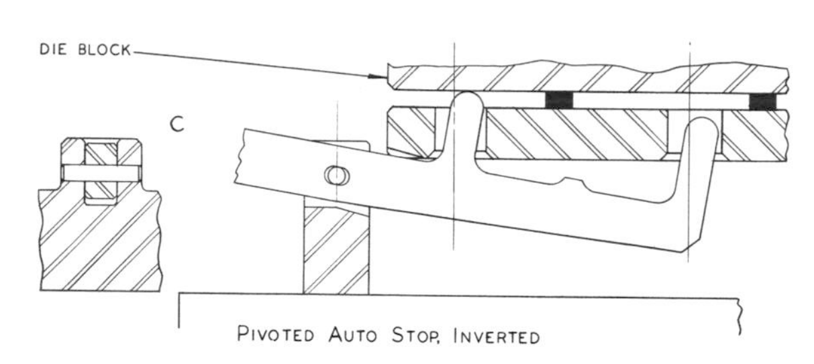

Figure 14·23 Inverted auto stop with integral actuator.

E, the actuating nib extends through the stripper a distance *Y* in a manner similar to the nose of the lever (distance *N*). The actuating nib must be located outside the area covered by the stock strip. A stop of this type could not function if it were mounted in the stripper. Instead, a separate pivot block must be provided in a manner similar to that illustrated.

To make this kind of stop lever, it is necessary to determine the height *Y* of the actuating nib. This height is proportionally related to both *T* and *N*. For practical purposes when making stops, the following formula may be used:

$$Y = \frac{EN}{F} + \frac{ET}{F}$$

EXAMPLE:

$T = \frac{1}{16}$ in. $\quad N = \frac{1}{8}$ in.
$E = 1$ in. $\quad F = 4$ in.

Substituting,

$$Y = \frac{1(\frac{1}{8})}{4} + \frac{1(\frac{1}{16})}{4}$$

$$= \frac{1}{32} + \frac{1}{64}$$

$$= \frac{3}{64}$$

$$N - Y = \frac{5}{64} \text{ in.}$$

Therefore, in this instance, the height of the actuating nib should be 5/64 in. less than nose height *N*.

In the interests of simplicity, certain factors have been disregarded in the derivation of this and other formulas given in this section. The formulas as given are entirely adequate for the intended purpose, which is the practical determination of sizes and proportions for stop construction.

Inverted auto stop with overhead actuator. The stop represented in Fig. 14·24 is similar to the previous stop in that the actuating point lies between the fulcrum and the nose. This stop differs in that it is actuated by a separate tripper. The tripper is a part of, and moves with, the upper punch-holder assembly.

In the case illustrated, the tripper is threaded and screwed into the die block. The tripper is made of drill rod, heat-treated. A cross-drilled hole is provided, which permits a pin to be inserted in lieu of a wrench for adjusting the tripper height. The tripper is safetied by a setscrew from one side. A brass plug is used between the setscrew and the tripper to prevent thread damage. When compared to the previous stop, this stop offers the advantage of an adjustable tripper. This stop, too, requires a separate pivot block.

TRIM STOPS

In cases where the stock strip is trimmed in the die, the resulting offset can be used for stopping purposes. When a stop is installed in conjunction

Figure 14·24 Inverted auto stop actuated by upper mounted tripper.

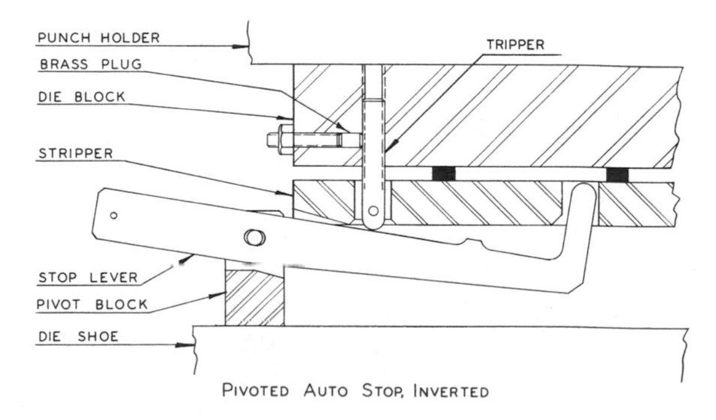

with a trimming operation, it is called a trim stop. An example of a trim-stop arrangement is illustrated in Fig. 14·25. For clarity, the stripper is not shown.

The illustrated die is equipped with a pusher to aid in positioning the stock strip against the back gage. Pushers facilitate proper starting of the stock strip. They are a definite asset on dies in which trim stops are incorporated.

Operating Sequence, Trim Stop. Leading end of stock strip is fed into the stock channel until its front corner contacts the stop. Stock strip is now at start position in the illustration. At the press stroke, the trimming punch trims the stock strip to the required width in the manner shown. Stock strip is then advanced until the offset strikes against the stop insert. This cycle is repeated until the stock strip is consumed.

The trim punch produces a notch along the front edge of the stock strip. Depth of the notch is the amount of offset. It is equal to the difference between the original stock-strip width and the required trimmed width. The illustrated die is not equipped with pilots. Therefore, the registry position coincides with the stopping position. Because of this, the length of the notch along the stock strip must be precisely equal to the advance.

View *B* is an exaggerated view showing the relationship existing between die opening, punch, and stop. As stated, the trim punch produces the

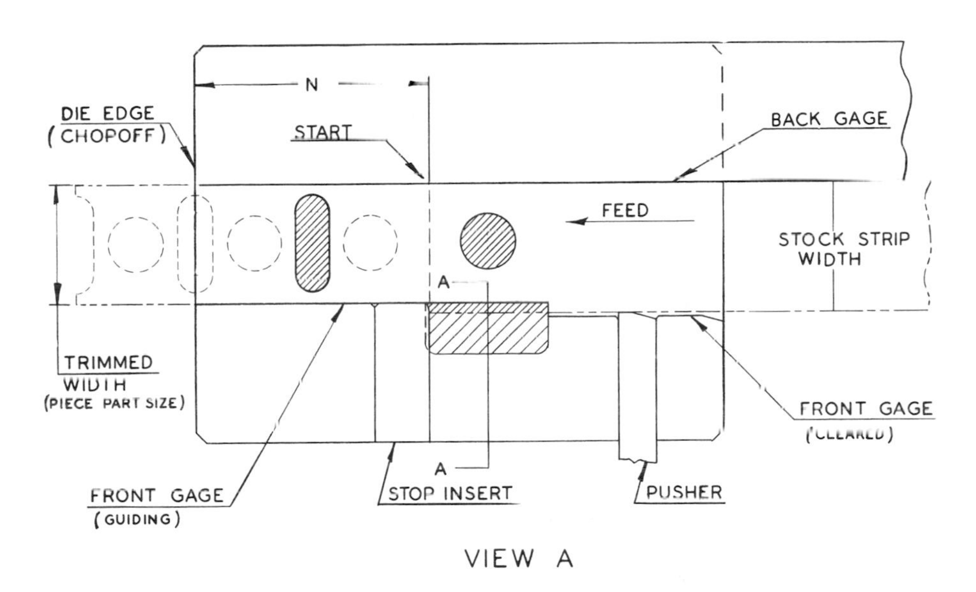

Figure **14·25** Trim-stop arrangement.

notch in the stock strip. Therefore, the notch is sized by the punch. The punch is made to the required dimensions, and cutting clearance is provided in the die opening as shown. The distance between the stop face and the far end of the punch is equal to the advance.

It is good practice to make the cutting length of the punch slightly longer than the advance in order to provide an overcut. One method of providing an overcut is that shown, where the cutting length of the punch is made equal to the advance plus overcut. The punch is then angled back through distance *D* in order to clear the stop insert. Distance *D* is usually made equal to, or slightly less than, the smallest anticipated offset.

The amount of overcut to use depends upon circumstances pertaining to the individual job. Overcuts are, generally, in proportion to the size of the work; that is, smaller overcuts are used for small work. Probably the most prominent factor influencing the amount of overcut is the size of the offset. A deeper offset permits the use of larger overcuts.

Overcuts of only 0.002 in. are sometimes used, especially on delicate work. Overcuts larger than 1/64 in. are seldom necessary. For the majority of applications, overcuts of 0.005 to 0.010 in. will be satisfactory. The presence of the overcut permits the stopping position to be altered slightly, if required. This is a very real advantage.

Extra clearance is provided at the stop end of the die opening, as noted. This clearance serves two purposes:

1. Needless fitting of die opening to punch is eliminated.
2. If necessary, the advance distance can be increased a slight amount without altering the die opening.

Keep in mind that the proportions shown in view *B* are exaggerated for illustrative purposes.

Trim punches are commonly provided with heels (section *A-A*). The back of the heel is in sliding contact with the wall of the die opening. This enables the punch to resist possible deflection resulting from lateral thrust caused by the cutting action.

The trimming operation produces a scrap slug which is cut on two sides only. Since the slug is not tightly confined in the die opening, there is a decided tendency for the trim slugs to pop up from the die opening. Slug pulling is a serious hazard to successful die operation. The most common method of overcoming this slug-pulling tendency is to equip the punch with spring pins, as shown. If the trimmed scrap is quite narrow, the space available for spring pins is restricted. In some cases where the scrap is narrow, spring pins have been installed at an angle, as seen in view *C*.

The material trimmed from the stock strip becomes scrap. Because of this, there is understandably a strong tendency to make the offset too narrow in an attempt to conserve material. However, it is better to sacrifice a slight amount of material in order to be assured of trouble-free production. The offset should be large enough to permit practical installation of adequate spring pins in the trim punch.

Essentially, a trim-stop arrangement is a positive stop installed in conjunction with a trimming operation. The degree of feeding accuracy which can be obtained is comparable to any other positive stop.

Trim stop for piloted die. Trim stops can also be applied to dies which are equipped with pilots (see Fig. 14·26). As with any stop used in conjunction with pilots, the stop must permit the stock strip to be overfed in order to allow the pilots to register the stock strip. The principle is the same as that illustrated in Fig. 14·1, views *B* and *C*.

As shown in Fig. 14·26, the punch length is made equal to the advance distance plus the overfeed. Because the punch length is greater than the advance, overcutting is an inherent condition with an arrangement of this kind. The amount of overcut is equal to the amount of overfeed. Because of the difference between stopping position and registry position, the risk of slug pulling is even greater than it is in a die which is not equipped with pilots.

Trim stops summarized. Feed accuracy obtainable with a trim stop is comparable to the accuracy which can be obtained from any other type of positive stop. Slug pulling is a characteristic

Figure 14·26 Trim-stop arrangement for piloted stock strip.

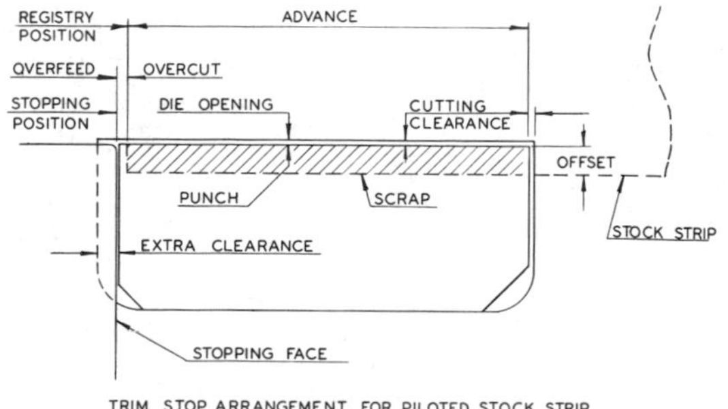

hazard of the trimming operation, and provision must be made to prevent it.

If the stock strip is not piloted, the stop must act as a true gage; stop position and registry position are the same. If the stock strip is piloted, it must be overfed in order to avoid interference with accurate registry by the pilots.

Trim stops are sometimes incorporated in dies where trimming is not required in order to secure the benefits accruing to a positive stop. This is also done on occasion to eliminate the need for a series of stops in a progressive die which has a great number of stations.

The offset in the stock strip is made by the punch. Therefore, the size of the offset is derived from the punch size.

The trim stop should be installed as part of the first die station. This eliminates the need for extra stops and thereby simplifies both construction and operation of the die. Normally, the trim stop is located along the front edge of the stock strip, as shown in the illustrations. This is true because, generally, the stock strip is meant to gage against the back gage. As the front of the strip is trimmed away, any variation in the original stock-strip width affects only the trimmed scrap and not the trimmed-strip width. If in an exceptional case the stock strip is gaged from the front gage, the trim stop is then located along the back of the strip.

The trim stops shown in this section are typical examples. Variations will be necessary to suit individual job requirements. However, the principles which are discussed in this section should be thoroughly studied and understood so that they can be successfully applied to different situations.

SIGHT STOPS

A sight stop is not, in reality, a stop. It is merely some visual reference with which the operator can line up the leading end of the stock strip when starting the strip through the die. It is not reasonable to expect more than nominal accuracy or consistency from sight stopping. In spite of this, the use of sight stops is not uncommon, since there are instances where sight stopping is satisfactory. In fact, there are conditions where visual reference is the only practical procedure. Parting dies and chopoff dies, for example, require the employment of sight stopping.

As in the case of chopoff dies, a die edge may be used for visual reference. Or the lead end of the stock strip may be sighted in relation to a die opening, as with a parting die. If these methods are not applicable, a sight line may be inscribed at the required location. In addition to being properly located, sight lines must be clearly visible to the operator when he is running the die.

CHAPTER 15

STOCK MATERIAL UTILIZATION AND STRIP LAYOUTS

UNIT STOCK

The term unit stock is applied to pieces of stock material which are fed individually into dies for processing. In the case of a large stamping, a standard mill-size sheet might be practical as unit stock. In most cases, however, it is necessary to cut the sheets into the required units with squaring shears. In some instances, savings in stock material costs are effected by shearing unit stock from mill ends or even from scrap stock material. At times, unit stock may be produced from strip stock by means of a chopoff die.

Two examples of unit stock are pictured in Fig. 15·1. The rectangular piece of unit stock *A* is fed into a piercing die to produce the piece part *B*. Disc *C* is the scrap slug produced by the piercing operation. In the case illustrated, slug *C*, which would ordinarily have been scrap, becomes the unit stock from which the hexagonal piece *D* is blanked, leaving only ring *E* as scrap.

STOCK STRIPS

Stock strips are fed into the die and advanced the required advance distance at each press stroke for a series of repetitive operations. The fundamental relationship of a stock strip to the stampings produced from it is indicated in Fig. 15·2.

Figure **15·1** Unit stock as related to stampings produced from it.

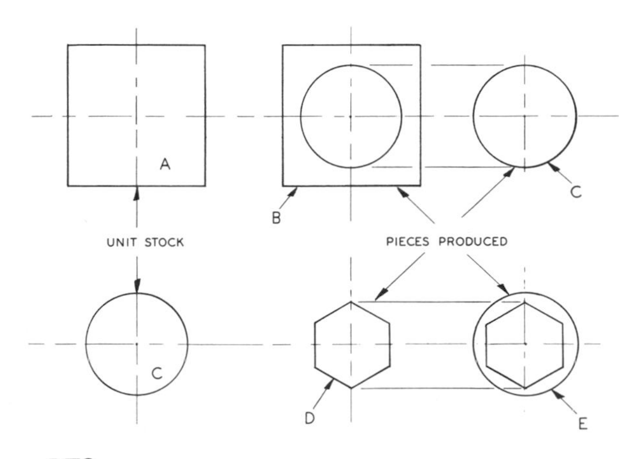

If a stock strip is cut from sheet stock, its length *L* may be equal to a standard sheet length, or it may be sheared to a specified length, if desired. Strip width *W* must, of course, be compatible with the required piece-part size. For strip stock furnished in coil, strip length is a matter of agreement between producer and consumer. Coiled stock is considered in terms of weight rather than length. For example, with cold-rolled strip steel, the weight of standard coils is considered to be 200 lb per in. of width, min. per coil. Strip stock is also furnished straightened and cut to specified lengths.

Certain conditions pertaining to strip stock, and of possible concern to the diemaker, are illustrated in Figs. 15·3 to 15·6.

Figure 15·3: Edgewise Curvature. (Also called camber.) Lateral departure of the edge from a straight line, which may be unidirectional or reversing; if reversing, it is sometimes called snake, or snaky.

Figure 15·4: Longitudinal Bow. (Also called longitudinal curl.) A curved departure from longitudinal flatness. With coiled stock, this condition is unidirectional and is usually attributable to coil set. Stock straighteners are commonly employed to correct this condition.

Figure 15·5: Crown. Variation in thickness across the strip from edge to center or from edge to edge.

Figure 15·6: Dish. The transverse departure of the concave surface from a straight line which touches the two edges.

Grain direction in stock strips. Normal grain direction is shown in Fig. 15·7, view *A*. Grain direction is parallel to the length of the strip. This is normal for all materials, both sheet and in coil. If necessary, sheet stock can be sheared across

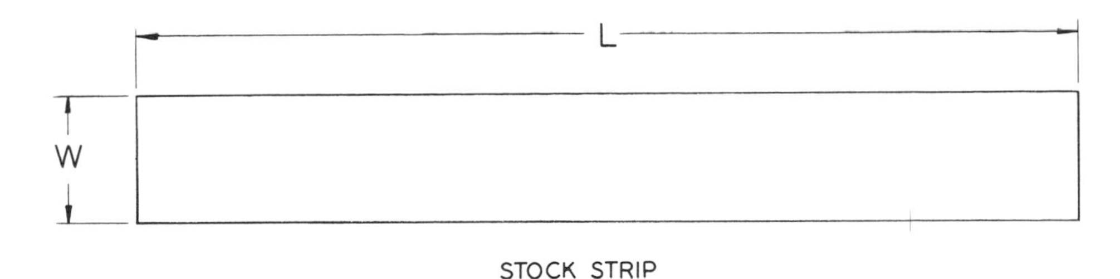

Figure 15·2 Strip stock in relation to stampings produced from it.

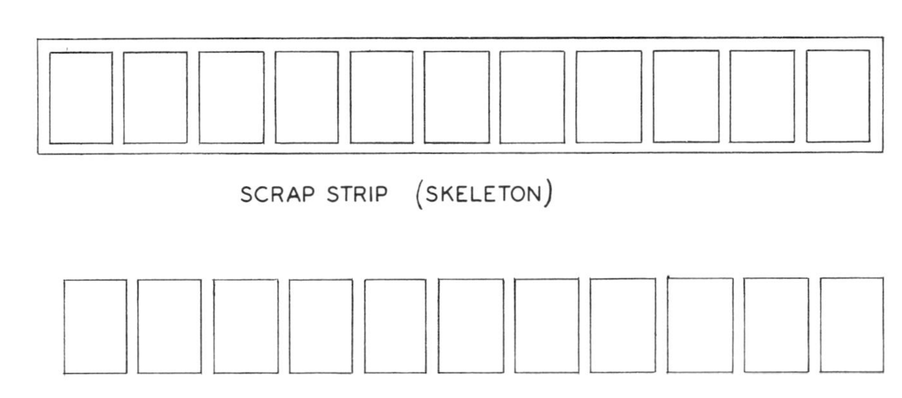

the grain as shown in view *B*. Instances where this grain direction would be advantageous are very rare and should be avoided if possible.

Scrap strip terminology. Terms commonly used to describe die-processed stock strips are illustrated in Fig. 15·8 (see also Chap. 1, Fig. 1·1).

Material conservation. Piece-part contours are highly individualized. However, in a general way quite a number of piece parts will conform to the basic contours shown in Fig. 15·9. This drawing illustrates, by comparison, how relative material consumption for various shapes is affected by the manner in which the blank is positioned in the stock strip.

Material costs are an important aspect of stamping production. It follows, therefore, that conservation of stock material is an important consideration in diemaking. The importance of stock saving is in direct proportion to volume of production: As production requirements increase, so does the importance of stock material conservation. It must be remembered, however, that stock saving is secondary to the satisfactory production of an acceptable piece part.

BLANK (PIECE-PART) ORIENTATION FACTORS

Grain direction (see Figs. 3·55, 3·56, and 15·7) is the most important single factor in determining the relationship of a blank or piece part to the stock strip. If a specific grain direction is necessary, the piece part must be oriented to agree with this requirement. Also, but less frequently, other piece-part conditions (such as burr-side requirements) may affect orientation in the

Figure 15·3 Edgewise curvature (camber).

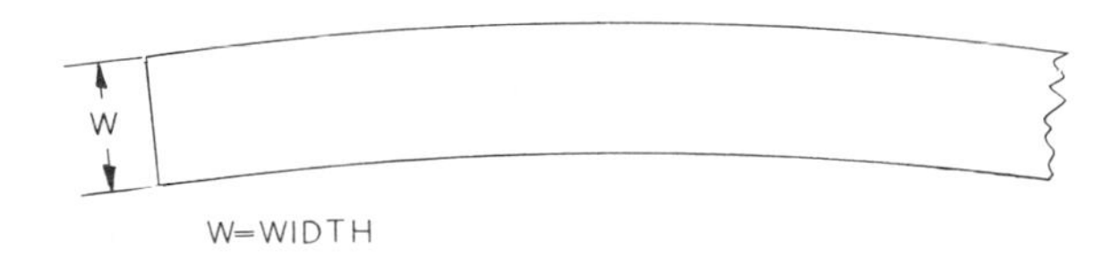

Figure 15·4 Longitudinal bow.

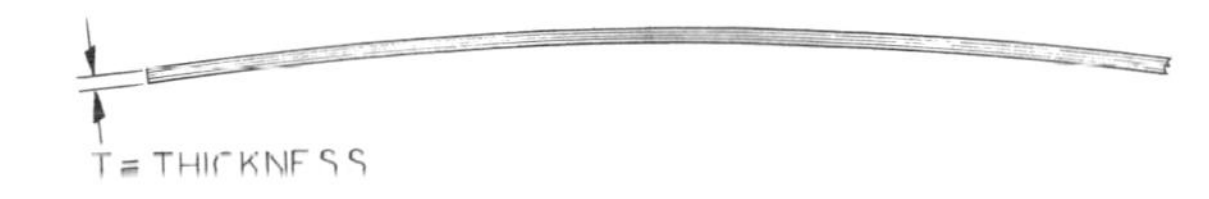

Figure 15·5 Crown = a − b.

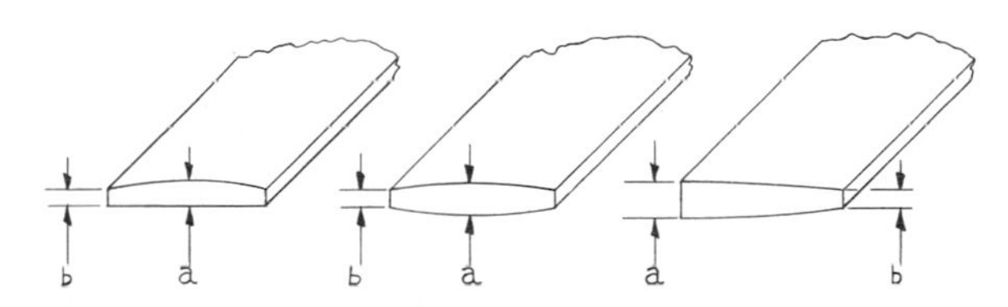

Figure 15·6 Dish.

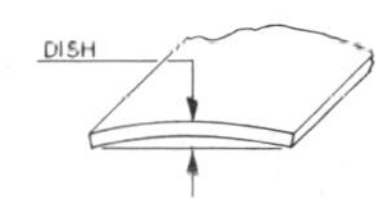

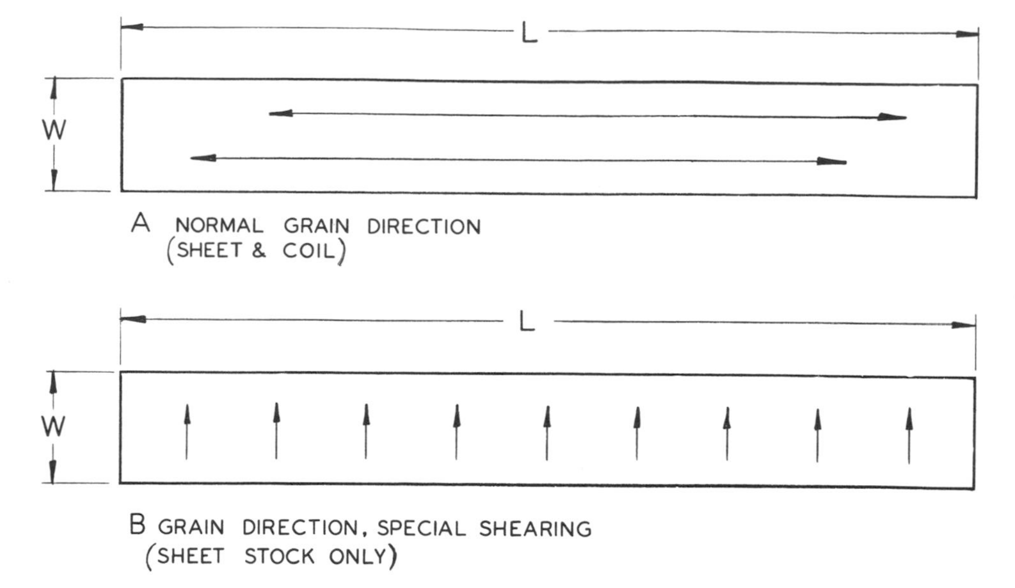

Figure 15·7 Grain direction in stock strips.

stock strip. Assuming that factors which affect the quality of the piece part are satisfied, die function, operating ease, economy of stock material, and die construction cost are the factors which influence a strip layout. These factors are closely related to production requirements. Where production is low, operating ease and economy of stock material may be sacrificed in favor of lower die cost. For high production, die cost becomes less important, while operating ease and economy of stock material assume greater significance.

Wide run versus narrow run. Blanks positioned as shown in Fig. 15·10, view *A*, are said to be run the wide way. Those in view *B* are described as running the narrow way. In the absence of other circumstances which would dictate orientation, the wide run is generally more desirable than the narrow run. There are three reasons for this:

1. The shorter advance distance required for the wide run tends to facilitate feeding.
2. With the wide run, more pieces will be produced from any given strip length. Therefore, fewer strips are required to produce a given number of piece parts. This expedites production, since less time is consumed in handling stock strips and in starting the strips through the die.
3. For operations which entail back and front scrap webs (see Fig. 15·8), the wide-run strip requires less material per piece produced than the narrow run. The amount of material saved is in proportion to the difference between the length of the blank and its width. The greater the difference, the greater the stock saving.

The first two reasons cited above are valid for most situations. The third is valid only for stock strips which have both front and back scrap webs.

Strip reaction to stopping. Ease of die operation can be influenced by reaction of the stock strip to stopping. This is readily apparent in Fig. 15·11. View *A* can be considered the optimum condition for maximum operating efficiency. Here, the advantages of the wide-run principle are present. In addition, orientation of the blank is such that the most favorable stopping condition expedites feeding the strip through the die. In the vast majority of cases it is advantageous to feed the stock strip with its back edge in contact with the back gage. The necessary feeding clearance is provided at the front gage. Therefore, the most desirable stopping effect is one which tends to displace the stock strip toward the back gage rather than the front. In view *A*, as the leading edge of the scrap bridge contacts the stop, its angle of attack tends to wedge the strip against the back gage.

Suppose, however, that the piece part has a grain direction requirement which requires the blank to be run the narrow way. Rotating the blank position horizontally within the plane of

Figure 15·8 Scrap-strip terms.

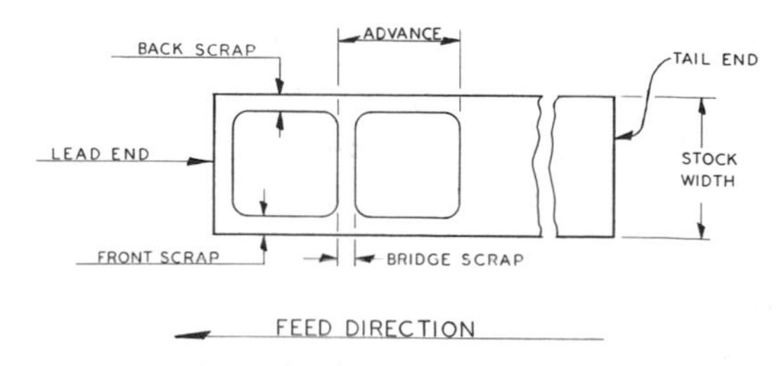

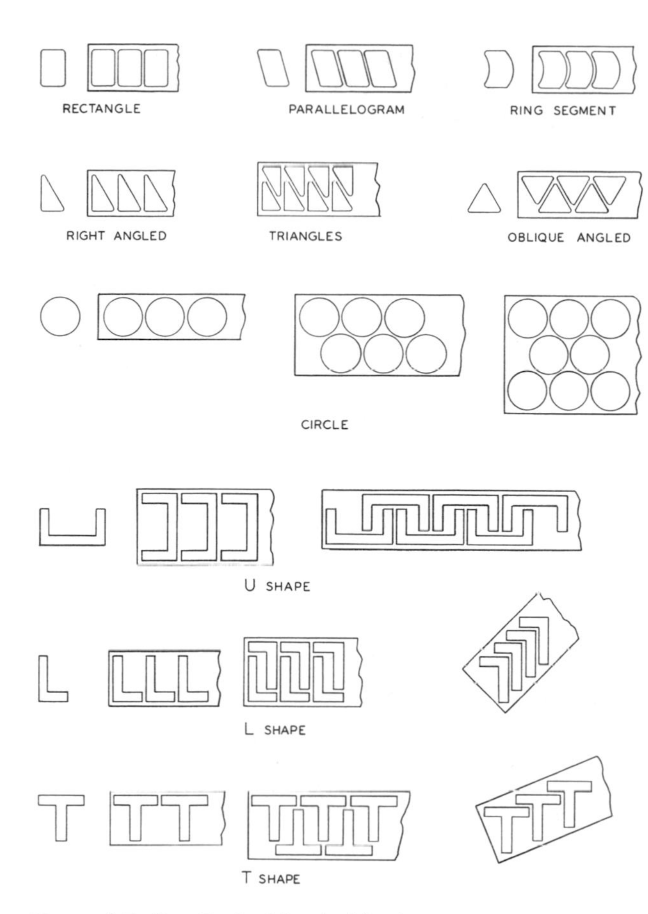

Figure 15·9 Typical basic blank contours.

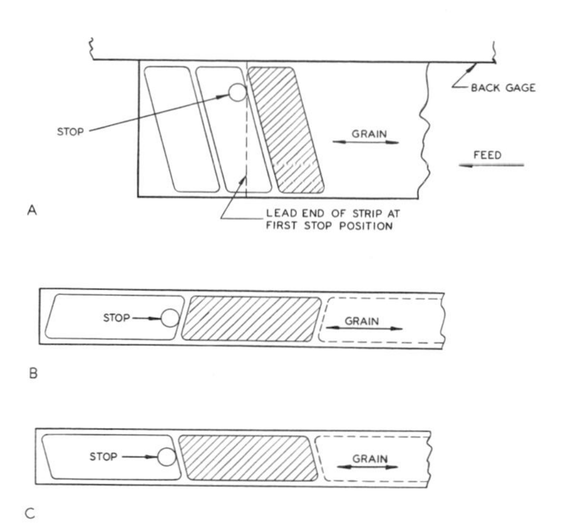

Figure 15·11 Strip orientation of blank: stop and burr-side considerations.

the strip produces the condition shown in view *B*. Here, the angle of the scrap bridge as it strikes against the stop tends to displace the strip toward the front gage. Since this is a less desirable stopping condition, it would be better to change the blank orientation further by inverting the blank as shown in view *C*.

Caution! Inverting the blank position will also invert the burr side with respect to the blank contour. If the piece part has a definite burr-side requirement, this would take precedence over the stopping condition. Keep in mind also that inverting the die construction may provide a satisfactory solution to burr-side problems.

Projecting contours. The relationship of projecting contours to feeding is an important consideration. Projecting prongs, either on the scrap or the piece part, are a potential source of interference with feeding. Whenever practical, strip orientation of the piece part should be such that any prongs or projections tend to trail the feed direction. A trailing scrap projection is pictured in Fig. 15·12, view *A*.

In view *B*, the projecting prong is leading the feed. This should be avoided if possible. However, especially with more complex piece parts, it may not always be practical to avoid leading scrap projections. In such cases, be certain that strategically located stock lifters (or the equivalent) are provided to facilitate feeding. When doubt exists as to the necessity of lifters, it is a good idea (be-

Figure 15·10 Wide run in contrast to narrow run.

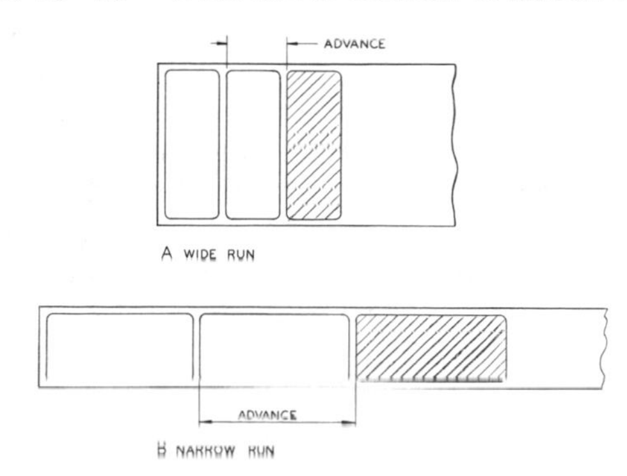

Figure 15·12 Feeding projecting contours.

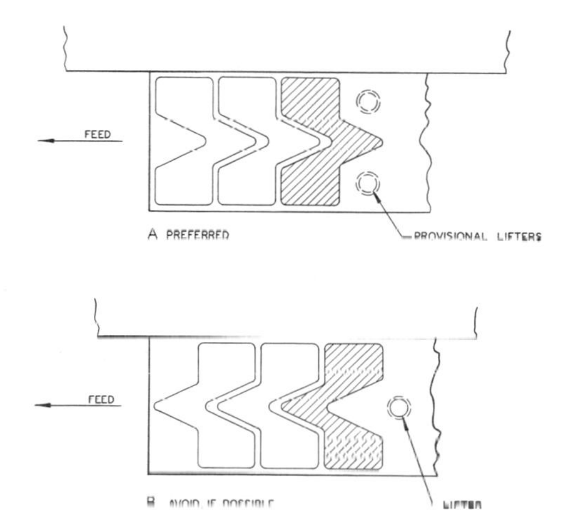

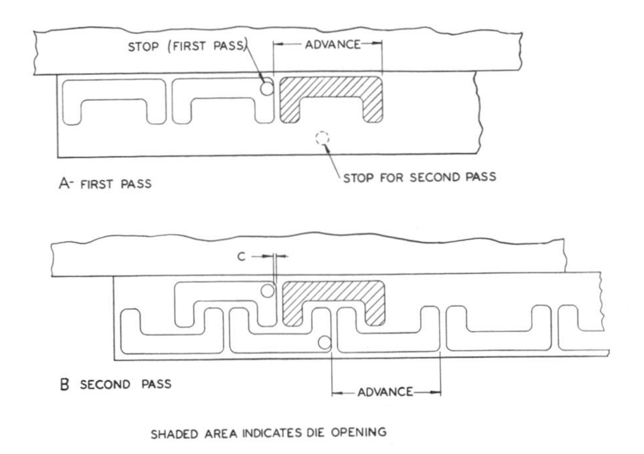

Figure **15 · 13** Interlaid blanks—two-pass die.

fore heat-treating) to provide effectively located lifter openings in the appropriate die members if the members require hardening. This will preclude the need for hard-drilling or for annealing hardened die components in the event that lifters are subsequently proved necessary.

INTERLAID BLANK ORIENTATION

The contour of some blanks may be such that interlaid orientation in the strip can produce a considerable saving in stock material. The blank contour shown in Figs. 15 · 13 and 15 · 14 is an excellent example of such a condition.

Two-pass method. One procedure for economical strip orientation of such blanks is the two-pass method shown in Fig. 15 · 13. With this procedure, interlaying is achieved by feeding the stock strip through the die twice, in the manner illustrated. A series of strips is fed through the die for the first pass. The optimum number of strips varies according to individual job conditions and production requirements. The partially skeletonized strips are then turned over or rotated end for end and fed through the die for the second pass, which completes the procedure.

In design and construction, a two-pass die is the same as a normal one-pass die except for three factors:

1. A two-pass die often requires a wider stock-strip channel or tunnel to accommodate a wider stock strip.
2. In most cases, it is best to install an additional stop which operates from the openings made in the stock strip as a result of the first pass. This associates the second pass with the first pass, maintaining a more accurate relationship between the two. The extra stop is especially desirable where blank contours are closely interlaid, because it tends to prevent overlapping of the second-pass blanks into the openings left by the first pass.

   Because of interference, solid stops are obviously not suited to these dual stop applications. Spring-pin stops, however, lend themselves nicely to this method when applied to inverted dies and compound dies. For blank-through dies, swinging lever stops are also applicable. Ideally, when dual stops are used, each stop should function only when it is needed. To accomplish this, provision can be made to lock out each stop during its idle period. Another means of avoiding possible interference is to locate the stops so as to provide a slight clearance as shown at *C* in view *B*. A clearance of 0.005 to 0.010 in. should be sufficient. If more clearance is deemed necessary, the stop should be locked out.
3. For multiple-pass operations, scrap bridges between blanks should be wider than normal. This is especially true for the scrap bridges created between the blanks in one pass and the blanks in the next pass. For a two-pass die the minimum scrap bridge between passes should be 50 to 100 per cent wider than normal.

Two-pass advantage and disadvantages. The advantage of the two-pass method is maximum stock material utilization combined with minimum die cost. The disadvantages can be listed as follows:

1. Extra stock handling and manipulation. From 10 to 20 per cent more production time will generally be required to produce blanks with this method as opposed to a one-pass die which produces one blank per press stroke. This must be balanced against the savings in stock material and die cost.
2. Cutting stresses and strains induced during the first pass may introduce edgewise curvature (camber) in the strip. If this condition becomes pronounced, it can cause serious difficulty in feeding the second pass. The tendency to camber is strongest where blank contours are such that the residual scrap skeleton is relatively fragile. It is not usually practical to use coiled stock in conjunction with the two-pass method. When the coil is rewound after the first pass, the strip tends to kink at the weak sections caused by the blank openings. In many cases, kinking can be pronounced enough to cause the coil to become bulky and unwieldy. The kinks are not readily eliminated by normal straightening procedures and remain a decided

handicap to successful feeding of the second pass.

3. The two-pass method should be restricted to single-stage dies; it is not feasible for multiple-station progressive dies. All in all, the disadvantages of the two-pass procedure are such that the method is not generally practical where high production is required.

Two-pass method summarized. This procedure lends itself to low- and medium-production applications where the blank contours are such that a considerable stock material saving is effected. With the exclusion of a few very specialized applications, the multiple-pass method is efficient only when used in conjunction with interlaid blank orientation. Also with a few specialized exceptions, multiple-pass dies employing more than two passes are not considered feasible. While it is true that the two-pass method is practical only for a relatively small percentage of applications, it can under proper circumstances be well worthwhile.

Gang dies (one pass). A gang die is essentially two (or more) dies combined into one. Since only one pass is required for feeding a gang die, it does not suffer from the disadvantages inherent in multiple-pass operations. Economy of stock material with gang dies is at least as good as with multiple-pass dies. In some instances, gang dies may be even more efficient in regard to stock material consumption, because the extra scrap-bridge width allowance can be less than for an equivalent multiple-pass operation. In most cases, gang dies producing two blanks per press stroke can be operated successfully using normal or up to normal plus 25 per cent scrap bridge widths. However, it should be kept in mind that gang dies which produce a greater number of blanks or pieces per press stroke may require wider scrap allowances.

In addition to economy of stock material, an obvious asset of gang dies is the production of more pieces per press stroke. This combination of factors, if wisely used, can be a decided asset where production requirements are great enough.

Figure 15·14 Interlaid blanks—gang die, one pass. Two blanks per stroke.

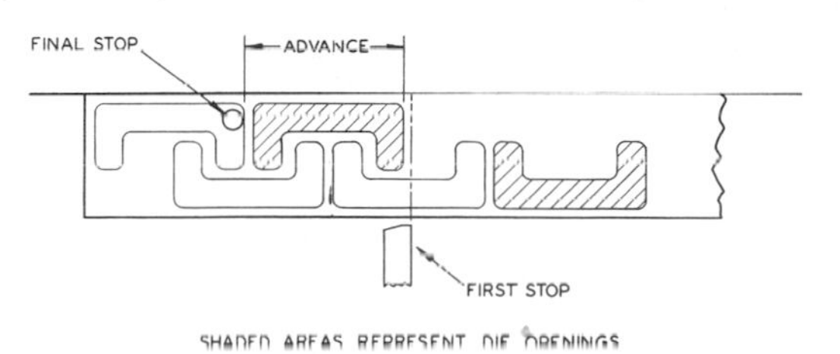

Die cost is, of course, greater for a gang die than for a die which produces one piece per stroke. Generally, the tool cost can be considered as approximately the same on a per-piece basis. That is, a two-gang blanking die will cost approximately twice as much as a single-blank die. Actually, for simple gang dies (round work, for example) the tool cost can be somewhat less per piece produced. However, for more complex piece parts, the tool cost per piece can be considerably greater. In fact, for really complex work involving other than cutting operations, gang dies are not generally practical from the standpoint of tool cost.

Gang dies summarized. In all respects except tool cost, gang dies may be considered superior to multiple-pass dies.

The use of gang dies is generally limited to high-production requirements. However, for simple dies where a considerable stock material economy is effected, gang dies can be practical for medium-production requirements.

Gang dies, unlike multiple-pass dies, are not restricted to single-stage operations. The principle is practical for many applications where multiple-station progressive operations are required, provided only that operations are simple enough to permit trouble-free operation and ease of maintenance.

Gang dies are not generally well suited to very complex work. This is not due merely to high initial tool cost. An unsuitable gang die can present many running problems and be impractical to maintain.

The pieces produced in gang dies are usually, but not necessarily, duplicates. On occasion, the gang-die principle may be advantageous for the simultaneous production of different piece parts.

The employment of the gang-die principle is not necessarily restricted to interlaid blank orientation.

Simple gang dies can, ordinarily, be operated at the same press speeds as the equivalent single die.

Tandem gang dies. A tandem gang die is one which produces two or more pieces from the same row simultaneously (at one press stroke). The pieces are usually, but not necessarily, duplicates. A tandem die is a special form of gang die. It may be single-row or multiple-row. The pattern in Fig. 15·15 is two-row, four blanks per stroke, producing duplicate pieces.

A characteristic of tandem strip layouts is the

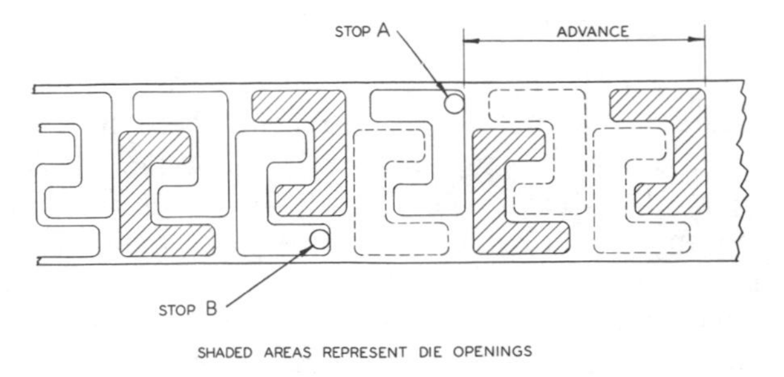

Figure 15·15 Gang die—tandem, one pass. Four blanks per stroke.

longer advance distance necessary to accomplish alternate blanking within the same row. The advance distance required in the illustrated instance is approximately twice the advance distance normally required for the equivalent double-row, two-blank gang die.

In some cases, the tandem principle is employed to achieve an interlaid stock-strip pattern. In other instances such as that in the illustration, the tandem principle is used simply to produce more pieces per press stroke.

Tandem dies are more complex (as is any gang die). Therefore, the principle is normally limited to applications where production requirements are high. Tandem dies and other gang dies (unlike multiple-pass dies) are best suited to the running of coiled stock. This fact, too, is in accord with high-production requirements.

Because of the relatively large die area required for tandem dies, the principle tends to be limited in application to dies which produce smaller piece parts. This is especially true if the dies are progressive in the sense that more than one operation is performed per piece part.

In keeping with high-production practices, tandem dies are usually operated in conjunction with automatic feeds, feeding coiled stock. When this is the case, stops other than a reference stop for the feed are often eliminated. However, stops may be required for one reason or another. For example, it may be desirable to make an occasional short run without using the automatic feed. When stops are required, the strip layout should be well considered, to provide optimum stopping conditions. Note that two stops are provided in the illustration to permit maximum stock-strip consumption. Actually, in this case stop *B* permits blanking one more piece per strip, provided the strips are of optimum length. This may or may not be worthwhile, depending on the particular job. As a rule, one stop only, located at *A* (not *B*), would be considered optimum for most similar applications. Keep in mind that if stops are located in blanked-out strip openings after the final blanking stations, they may interfere with feeding the required advance distance. This is generally true for tandem dies, owing to the fact that more than one blanked-out opening is included, in tandem, within the advance.

Any multiple-pass die or any gang die is, necessarily, more complicated than the equivalent single die. This increases the possibility of error or confusion in the design of the die. Therefore, when studying and analyzing the design before beginning construction, give especial attention to the following conditions:

1. Possible Overlapping of Blanks. Check that adequate scrap bridges exist between the blanked-out openings, particularly in the row-to-row relationship.

2. Stops. Check and be certain that stops will not catch in the wrong strip openings.

3. Grain Direction. Be sure that grain-direction requirements have not been overlooked in order to achieve interlaid orientation.

4. Surface Finish of Stock Strip. If one surface of the stock strip is plated, polished, or otherwise different from the other side, double-check to be sure that the blanks are oriented properly with respect to the surface finish. This point is emphasized here because two-pass dies especially are susceptible to error in this respect.

Interlaid blank orientation summarized. This procedure is, simply, a means of securing maximum economy of stock material. Either gang dies or the multiple-pass (usually two-pass) method may be employed to achieve interlaid blank orientation. The use of interlaying should be restricted to piece parts whose contours make the procedure worthwhile.

Interlaid orientation imposes rather strict limitations on grain-direction possibilities. In fact, piece-part grain-direction requirements often prohibit the use of interlaid strip patterns even though the blanks are otherwise ideally suited to interlaying.

INTERPOSED BLANK ORIENTATION

Blank-to-strip orientation of the type pictured in Fig. 15·16 is called interposed orientation. Interposition is distinguished from interlayment as follows: With interlaid orientation, blank contours have different relative strip locations, as shown in Figs. 15·13 to 15·15. Interposed blank contours, however, have the same relative position in the strip. Interposal occurs when an area of one blank extends into what would otherwise be a scrap area subtended by the contour of the

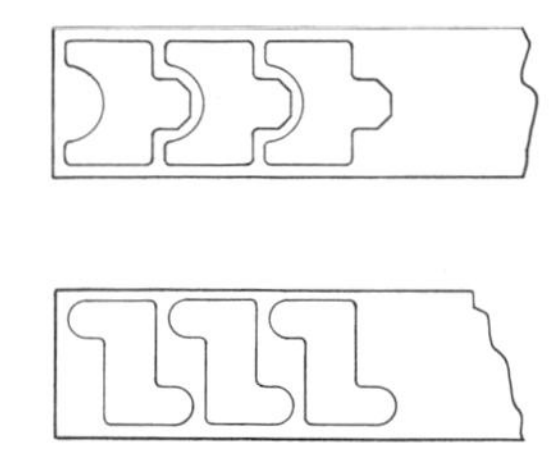

Figure **15·16** Interposed orientation of blanks.

adjacent blank. From Fig. 15·16, it is readily apparent that interposed orientation can be a practical and effective way to achieve efficient utilization of stock material.

SCRAP-WEB ALLOWANCES

Optimum scrap-web sizes vary in accordance with the following:

1. Thickness of the Stock Material. Thicker stock requires wider scrap webs. Therefore, a proportionate relationship exists between the width of the scrap web and the stock thickness. For this reason, the width of the scrap web is customarily expressed in terms of T, where T is the stock material thickness. For thinner gages of stock material, however, it is not practical to attempt to maintain a proportional association between the stock material thickness and the width of the scrap web. Instead, a minimum value is assigned for scrap-web widths in accordance with size, shape, and operational requirements.

2. Length of Scrap Web. Long scrap webs must be made wider than short scrap webs. This is necessary in order to maintain adequate web strength. Since large blanks produce longer scrap webs, it follows that large blanks require wider scrap webs.

3. Shape. The strength of a given scrap web is associated with its contour. Therefore, web contour must be considered in order to determine optimum width for a given scrap web.

4. Type of Operation. Some operational procedures require stronger scrap webs than others. Width requirements of scrap webs are narrowest for single-station dies such as plain blanking dies, compound dies, etc. Wider (stronger) scrap webs are necessary for progressive dies. For gang dies, scrap webs can normally be made the same width as would be required for the equivalent progressive die. Two-pass dies require wider scrap webs than one-pass dies: generally 50 per cent wider.

In the course of time, it has been ascertained that certain different proportions of stock material thickness T may be used to determine adequate scrap widths for various typical blank contours. These proportions are indicated as G and H in Figs. 15·17 and 15·18. It is obvious that this procedure is not suitable for the thinner gages of stock material because the resulting scrap webs would be too narrow. Therefore, the smallest feasible scrap width for each condition is also specified. Scrap webs should not be made narrower than these smallest feasible widths.

The allowances illustrated are intended as a guide to determining practical minimum scrap-web allowances. Each figure illustrates a characteristic contour group and accounts for stock material thickness, blank size, single-station operations, and progressive operations. By judicious association, they can be used to determine optimum scrap-web requirements for most applications.

Figure 15·17 indicates appropriate minimum allowances for scrap webs created by opposing curved edges or curved edges adjacent to straight edges. These contours tend to produce comparatively strong scrap skeletons, permitting the use of relatively narrow scrap webs. The different contours shown in views A, B, and C are the same in principle as far as scrap webs are concerned. Therefore, the allowance is the same in each case.

EXCEPTION:

When radii R (view B) are less than $2T$ or $\frac{1}{16}$ in., whichever is larger, the scrap width should be treated the same as for adjacent sharp corners (see Fig. 15·19).

Figure 15·18 indicates proportions which are suitable for scrap webs whose edges are parallel. These contours tend to leave weaker scrap skeletons, which accounts for the greater scrap-web allowance. For gang dies, use allowance indicated for progressive dies. For gang dies with tandem openings, add 25 per cent. For two-pass dies, add 50 per cent.

Since adjacent sharp corners (Fig. 15·19) create a focus for possible origin of fractures, they necessitate wider scrap webs. To provide a safety factor, scrap webs of this kind are made 25 to 50 per cent wider than would otherwise be required.

WIDTH OF STOCK STRIP

To determine the optimum stock width W (Fig. 15·20) it is necessary first to determine the overall width of the blank or blanks. This width is indicated at F (views A, B, and C). To F add both front and back scrap H. Using the designations shown,

$$W = F + H + H$$

This sum is the minimum required stock-strip width. By fortunate chance, the sum $F + H + H$

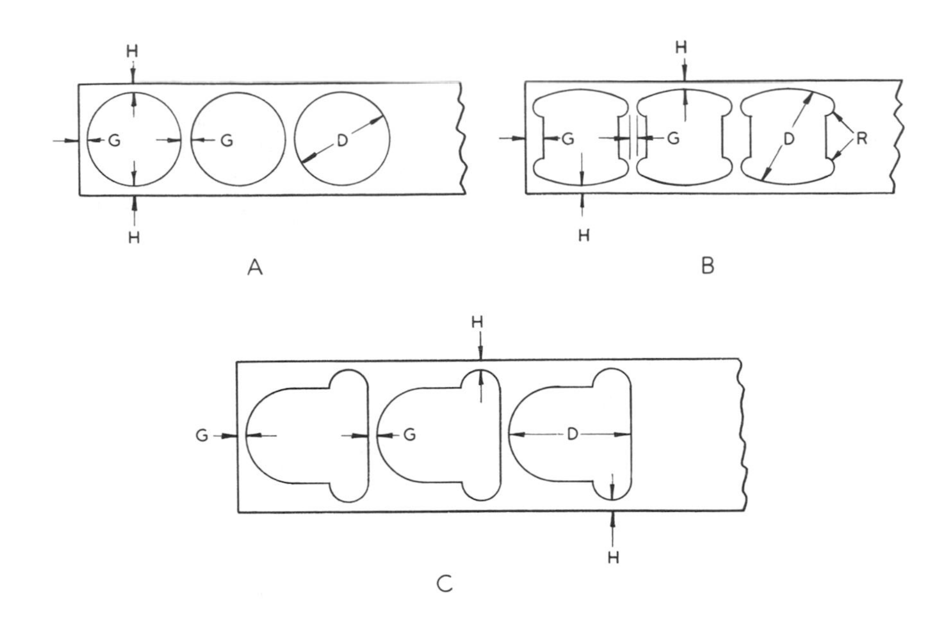

ONE-PASS SINGLE STATION DIES		
T= STOCK THICKNESS		
D (INCHES)	G OR H	SMALLEST G OR H (INCHES)
TO 1	$\frac{3}{4}T$	$\frac{1}{32}$
1-3	T	$\frac{3}{64}$
3-6	$1\frac{1}{4}\ T$	$\frac{3}{32}$
6-10	$1\frac{1}{4}\ T$	$\frac{1}{8}$
10-15	$1\frac{1}{2}\ T$	$\frac{1}{8}$

ONE-PASS PROGRESSIVE DIES		
T= STOCK THICKNESS		
D (INCHES)	G OR H	SMALLEST G OR H (INCHES)
TO 1	T	$\frac{3}{64}$
1-3	$1\frac{1}{4}\ T$	$\frac{1}{16}$
3-6	$1\frac{1}{2}\ T$	$\frac{3}{32}$
6-10	$1\frac{1}{2}\ T$	$\frac{1}{8}$
10-15	$1\frac{3}{4}\ T$	$\frac{5}{32}$

Figure 15·17 Scrap-web allowance between opposed curves or between curves adjacent to straight lines.

may equal a standard strip width. If not, either of two procedures may be followed:

1. Least common practice: Stock is sheared or slit especially to this width.
2. Most common practice: The next wider commercially available stock-strip width is used. It should be mentioned that this practice permits a larger tolerance on the stock-strip width dimension.

ADVANCE DISTANCE

In Fig. 15·20, J represents the overall blank size parallel to the feed direction. E indicates the blank dimension to which the scrap allowance or allowances G must be added in order to determine the required advance distance A.

Adjacent Blanks. Where strip orientation is such that blanks are adjacent (view A), E is equal to J. Therefore, with this kind of orientation

$$A = E + G = J + G$$

Interposed Blanks. With interposal (view B), E is less than J. The advance distance A is not equal to $J + G$. The advance distance

$$A = E + G \qquad \text{only}$$

The advance distance may be measured from any given point in one station to the equivalent point in the next station ($A' = A$). Dimension $K = J - A$. This dimension may be of occasional value if it is desired to know the position of the strip tail end in relation to the blanking position. Strip orientation of the blanks must be decided before the E dimension, and ultimately the advance distance, can be determined.

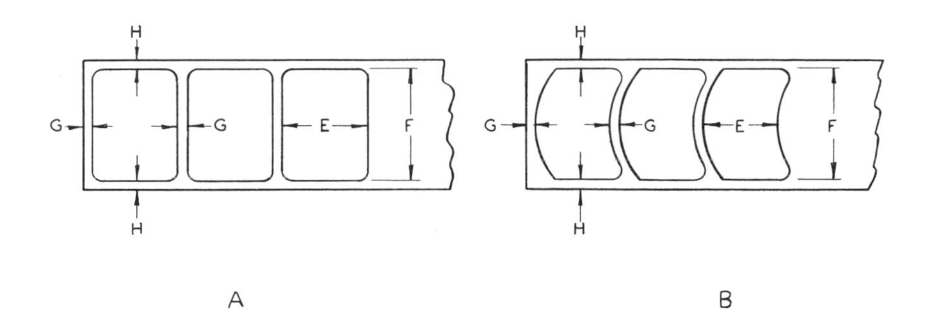

ONE-PASS SINGLE STATION DIES		
T= STOCK THICKNESS		
E OR F (INCHES)	G	SMALLEST G (INCHES)
0-2	T	$\frac{3}{64}$
2-6	$1\frac{1}{4}$ T	$\frac{1}{16}$
6-10	$1\frac{1}{2}$ T	$\frac{5}{64}$
1-15	$1\frac{3}{4}$ T	$\frac{3}{32}$

ONE-PASS PROGRESSIVE DIES		
T= STOCK THICKNESS		
E OR F (INCHES)	G	SMALLEST G (INCHES)
0-2	$1\frac{1}{4}$ T	$\frac{1}{16}$
2-6	$1\frac{1}{2}$ T	$\frac{5}{64}$
6-10	$1\frac{3}{4}$ T	$\frac{3}{32}$
10-15	2T	$\frac{1}{8}$

E OR F	H	SMALLEST H
0-2	$1\frac{1}{4}$ T	$\frac{1}{16}$
2-6	$1\frac{1}{2}$ T	$\frac{5}{64}$
6-10	$1\frac{3}{4}$ T	$\frac{3}{32}$
10-15	2T	$\frac{1}{8}$

E OR F	H	SMALLEST H
0-2	$1\frac{1}{2}$ T	$\frac{1}{16}$
2-6	$1\frac{1}{2}$ T	$\frac{3}{32}$
6-10	$1\frac{3}{4}$ T	$\frac{1}{8}$
10-15	2T	$\frac{3}{16}$

Figure **15·18** Scrap-web allowance for webs with edges parallel.

Interlaid Blanks. In many instances, interlaid blank contours will require that more than one scrap-bridge dimension G be added into the advance distance. The advance distance A in view C includes two scrap bridges. Here, the advance distance is determined

$$A = M + G + E + G + P$$

This is a procedure which can, in principle, be applied to determine the required advance distance for interlaid orientations. Actually, each job is an individual case which must be analyzed and considered according to its peculiar needs. In view C, for example, the advance distance is arbitrarily, but logically, given as a referent of the center line of the E dimension. Thus,

$$M = P = E/2$$

Therefore, in this special instance,

$$A = 2E + 2G$$

When the various required dimensions are added together to determine the advance, the sum often is a nonstandard dimension. It is then customary to round off this figure by raising it to

Figure **15·19** Scrap-web allowance between adjacent sharp corners or between sharp corner and straight edge.

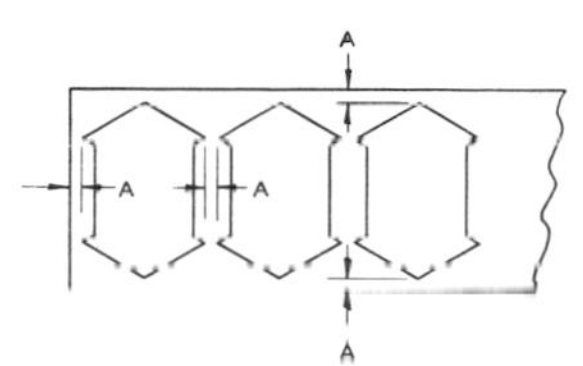

the next higher nominal dimension. One practice, which seems to be rather widely accepted, is to raise the calculated required advance dimension to the next higher multiple of 0.010 in. for advances less than 4 in. For advances of 4 to 8 in., raise the figure to the next higher multiple of 1/32 in. For advances of 8 to 16 in., raise to the next higher multiple of 1/16 in. When the advance is more than 16 in., raise the figure to the next higher multiple of 1/8 in.

This rounding off procedure is intended as a guide to aid in determining a suitable advance dimension. Use of the procedure should be tempered by logic and experience.

STRIP LAYOUT PROCEDURE

To determine any advance distance or any stock-strip width, the blank orientation with respect to the strip must be known. Since the importance of securing optimum orientation cannot be denied, it is axiomatic that this phase of any job must be given thorough consideration.

For simple blanks, the most desirable strip position is usually rather obvious. With complex piece parts, however, considerable work may be entailed in determining optimum orientation. Because of the number of different factors which must be taken into account, the most practical way to determine blank orientation involves cut-and-try methods. To follow this procedure:

1. Determine the flat-blank sizes and contours, accurately performing and recording all necessary calculations including bend allowances, etc.
2. Make a template of the developed flat blank. For this purpose a template made of lightweight cardboard (card stock) will do nicely. In the trade parlance, such templates are called "paper dolls."

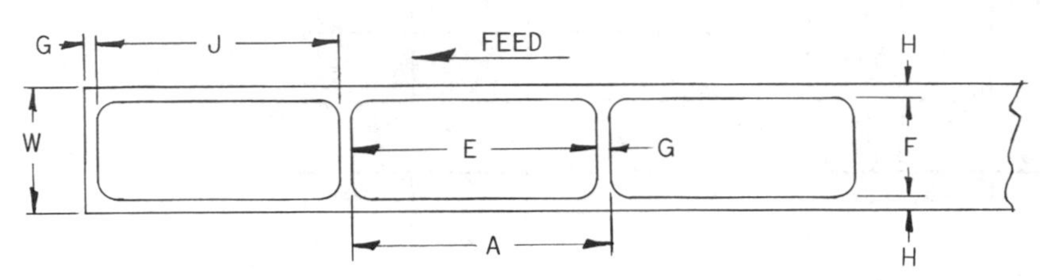

VIEW A: ADJACENT BLANKS

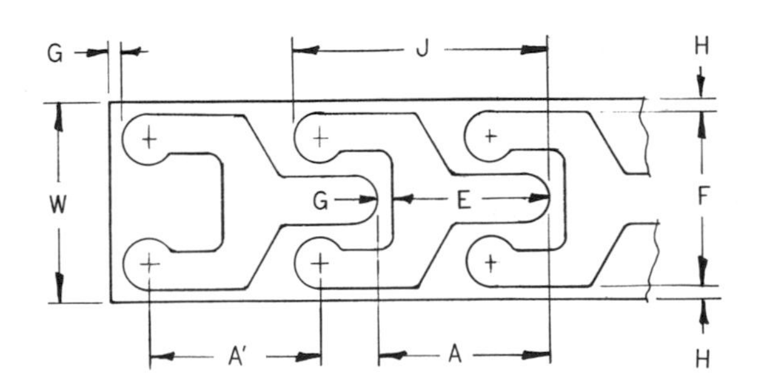

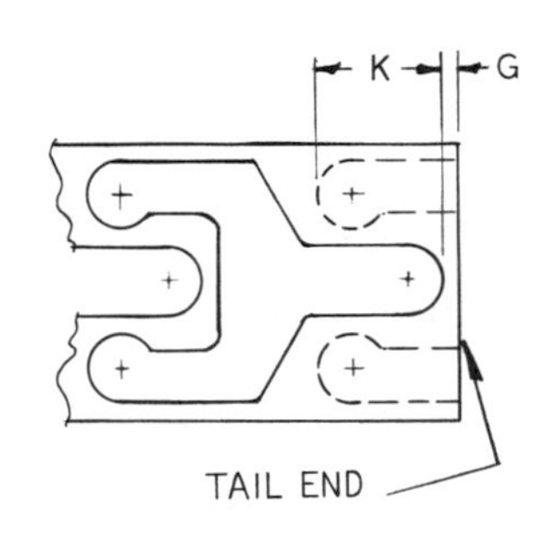

VIEW B: INTERPOSED BLANKS

Figure 15·20 Stock-strip layouts.

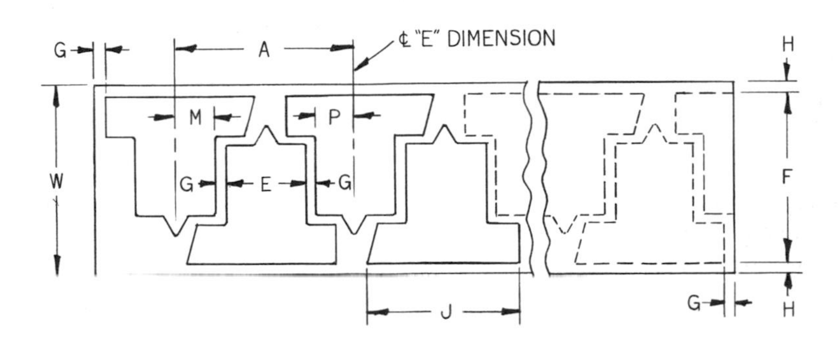

VIEW C: INTERLAID BLANKS

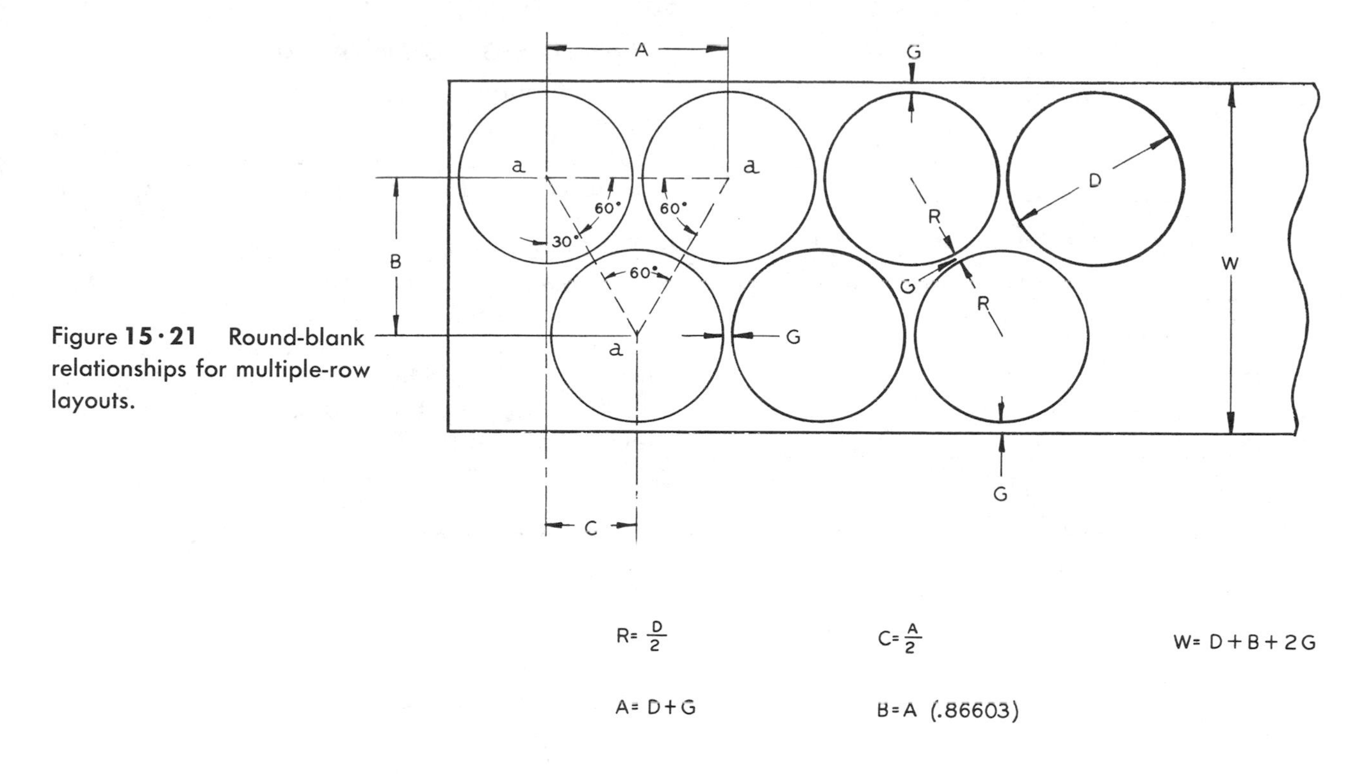

Figure **15·21** Round-blank relationships for multiple-row layouts.

3. Using the template as a pattern, draw the outline of the blank on a piece of paper.
4. Proceeding by visual reference in a trial-and-error fashion, lay the template in various logical positions alongside the pictured outline. When the most favorable relationship is found, trace around the template again to record the relationship.

Repeat steps 3 and 4 as often as necessary. Keep in mind that, excepting tandem-die layouts, all blanks in the same row must have the same relative strip position. Do not reverse the template or turn it over unless this is required for a two-pass die or a gang-die layout. When making a template as described in step 2, it is a good idea to mark the template for burr side and/or grain direction if, and as, required. On occasion, two or more templates may be helpful.

When the relative blank positions have been decided, check for the following:

1. Grain direction
2. Burr side
3. Relationship to a particular side of the stock strip, if required (polished or plated stock, etc.)
4. Die operation
5. Possible undesirable die construction due to the blank orientation

MULTIPLE-ROW ORIENTATION OF ROUND BLANKS

Round blanks are a special case, illustrated in Fig. 15·21. Optimum blank positions, advance distance, and stock-strip width all can be determined by substituting the required values in the relationships shown. These relationships are derived from the fact that the triangle *a-a-a*, created by the centers of three mutually adjacent blanks, is an equilateral triangle.

The relationships shown can be used for layouts having any number of rows. Just remember to add in another *B* dimension for each additional row. That is, for three rows

$$W = D + 2B + 2G$$

for four rows

$$W = D + 3B + 2G \qquad \text{etc.}$$

PARTING OPERATIONS

In relation to scrap allowance, these operations are a special case, illustrated in Fig. 15·22. For all practical purposes, punch strength is the factor which dictates the scrap allowance *G*. Because there can be so many different highly individual situations, it is not feasible to set up a definite rule for determining *G*. However, the illustration depicts two very general situations. These, with their accompanying tables, may be used as an associative means of deciding approximate minimum scrap allowance for parting operations. The figures given assume that the cutoff punches will be heeled into the die blocks. *Note:* Conditions which are not as ideal as those illustrated will require a larger scrap allowance.

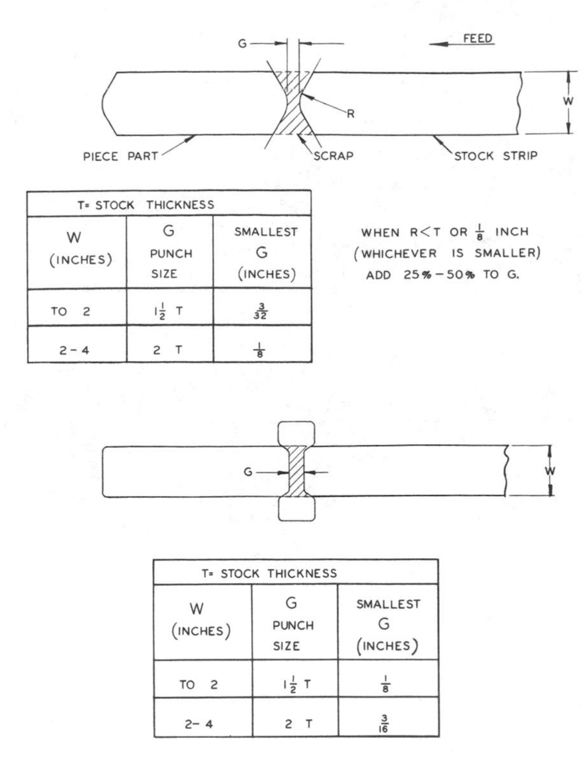

T = STOCK THICKNESS

W (INCHES)	G PUNCH SIZE	SMALLEST G (INCHES)
TO 2	$1\frac{1}{2}$ T	$\frac{3}{32}$
2–4	2 T	$\frac{1}{8}$

T = STOCK THICKNESS

W (INCHES)	G PUNCH SIZE	SMALLEST G (INCHES)
TO 2	$1\frac{1}{2}$ T	$\frac{1}{8}$
2–4	2 T	$\frac{3}{16}$

Figure **15·22** Parting scrap allowances.

SCRAPLESS STOCK-STRIP LAYOUTS

By definition, a scrapless die is one which produces piece parts (or blanks) without producing scrap of any kind. Possibilities for truly scrapless operations are definitely limited. The most common application requires a chopoff die and is pictured in view *A*, Fig. 15·23. Here, the advance distance *A* is the distance from the punch cutting edge to the stop. Length *L* of the blank is equal to the distance from the die cutting edge to the stop, or $A + C$, where $C =$ cutting clearance. Blank width *W* is the same as the width of the stock strip, as furnished. Variations in stock-strip width will cause like variations in blank width.

View *B* is a tandem type of strip layout. Two pieces are produced at each press stroke. One piece (No. 1) is produced by the chopoff method and is ejected to the rear. The other piece (No. 2) is blanked through the die. As shown, the advance distance *A* is measured from the stop to the more remote cutting edge of the punch. Two blanks are included within the advance distance. Overall width of the blanks is the same as the stock-strip width *W*. Left-to-right dimensions of blank No. 1 are controlled by the relationship of the stop to the near cutting edge of the punch. Left-to-right dimensions of blank No. 2 are derived from the die opening, as with a conventional blanking die.

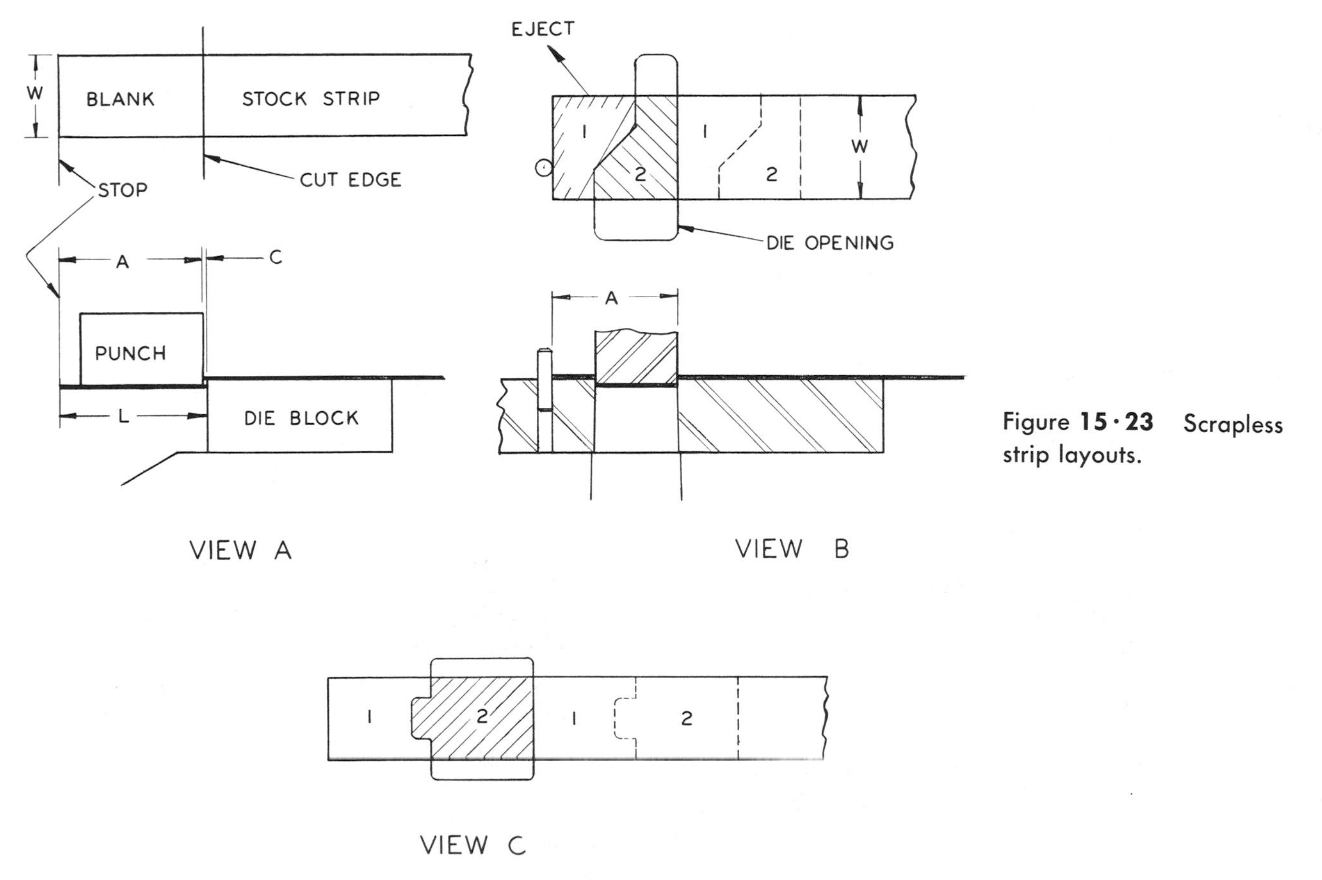

Figure **15·23** Scrapless strip layouts.

View C depicts the production of dissimilar blanks by the scrapless method. Note the resemblance in principle between this view and the parting operations in Fig. 15·22.

Semiscrapless stock strips. This term is most commonly used to describe strips where the blank orientation is such that there is no front scrap, back scrap, or scrap bridge. As pictured in Fig. 15·24, scrap may be present in the form of slugs which are a result of piercing, notching, and/or trimming operations. Unlike scrapless dies, which are inherently limited in scope, semiscrapless procedures may be adapted to a wide range of applications.

DETERMINING STOCK MATERIAL REQUIREMENTS

To determine accurately the amount of material required to produce a given blank, it is necessary to know the advance distance and the stock-strip width. In many cases it will be necessary to make one or more strip layouts in order to ascertain these dimensions. Once these dimensions have been determined (see Fig. 15·25), then the area required per blank is equal to the advance times the width divided by the number of passes or by the number of blanks per stroke, whichever is appropriate. This may be expressed as

$$M = \frac{AW}{N}$$

Figure **15·24** Semiscrapless strip layouts.

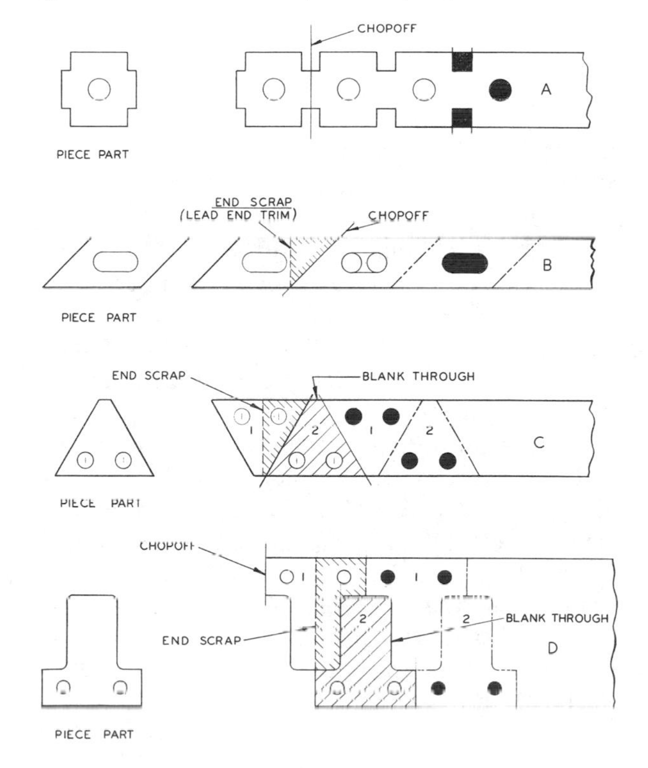

where M = area of material required per blank
A = advance distance
W = stock-strip width
N = no. passes (multiple-pass operations)
or N = no. blanks per stroke (gang dies)

To calculate minimum strip length L for a given number n of blanks,

single-row strip $$L = \left(\frac{A}{N}n\right) + G$$

multiple-row strip $$L = \left(\frac{A}{N}n\right) + \frac{A}{2} + G$$

To calculate weight of material required per blank,

$$P = MTC$$

where P = weight of material required per blank
M = area required per blank
T = stock material thickness
C = unit weight of material

For most purposes, the following unit weights may be used:

Aluminum	0.095 lb per cu in.
Brass	0.310 lb per cu in.
Copper	0.320 lb per cu in.
Steel	0.283 lb per cu in.

Different alloys have different unit weights. If a more accurate weight is desired for a specific material, it can be secured from the material suppliers.

Figure **15·25** Relationships for calculating area of material required per blank.

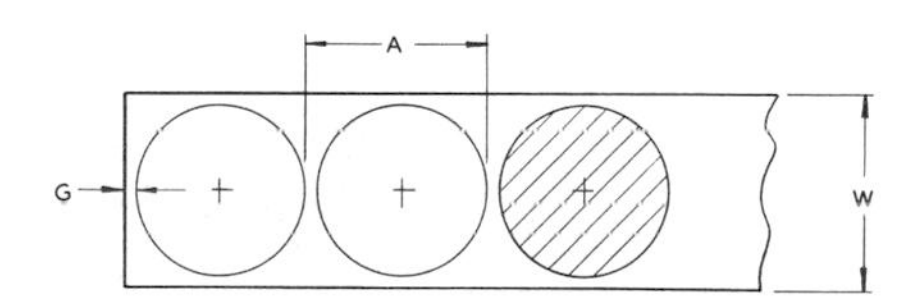

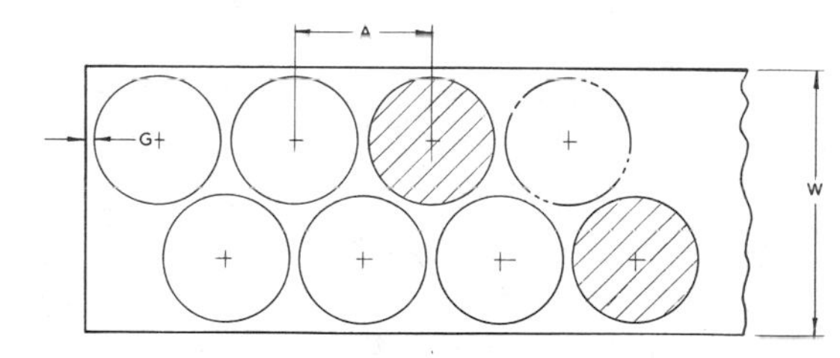

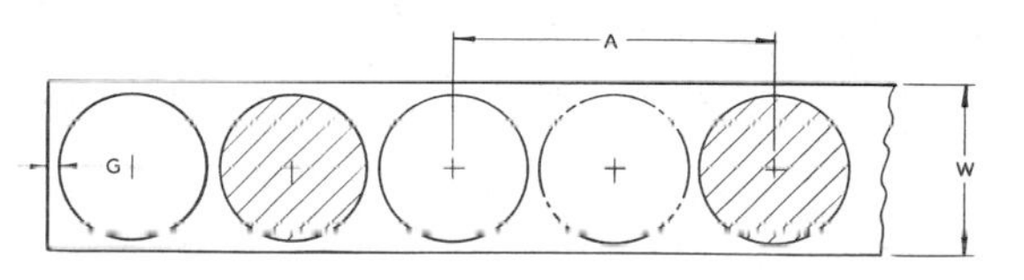

COMPARATIVE STRIP LAYOUTS

A piece part is sketched in Fig. 15·26. Accompanying the piece part is a sketch showing the required flat blank.

It is often necessary to make two or more strip layouts in order to determine, by comparison, the optimum method for a given job. In the case of the illustrated piece part, five comparison layouts are shown in Fig. 15·27. So many layouts would not normally be required. They are used here in order to provide a wide range for comparison and to accentuate the fact that a number of different orientations may be possible for a given blank. Also emphasized is the fact that variations in orientation can have a pronounced effect upon the amount of stock material required per blank.

Strip A: Parting Method. Advance distance was determined in the following manner: Consulting Fig. 15·22 yielded a minimum G dimension of $1\frac{1}{2}T$, or $\frac{3}{32}$ in. However, since the punch configuration required is less favorable than that shown, the next higher minimum allowance, $2T$ or $\frac{1}{8}$ in., was chosen. This yielded a tentative G dimension of 0.125 in. Adding this to the blank length, 2.972 in., gave a minimum advance distance of 3.097 in. Rounding this off to the next higher multiple of 0.010 in. resulted in the advance distance of 3.1 in., as shown. Therefore, the final G dimension is the difference between 3.1 and 2.972 in.

$$3.1 - 2.972 = 0.128 \text{ in., punch size}$$

Applying the formula $AW/N = M$ results in $(3.1 \times 1.000)/1 = 3.1$ sq in. of stock material, per blank, required for this method.

By examination, the strip layout reveals that this would not be a practical method for the following reasons:

Figure **15·26** Piece part and developed flat blank.

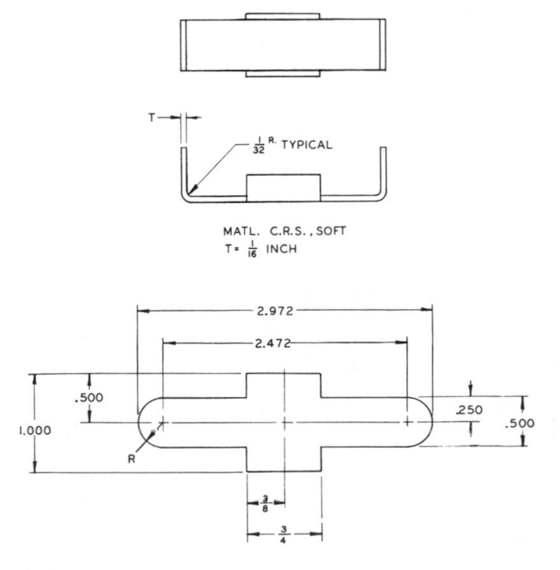

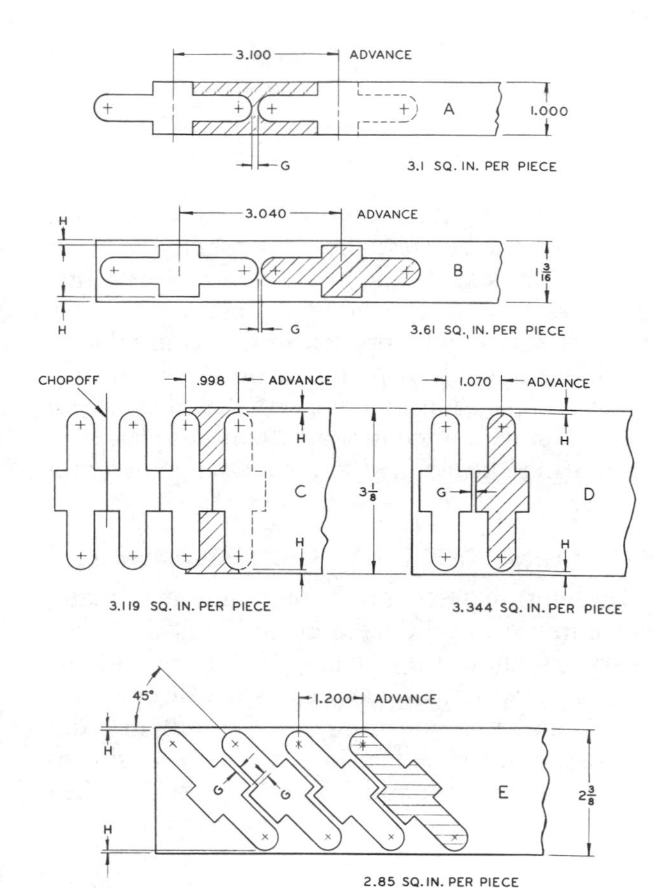

Figure **15·27** Comparative strip layouts.

1. The 1.000 dimension depends upon the stock-strip width.
2. This layout has the weakest die proportions of all the layouts shown, because of the relatively long projections required.

In short, this blank is not well suited to production by the parting method.

Strip B: Blanking Die, Narrow Run. To find stock width: From Fig. 15·18, $H = 1\frac{1}{2}T$, which is $\frac{3}{32}$ in. Then, $2H$ + blank width = $1\frac{3}{16}$ in. stock width. To find advance distance: From Fig. 15·17, $G = T$, which in this case is $\frac{1}{16}$ in. Blank length 2.972 + 0.0625 = 3.0345. This is rounded off to 3.040. Stock material requirement per blank is 3.61 sq in.

With this method the die will be structurally sound and economical to make. The blanks will be dimensionally accurate. The major objection is the comparatively high stock material consumption. The fact that the blank is run the long way may be a disadvantage as far as feeding is concerned. Unless special considerations were present which are not evident here, this layout is inferior to strips *D* and *E*.

Strip C: Notch and Chopoff Die. Referring to

Fig. 15·17, $H = T$. Then, for strip C, $H = 1/16$ in. Thus, minimum stock width is $0.0625 + 2.972 + 0.0625 = 3.097$ in. Raising this figure to the next higher nominal size gives a stock width of 3⅛ in. The advance distance is based upon a cutting clearance of 0.002 in. for the chopoff. (To avoid confusion, the possibility of skidding is being ignored here.) Stock material consumption in this case is 3.119 sq in. per blank.

This die would be stronger than the parting die, but not as strong as the blanking dies. Diemaking costs will be relatively high. It would be hazardous and time-consuming to attempt to feed the tail of the strip through the die, which entails considerable waste for any stock other than coil stock. The 1.000-in. blank dimension will be more accurate than with the parting die, but not as accurate as with the blanking dies. A matched cut is required for the rounded ends of the blank, which makes the die construction more exacting. The matched cut may, at times, be objectionable from the standpoint of piece-part appearance or function. This would not be a desirable method for producing this particular blank.

Strip D: Blanking Die, Wide Run. Stock width is same as previous strip. To find advance distance: For a tentative G allowance, T was chosen from Fig. 15·18, because in this case the adjacent scrap-bridge edges are only ¾ in. long. Since $T = 1/16$ in., 1.000 in. + 0.0625 in. = 1.0625 in. min. advance. This is rounded off to 1.070 in., which results in a stock material requirement of 3.344 sq in. per blank. Comparing this layout to strip B emphasizes the normal difference in stock material consumption for these two methods.

This die would be structurally sound and economical to make. Diemaking costs would be approximately the same as for layout B. However, in addition to low initial die cost, the die in layout D offers more efficient stock material utilization and ease of operation.

Strip E: Blanking Die, Angled Positioning. Here, positioning the blank at an angle produces an interpositioned blank orientation. This results in a considerable saving in stock material. Like dies B and D, this die would be structurally sound. However, a die of this kind will generally cost more to design and build than the dies represented by strips B and D. The greater die cost must be balanced against the material saving in order to determine the best method for a given situation.

In the illustrated case, the die required is a simple single-station blanking die. The additional die cost need not be large: 10 to 15 per cent more than for strip D would be average, if the die were properly designed. Caution! This would not hold true for a complex progressive die. In the latter case, initial tool costs have been known to increase 50 per cent and more as a result of angular blank orientation. Such an instance necessitates very high-production requirements in order to absorb the added die cost. Maintenance costs will also be proportionately higher for an angulated die, if the die is at all complex.

Deciding which layout to use. Actual determination of the optimum method for a given case is the function of the tool engineer or designer. As far as a choice between the layouts shown in Fig. 15·27 is concerned, the selection would be either strip D or strip E, depending upon the difference in tool cost balanced against the difference in cost of stock for the production required. This assumes, of course, that there are no factors involved other than those which have been here presented.

To choose between strips D and E, the diemaker must calculate the actual saving in stock material. To do this, simply determine the weight of the difference in material consumption and multiply by the cost per unit of weight.

EXAMPLE:

Strip D requires 3.344 sq in. per blank.
Strip E requires 2.85 sq in. per blank.
3.344 − 2.85 = 0.494 sq in. difference.
Using the formula given earlier in this section,

$$P = MTC$$

In this case
P = weight, lb
M = 0.494 sq in.
T = 1/16 in.
C = 0.283 lb per cu in.

Then $P = (0.494)\,(1/16)\,(0.283)$
$= 0.00874$ lb difference per blank.

For an accurate determination, P must be multiplied by the exact cost per pound. However, as an example, arbitrarily assume a price of 20 cents per pound. Then

20(0.00874) = 0.17480 cents per blank, difference

Rounding this figure off to 0.175 cents, the difference would be $1.75 per thousand blanks. Multiply $1.75 by the production requirement in thousands and compare the result to the estimated difference in die cost.

The kind of stock material can make significant changes in the die-cost to material-cost ratio.

EXAMPLE:

Suppose the stock material to be a copper alloy

whose weight per cubic inch is 0.31 lb and whose price is 90 cents per pound. Then

$$\begin{aligned} P &= MTC \\ &= (0.494)\,(1/16)\,(0.31) \\ &= 0.0096 \text{ lb, difference} \\ (0.0096)90 &= 0.864 \text{ cents per blank, difference} \end{aligned}$$

which yields a difference of \$8.64 per thousand blanks. In this case a far lower production requirement can justify increased die costs to effect savings in stock material consumption.

It is true that the average diemaker may not be required to analyze production costs. However, it cannot be denied that comprehension of the general relationship between die construction and production costs is a decided asset to any diemaker, especially for the diemaker who hopes to be selected for advancement to more demanding positions, such as supervisor, master mechanic, etc.

DERIVATION OF ANGLED BLANK ORIENTATION

If the blank position angle is dictated by a grain-direction requirement of the piece part, the angle is considered to be predetermined and the stock-strip layout is made to conform. On occasion, if the grain requirement is unfavorable because of die construction or stock material consumption considerations, it may be possible to secure a change in the piece-part specifications. This possibility should be investigated and disposed of at the time the strip layouts are in process.

Assuming that the blanking angle is not predetermined, then the angle should be made such that it achieves maximum economy of stock material. To derive the most effective angle for stock material economy, place two blanks in their most favorable proximal relationship, keeping their equivalent axes parallel. Then, connect any convenient point on one blank to the equivalent point on the next blank. The line of connection will be parallel to the feed direction. Length of this line will be equal to the advance distance. This procedure is exemplified in Fig. 15·28, which illustrates the derivation of the blanking angle for strip layout *E*, Fig. 15·27.

PROCEDURE:

(Fig. 15·28, view *A*) Locate blank in first position. The first position should permit positive and easy reference to a predetermined condition. Here, axis *bb* is positioned horizontally to facilitate reference.

Determine second position by trial and error, keeping blank axes parallel. The second position should produce the optimum proximal relationship to the first position. In this instance, the relationship is determined by dimensions *X* and *Y*.

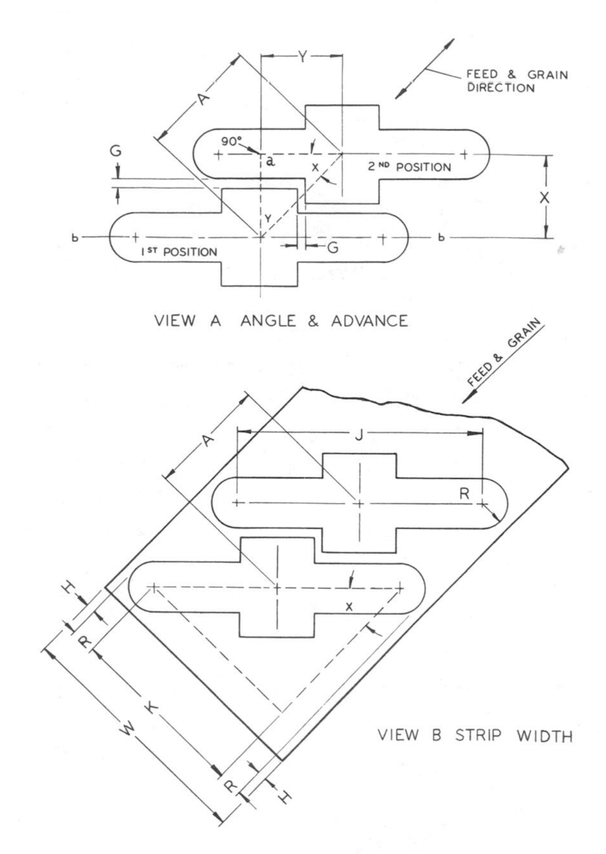

Figure **15·28** Derivation of angled strip layout.

To calculate advance distance and blanking angle (dimensions are inches)

$$X = 0.500 + G + 0.250$$
$$Y = 3/8 + G + 3/8$$

Because of the zigzag nature of the scrap bridge, a tentative allowance of $G = 1\tfrac{1}{2}T$ is chosen. Therefore,

$$\begin{aligned} X &= 0.500 + 0.094 + 0.250 = 0.844 \\ Y &= 0.375 + 0.094 + 0.375 = 0.844 \\ X &= Y \quad \text{in this case} \end{aligned}$$

Since $X = Y$, and angle $a = 90°$, then angle $x = 45°$. It follows that

$$\begin{aligned} A &= Y(\sec x) \\ &= (0.844)\,(1.4142) \\ &= 1.1936, \text{ min. required advance} \end{aligned}$$

Rounding off,

$$A = 1.200$$

Since $X = A(\sin x)$ and $Y = A(\cos x)$,

$$\begin{aligned} X \text{ or } Y &= 1.2(0.70711) \\ &= 0.8485, \text{ to suit adjusted advance} \end{aligned}$$

This results in an actual final scrap allowance of G equal to 0.0985 in.

Stock-strip width

$$H - T - 0.0625, \text{ tentatively}$$

$$J(\sin x) = K$$

$$2.472(0.70711) = 1.747976$$

Rounding off,

$$K = 1.748$$

$$H + R + K + R + H = W \quad \text{min.}$$

$$0.0625 + 0.250 + 1.748 + 0.250 + 0.0625 = 2.373, \text{ min.}$$

$$W = 2\tfrac{3}{8} \begin{array}{l} +\tfrac{1}{32} \\ -0 \end{array} \quad \text{in.}$$

Note the tolerance for W of plus $\frac{1}{32}$ in., minus 0. The tolerance was assigned in this manner in order to avoid strip widths which would be less than the minimum calculated dimension for W.

COMPARISON LAYOUTS: TANDEM, TWO-ROW, AND ANGLED

Figure 15·29 is not meant to convey the impression that there is any one method which will be most efficient in all cases. The illustration is meant to emphasize the fact that due consideration must be given to each and every individual case.

View A. The finished piece part is shown accompanied by its flat blank layout, which includes the necessary dimensions. The grain direction is specified 0 to 45°, indicating that, to be satisfactory, the bend must be made only within this given angular range. The presence of the grain specification restricts blank orientation possibilities to those which will satisfy the specification.

View B: Tandem Layout. It is obvious, by examination, that this strip is not truly efficient as far as stock material consumption is concerned. The large area of wasted material is apparent at a glance.

Dimensions are inches. Because of tandem orientation, minimum $G = 1\frac{1}{2}T$; minimum $H = 1\frac{3}{4}T$.

$$\begin{aligned} \text{Width } W &= H + 1.374 + H \\ &= 0.109 + 1.374 + 0.109 \\ &= 1.592, \text{ min.} \\ &= 1\tfrac{5}{8} \text{ in. standard stock-strip width} \end{aligned}$$

Advance

$$\begin{aligned} A &= 0.218 + G + 1.061 + G + 0.218 \\ &= 0.218 + 0.094 + 1.061 + 0.094 + 0.218 \\ &= 1.685, \text{ min.} \\ &= 1.690, \text{ adjusted} \end{aligned}$$

Material requirement, square inches per blank,

$$M = \frac{AW}{N}$$

where M = area material required per blank
N = no. blanks per stroke

$$\begin{aligned} M &= \frac{(1.69)1.625}{2} \\ &= 1.373 \text{ sq in. per blank} \end{aligned}$$

View C: Two-row Gang Layout. $G = 1\frac{1}{2}T$, $H = 1\frac{3}{4}T$, minimums.

Width

$$\begin{aligned} W &= H + 1.374 + G + 0.312 + H \\ &= 0.109 + 1.374 + 0.094 + 0.312 + 0.109 \\ &= 1.998, \text{ min.} \\ &= 2 \text{ in. standard stock-strip width} \end{aligned}$$

Advance

$$\begin{aligned} A &= 0.218 + G + 0.156 + 0.687 \\ &= 0.218 + 0.094 + 0.156 + 0.687 \\ &= 1.155, \text{ min.} \\ &= 1.160, \text{ adjusted} \end{aligned}$$

Material requirement

$$\begin{aligned} M &= \frac{(1.16)2}{2} \\ &= 1.16 \text{ sq in. per blank} \end{aligned}$$

Figure **15·29** Comparison layouts.

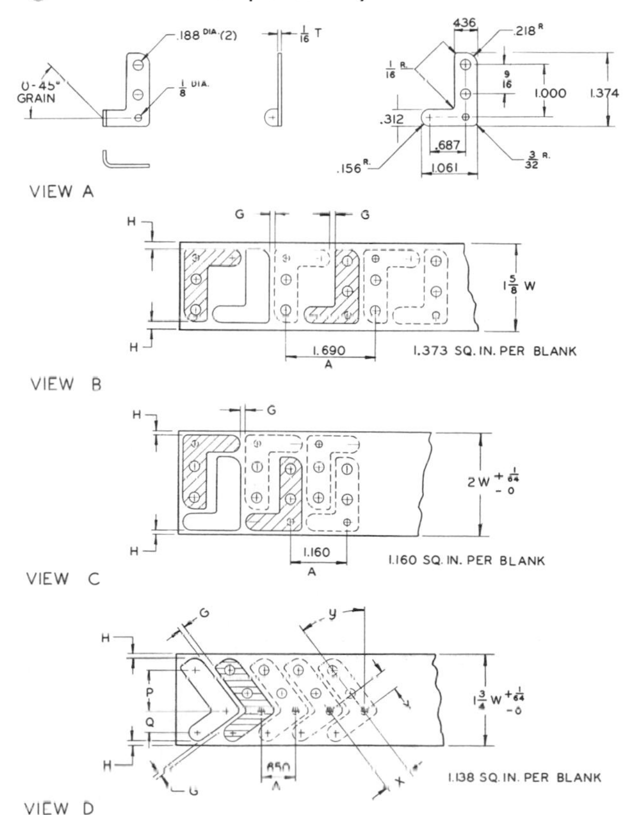

View D: Angled Layout. $G = 1\frac{1}{4}T$, $H = 1\frac{1}{4}T$, minimums. Determine X, Y, and angle y, then advance distance A, and finally stock width W.

$$\begin{aligned} X &= 0.218 + G + 0.218 \\ &= 0.218 + 0.078 + 0.218 \\ &= 0.514, \text{ min.} \end{aligned}$$

$$\begin{aligned} Y &= 0.156 + G + 0.156 \\ &= 0.156 + 0.078 + 0.156 \\ &= 0.390, \text{ min.} \end{aligned}$$

Angle y

$$\text{Tan } y = \frac{Y}{X} = 0.7587548$$

$$\begin{aligned} y &= 37°11'23'' \\ &= 37° \qquad \text{rounded off} \end{aligned}$$

Advance

$$\begin{aligned} A &= X(\sec 37°) \\ &= 0.6436 \\ &= 0.650, \text{ adjusted} \end{aligned}$$

Strip width W

$$\begin{aligned} P &= (\cos 37°)1.000 \\ &= 0.7986, \text{ round off to } 0.799 \\ Q &= (\sin 37°)0.687 \\ &= 0.4135, \text{ round off to } 0.414 \end{aligned}$$

Figure **15·30** Suggested blanks for strip layout practice problems.

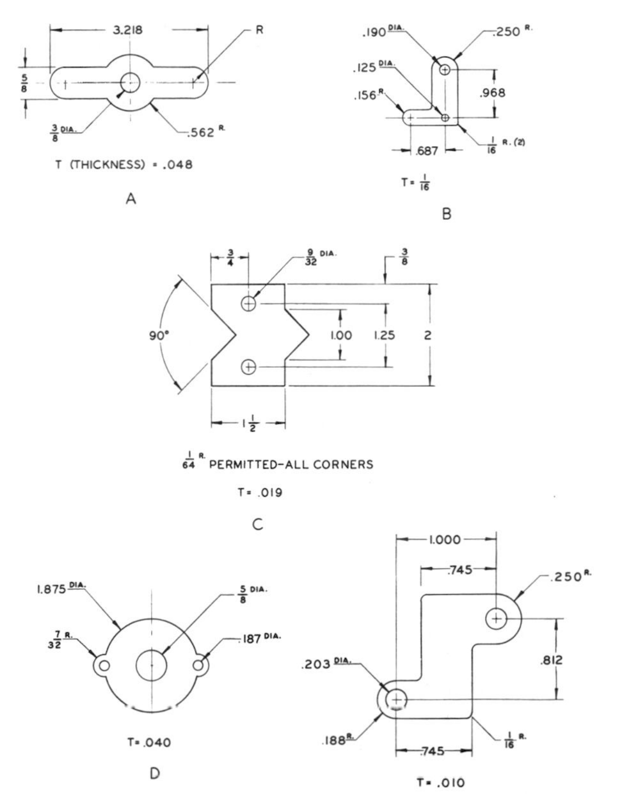

Width

$$\begin{aligned} W &= H + 0.218 + P + Q + 0.156 + H \\ &= 0.078 + 0.218 + 0.799 + 0.414 + 0.156 + H \\ &= 1.743, \text{ min.} \\ &= 1\tfrac{3}{4} \text{ in.} \end{aligned}$$

Material requirement

$$\begin{aligned} M &= \frac{(1.75)0.65}{1} \\ &= 1.138 \text{ sq in. per blank.} \end{aligned}$$

PRACTICE LAYOUTS

The know-how of strip layouts can only be acquired through familiarity resulting from practice. As a start in this direction, make comparison strip layouts of the piece parts illustrated in Figs. 15·30 and 15·31. Follow through completely on each layout, determining advance distance, stock width, and material consumption. When the layouts are complete, consider each method from the standpoint of die construction, die maintenance, and facility of die operation in production. Assign different production quantity requirements (for example, 10,000 parts, 100,000 parts and 1,000,000 parts) to see how the different requirements might affect the choice of strip layouts. Do not jump to conclusions; be sure to make thorough comparisons.

The best way to acquire judgment and familiarity with blank orientation possibilities is to secure piece-part drawings or stampings and make comparison strip layouts from them. If possible, compare your results to the actual production procedures and/or engage in discussions concerning the subject of strip layouts.

Figure **15·31** Suggested piece part for strip layout problem.

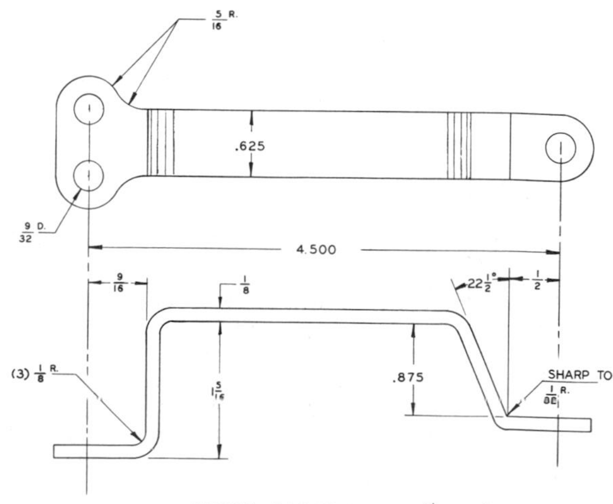

CHAPTER

16

DIE SETS

While the exact date of manufacture of the first die set is not a matter of record, the year 1920 is approximately correct. Die sets were introduced by mass production manufacturers such as automotive, appliance, and business machine companies.

The die set is one of the basic tools of the stamping industry, and while not a glamour product, such as numerical controls, electrical discharge machines, and exotic metals, it is as vital to metalworking as the tool bit, drill, reamer, and grinding wheel.

The industry which manufactures die sets commercially comprises six national and eight to ten regional companies whose overall volume is relatively small in contrast to the importance of its product to metalworking mass production. The die-set industry operates through assembly branches or distributors on a national scale or locally with independent special die-set manufacturers on a small scale. It offers a tremendous variety of sizes and types of die sets, which are assembled to order as well as offered off the shelf like cutting tools or abrasives.

One of the best sources of die-set data is a die-set supplier's catalog. If possible, secure one of these catalogs to use for reference in conjunction with this discussion of die sets.

ROLE OF DIE SETS IN INDUSTRY

Definition and description of a die set. A die set can be defined as a subpress unit consisting of a lower shoe and an upper shoe, together with the guideposts and bushings by means of which the shoes are aligned.

Figure 16 • 2 pictures one of the most commonly used die-set types: a standardized back-post die set. Its salient features are enumerated in the Fig. 16 • 3 legend. The punch and die components are mounted on the inner surfaces of the die set to complete the finished punch-press tool. Occasionally, a component is secured to a sidewall of one or both of the shoes. Such cases usually require special machining of the sidewall to provide a mounting surface. The exploded drawing (Fig. 16 • 3) depicts the individual components of the die set.

Purpose of die sets. The purpose of a die set is to unitize the entire die assembly. Some of the advantages realized by assembling die components to a properly selected die set are:

1. *Accuracy of Setup.* The die can be installed in the press as a self-contained unit, assuring proper alignment of the various punch and die members.

2. *Improved Piece-part Quality.* The quality of the work produced is enhanced by the assured setup accuracy.

3. *Increased Die Life.* This is a result of proper alignment.

4. *Minimum Setup Time.* Setup time is kept to the minimum because the die is installed as a unit.

5. *Facilitation of Maintenance.* Die components can be removed and reassembled without disturbing their relationship to each other. Cutting components can be sharpened in assembly, as units, without removing them from the die set. This can be a distinct advantage over removing the components and sharpening them as separate pieces.

6. *Alignment of Punch and Die Members.* A die set can be a means of keeping the punch and die members properly aligned during the working process. However, a die set cannot be expected to compensate for a punch press which is not in good condition. Neither should a die set be expected to operate satisfactorily if heavy, unbalanced work forces exist. Such loads should be compensated for in the design of the die; they should not be transferred to the guideposts and bushings of the die set.

7. *Facilitation of Storage.* On completion of the production run, the die can be stored as a unit ready to be placed in production again immedi-

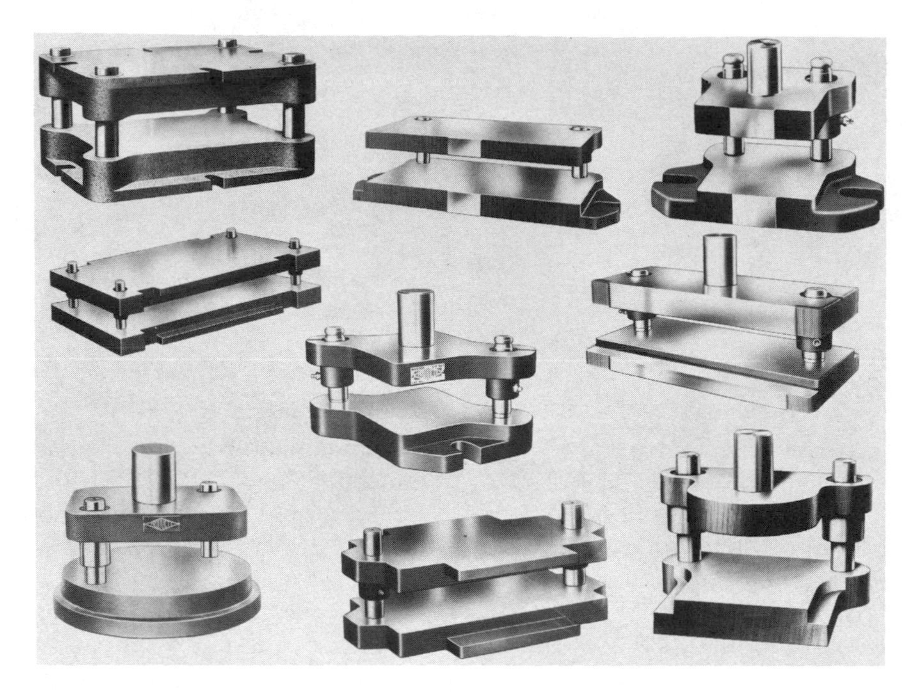

Figure 16·1 Some typical standardized die sets.

ately (this is discussed under "Die Inventory Control" below).

Size range of die sets. Possible die-set sizes are limited only by the facilities of the manufacturer. Die sets are made as small as 2 in. by 2 in. They can also be made in sizes up to 15 ft by 8 ft and larger, if necessary. The photograph (Fig. 16·4) gives an idea of the size range of die sets.

Industry uses and applications

1. Primarily for metalworking in "captive" press rooms and contract job shops.
2. Also used in plastics, die casting, paper and fiber, and other industries.

TERMINOLOGY PERTINENT TO DIE SETS

The following terms are either directly pertinent or closely related to die sets:

die shoe: The die-set base is called the die shoe (or die holder). This remains true even though punches are sometimes mounted on this lower shoe. The great majority of standardized die sets have the guideposts mounted in the die shoe.

punch holder: The die-set top member is called the punch holder (or punch shoe). This remains true even though die blocks are sometimes mounted to this top member. The great majority of standardized die sets have the guide bushings mounted in the punch holder.

shank: Most punch holders in the smaller sizes are made with a shank which fits the clamping hole in the lower end of the punch press ram. The shank is used to center the die set in the press and to secure the punch holder to the ram. These shanks are generally an integral part of the punch holder. For semisteel punch holders, the shank is cast integrally as part of the punch holder. For steel punch holders, the shank is welded in place.

Instead of having an integral shank, the punch holder may be provided with an inserted shank. Standard inserted shanks are commercially available. These are screwed into a threaded hole which is provided in the

Figure 16·2 Typical standardized back-post die set.

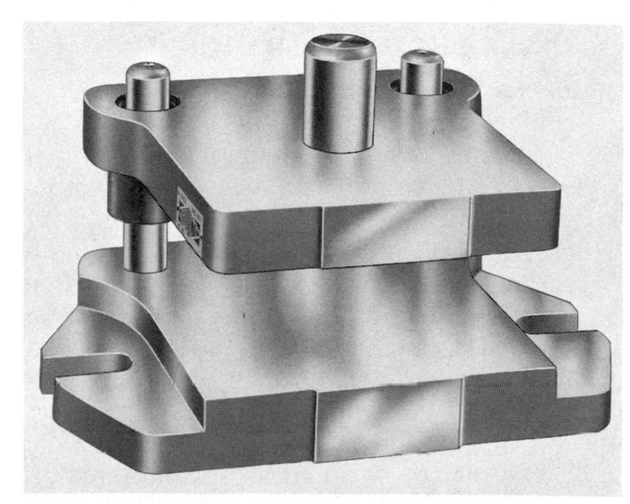

punch holder. Inserted shanks should be keyed to prevent possible rotation.

Shanks are not ordinarily provided on larger die sets. If a shank is provided, it is employed only to center the die in the press, not to secure the punch holder to the ram. Instead, the punch holder is bolted or clamped to the ram for security.

guideposts: Guideposts (also called leader pins or guide pins) are cylindrical pins which provide a means of alignment for the die set.

guidepost bushings: These are installed in the opposing shoe and engage the guideposts with a close sliding fit. The guideposts and bushings, acting together, align the die set.

flange: A ledge which is flush with the bottom surface of a die shoe or the top surface of a punch holder. The ledge extends beyond the die area to provide a means for clamping the shoe member to the bolster plate or press ram, whichever is appropriate. Flanges are more commonly provided on die shoes than on punch holders.

die area: The area available on the top surface of the die shoe and the lower surface of the punch holder for the mounting of punch and die components. The die-set guideposts and bushings are normally located outside this area.

shut height of die: The distance from the bottom of the die shoe to the top of the punch holder when the die is in its closed working position. Overall height of the guideposts must always be an adequate amount less than the shut height in order to ensure that the ram will not strike against the ends of the guideposts. For cutting dies, if possible, the guideposts should be short enough to accommodate the total amount that the shut height will be lowered because of sharpening.

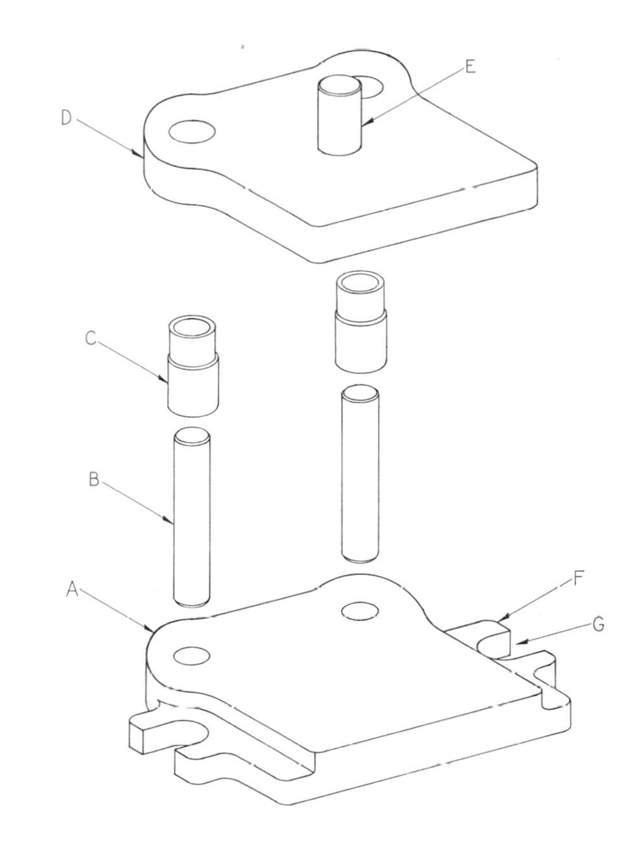

A—Die shoe. Die shoes are also referred to as die holders.

B—Guidepost (sometimes called leader pin and sometimes called guide pin).

C—Guidepost bushing.

D—Punch shoe. Punch shoes are also referred to as punch holders.

E—Shank.

F—Flange.

G—Bolt slot.

Figure **16·3** Components of typical die set.

Figure **16·4** For size comparison, a small back-post die set is shown placed on top of a large four-post die set. The fully extended 8-ft tape rule serves to indicate the actual sizes of these die sets.

MATERIALS FROM WHICH DIE SETS ARE MADE

1. Semisteel—a gray iron or cast iron containing 10 to 25 per cent steel.
2. Hot-rolled boiler plate—1018 to 1026 SAE steel.
3. A combination of platens, one a casting and the other steel.
4. Aluminum, magnesium, or special alloys as well as soft, semihardened, or hardened tool steels.

IMPROVEMENTS IN DIE-SET DESIGN THROUGH THE YEARS

In the early days of mass-produced stamped parts, tolerances and alignment requirements for dies making crude blanks and forms were not critical. The early die set was an open set which had no built-in guidance or alignment means and thus depended on the accuracy of the press itself to maintain alignment. It was necessary either to have accurate presses for close-tolerance work, or to compromise on the quality and uniformity of the parts produced. Through an evolutionary process the modern die set came into being.

The first die set to be self-aligning was made with two cast or steel ground plates which were strapped together and through-bored for two or more pins. Pins were pressed in one plate and were guided through the holes in the second plate, which were lapped for a sliding fit. Later, bosses were cast on the upper plates (cast sets) or were welded (in the case of steel sets) to give added length to the bored holes, thereby improving alignment. Sometimes oversize holes were made in the upper moving plate and the guide pins run through a poured babbit or other type of soft liner. This helped reduce friction on the guide pins and reduced scoring and galling, thus keeping the die set accurate and free-working longer.

The next improvement was the development of inserted bushings, usually pressed in with an interference fit and lapped to fit the guide pins. This, basically, is the die set now in use. With these improvements which made alignment of tooling the job of the die sets rather than the press, more critical sizes and tolerances could be held on the parts produced.

TYPES AND STYLES OF DIE SETS

Die sets are generally classed in two categories, catalog (standard sizes and shapes) and special.

Catalog sets. Advances in die-set design have brought about an almost endless choice of catalog types and styles within a limited size range (25 by 14 in. die space) which are produced by commercial manufacturers of die sets and stocked for fast assembly with pins and bushings. In addition to these catalog sets, die-set manufacturers can produce any practical type and size of die set the die designer specifies, usually at lower cost than would result from in-plant manufacture.

The most commonly used die-set styles are:

1. Back Post (Rear Pin). Two guideposts located toward the back of the set (see Figs. 16·5 and 16·6).

2. Center Post (Center Pin). Two guideposts at the left and right sides of the set on the shank center line (see Figs. 16·7 and 16·8). The two guideposts should differ in diameter in order to foolproof the set.

3. Diagonal Post (Diagonal Pin). Two guideposts, one located at the right rear corner and the other at the left front corner (see Fig. 16·9). These sets can be supplied with reversed guidepost locations. The two guidepost diameters must differ in order to foolproof this type of die set also.

4. Four Post (Four Pin). One guidepost located at each of the four corners of the die set (see Fig. 16·10). One guidepost center must be offset for foolproofing.

Special sets. A special die set is one which differs in any way from the standard catalog specifications. Special die sets are made to order. They may be very similar to catalog sets, or they may be radically different. The sets shown in Figs. 16·11 and 16·12 are special die sets.

Special die sets may have pockets, slots, or cutout areas. These may be roughed in or completely finished by the die-set manufacturer.

Figure **16·5** Back-post die set.

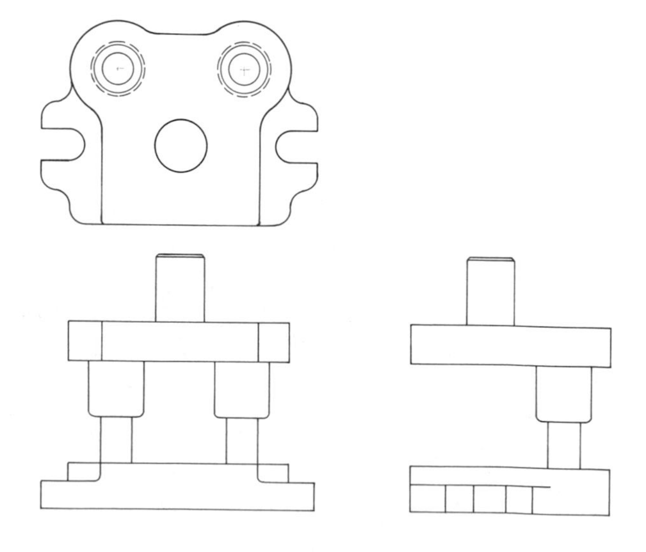

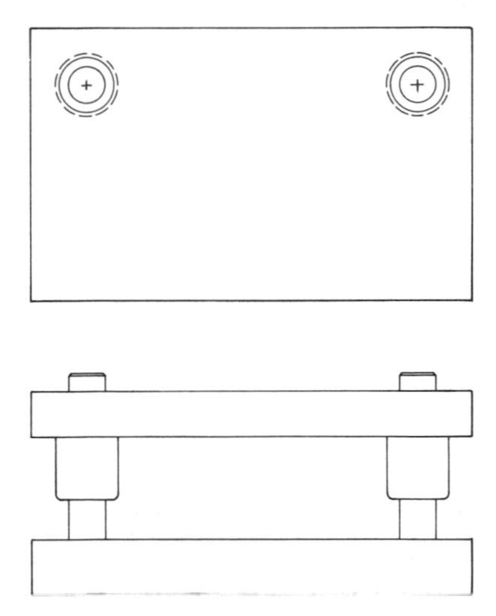

Figure 16·6 Back-post die set.

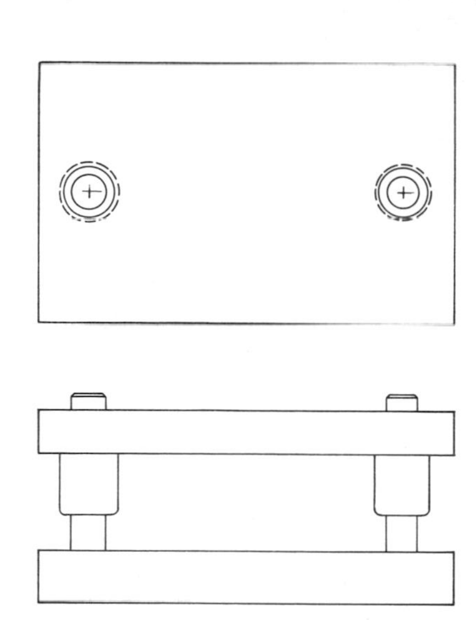

Figure 16·8 Center-post die set.

Die sets are thoroughly stress-relieved by the manufacturers before finishing and subpressing. Rough machining of deep pockets, cutouts, etc., should be done by the die-set manufacturer before the stress-relieving operation. If residual stresses are not removed, they will be gradually released in service. This can be the source of distortion and dimensional changes which can have serious consequences. The special set pictured in Fig. 16·13 will serve to emphasize the importance of proper procedures in the fabrication of die sets.

Types and styles of guideposts. An array of typical guideposts is pictured in Fig. 16·14. Increased press speeds, combined with longer production runs and the ever more exacting accuracy of dies, have instigated the development of a variety of different guideposts and bushings.

1. Commercial Guideposts. These guideposts (Fig. 16·15, view *A*) are normally centerless-ground and are often made with one diameter throughout their length. They are used when extra-close tolerance of fit between post and bushing is not essential. However, not all commercial posts are made this way; some are now being produced with the same configuration as the precision-grade guidepost, view *B*.

2. Precision Guideposts. Precision guideposts (press-fit type) are installed in the die set by being pressed into the appropriate shoe (usually the die shoe). They have a press-fit diameter which is slightly larger than the bearing diameter. They are normally ground on centers. The essential difference between precision grade and commercial grade is a matter of fit, tolerance, and finish. Precision posts are generally hard-chrome-plated and lapped, honed, or superfinished. Hard

Figure 16·7 Center-post die set.

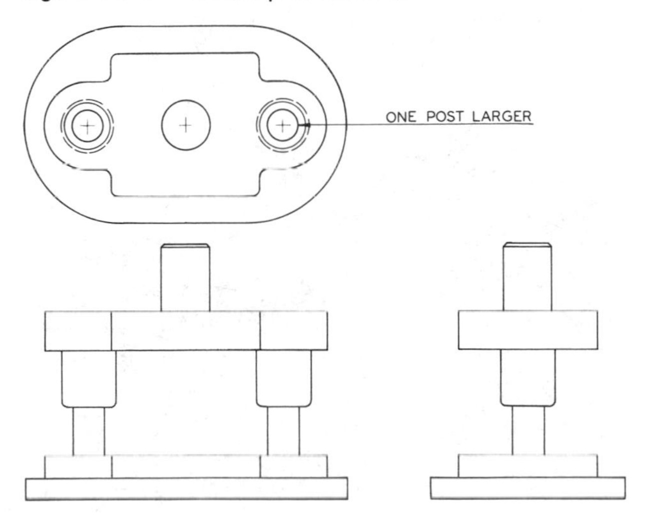

Figure 16·9 Diagonal-post die set.

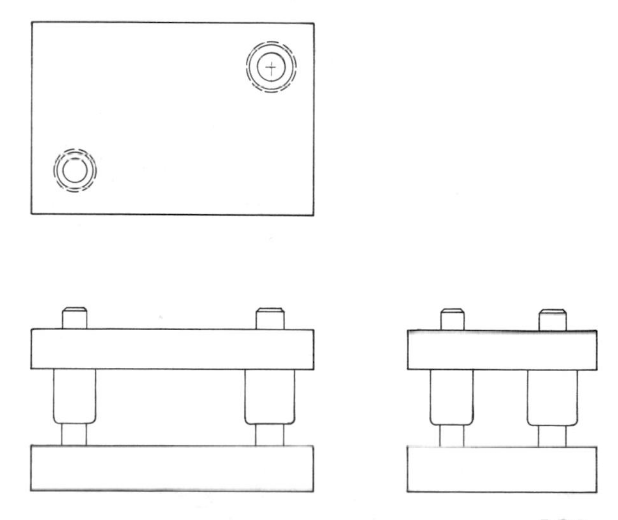

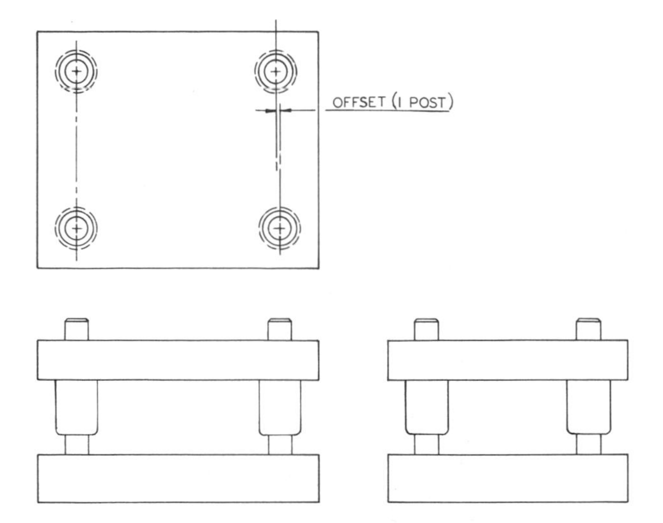

Figure 16·10 Four-post die set.

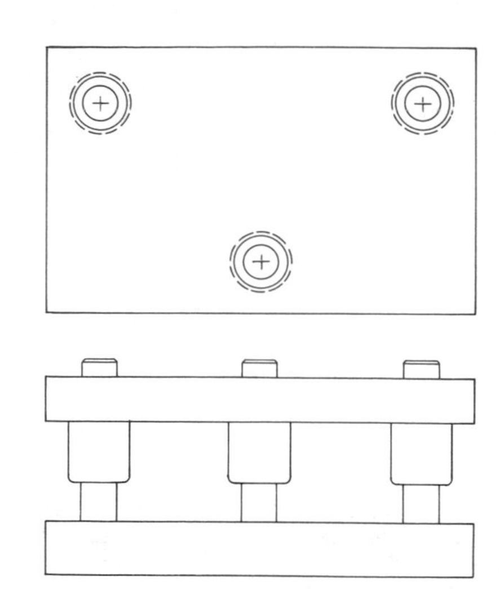

Figure 16·12 Special three-post die set.

chrome plating increases wear-resistance and reduces friction.

3. Shoulder Guideposts. As shown in view *C*, these pins (sometimes called bottle pins) have a larger press-fit diameter. This permits through-boring of the punch holder and die shoe with the same size hole for the press-fit diameter of the post and the guide bushing. The shoulder diameter is furnished with a grinding stock allowance for fitting to the bored hole.

4. Self-aligning Guideposts. These pins (Fig. 16·16) are made in precision grade. They were developed in order to eliminate jamming of the pins in the guide bushings during assembly and disassembly of the die set in the course of diemaking and maintenance. The first such pin was the Qwik-fit pin made by the Producto Machine Company. This uses a precision-ground ball and undercut, as pictured, on the lead end of the pin similar in principle to the go end of a go-no-go gage. The McVey pin, produced by Standard Die Set Division of Harsco Corporation, employs a similar shape to achieve the desired result, as does the Pilot pin offered by Danly Machine Specialties.

5. Removable Guideposts. These pins (Fig. 16·17) can be easily removed for replacement or for access to the dies for maintenance purposes, especially sharpening. Some typical removable guidepost descriptions follow.

a. Danly and Standard Die Set removable

Figure 16·11 Special two-post die set.

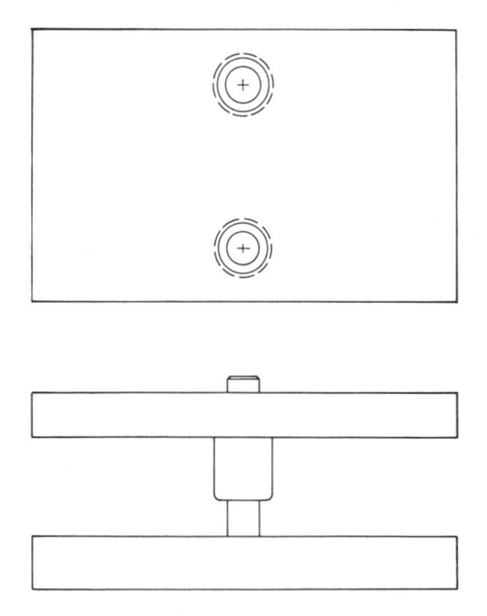

Figure 16·13 A large special set.

Figure 16 · 14 Some typical guideposts (guide pins).

guideposts (view *A*) have an axial hole through them, tapered at the press-fit end to engage a taper plug *D*. This end of the post is slotted to permit expansion. Driving the taper plug expands the post against the wall of the hole in the shoe. To remove the post, drive the taper plug back out.

b. Producto Machine Company makes the removable pin shown in view *B*. A socket cap screw *E* is used to advance the taper plug *F* for locking.

c. Union removable guideposts (view *C*) have a taper at the press-fit end fitted to the bushing *G*, which is press-fitted in the shoe. A socket cap screw *H* through the retaining cap *J* secures the post into the bushing.

Guide bushings for die sets. Bushings, which play an equally important part in die-set accuracy, are also available in a wide choice of

Figure 16 · 15 Three types of standard guideposts.

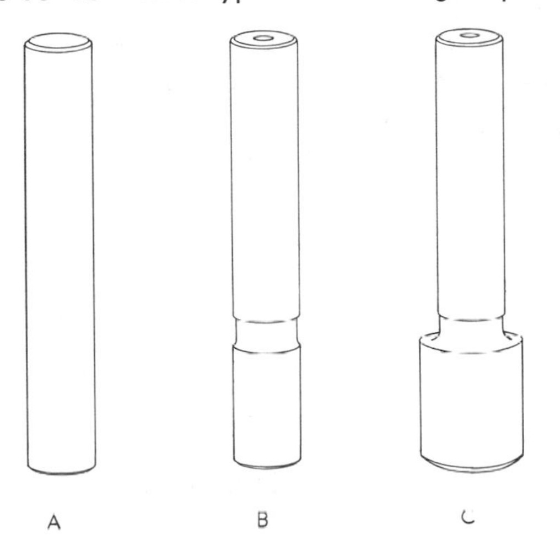

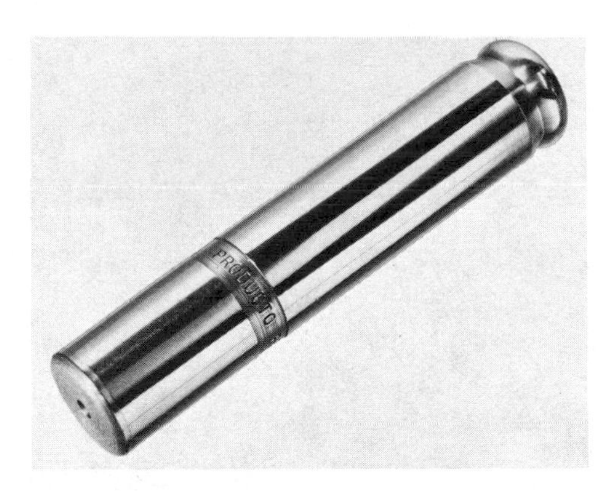

Figure 16 · 16 Self-aligning guidepost.

styles and materials (see Fig. 16 · 18). Some of the most commonly used bushings are:

1. "Precision Bushing" Types
- *a.* Steel, carburized and hardened, press-fit (PF) or demountable (D)
  1) Short straight (PF)
  2) Long straight (PF)
  3) Short shoulder (PF or D)
  4) Regular shoulder (PF or D)
  5) Medium shoulder (PF or D)
  6) Long shoulder (PF or D)
  7) Bronze plated (D)
  8) Special (PF or D)
- *b.* Bronze (solid), in Ampco grade #18, bearing bronze or commercial bronze
  1) Regular shoulder (PF or D)
  2) Medium shoulder (PF or D)
  3) Short shoulder (PF or D)
  4) Long shoulder (PF or D)
  5) Special (PF or D)
- *c.* Ball bearing
- *d.* Bronze plated

Figure 16 · 17 Removable guideposts.

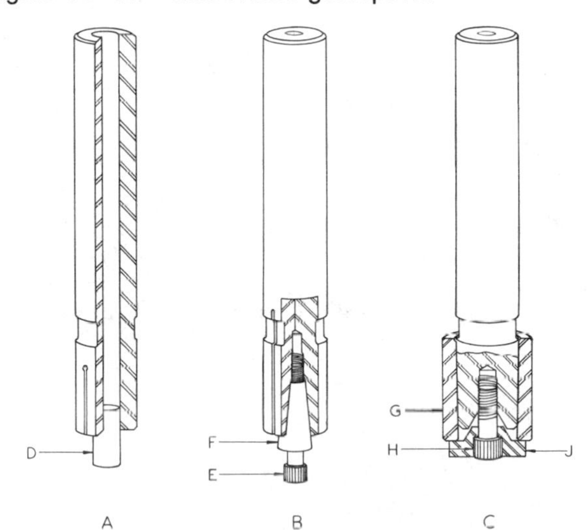

Figure **16·18** A group of typical guide bushings for die sets.

2. Commercial Bushings. These are not as closely fitted to guide pins but are available in all of the above types and materials except for the long straight style.

3. Bushings for Shoulder (Bottle) Guide Pins. These are also made in the styles mentioned above for commercial bushings, but they have grinding stock allowance on the press fit (or slip fit on demountable types) for fitting to bored holes.

Bosses for guidepost and/or bushing support. Refer to Fig. 16·19. Quite often bosses, either cast integral, demountable, or welded, are used to give added rigidity to the die set by supporting a greater length of the pressed-in portion of the guide pin. Also, when used to support the bushing, a boss places the bushing closer to the top of the die holder (further down the length of the pin) where less deflection of the pin is possible. This increases the rigidity and accuracy of the assembled set.

Figure **16·19** Die set with bosses for the guideposts and bushings.

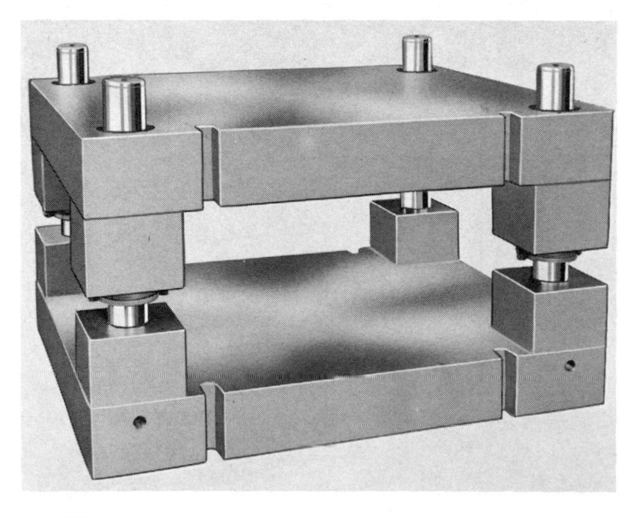

HOW DIE SETS ARE CLAMPED IN PUNCH PRESSES

The die set must also provide a method or means by which it can be fastened in place in the punch press. The catalog die sets commercially available usually are made with flanges on the die holders (lower plates) which contain bolt slots for easy mounting. The punch holders (upper plates) can be purchased with inserted, welded, or cast-integral shanks which are locked into the ram of the press. Special sets may also be made with shanks, but larger sets are usually bolted to an adapter plate on the ram of the press. Other means of fastening up the punch holders are keys, tee slots, or dovetails.

TOLERANCES ON DIE SET COMPONENTS

The die-set industry follows a set of general standards often better than those prescribed by ASA, which are obsolete and may soon be withdrawn from usage or revised. The standards of the die set industry govern thicknesses of plates, die areas, shank sizes, fit between pins and bushings in assembled sets, and parallelism between plates of assembled sets. Also covered are parallelism between top and bottom surfaces of each plate, perpendicularity of guide pins to working surface, and squareness and tolerance on diameters of shanks.

Fit between guidepost and bushing is loosely described in the standards, but industry standards as follows are generally the rule:

1. Super precision sets: 0.0002-in. total clearance (class 1 fit)
2. Precision sets: 0.0003 to 0.0005-in. total clearance (class 2 fit)
3. Commercial sets: 0.0006- to 0.0009-in. total clearance (class 3 fit)

Often greater clearance is required for heavy-duty blanking and forming dies or hot applications.

SURFACE FINISHES ON DIE SET COMPONENTS

Punch and Die Holders. Top and bottom surfaces of these parts are usually blanchard-ground, with some manufacturers furnishing surface-ground finish for ease of layout on the working areas. Where shanks are cast integral or welded, the shanks have a smooth turned finish.

Guide Pins. Most guide pins are ground with a 20-μ in. finish, but major manufacturers of die sets now improve this finish to 5 to 9μ in., which is ideal for wear and lubrication qualities required in the operating die set.

Bushings. Press-fit or slip-fit diameters on bushings are also ground with a fine finish, but only to simplify insertion into holes in the punch holders. The ID of the bushing is ground smooth but almost always honed or lapped to a smooth finish when fitted to the guide pins.

COMMERCIAL VERSUS PRECISION DIE SETS

The basic difference between commercial and precision die sets is the fit between pins and bushings in assembled sets. Tolerances on these fits have been previously outlined. The choice of fit is based on the accuracy which must be maintained in the part being produced or the allowable clearance between punches and dies. While part sizes may not necessarily require close tolerances, if the material being stamped is thin or difficult to cut clean, a very close fit of the punch and die must be maintained. Only with precision die sets can alignment of close-fitting punches and dies be held to prevent rubbing of one cutting edge against the other. In work where part size is not too critical or where punch and die clearance is great, a commercial set can be used with no resulting problems.

SHOULD DIE SETS BE MADE OR PURCHASED?

The die-set manufacturer in his range of catalog die sets offers one feature which makes it advantageous to buy die sets rather than to make your own. This feature is found in the close tolerances maintained between pin and bushing hole centers in punch and die holders. Interchangeability enables you to replace a punch or die holder quickly, should the need arise, with an off-the-shelf part which can be assembled to the mating part you already have tooled up. Should you ever have occasion to make your own set and later find it necessary to replace punch or die holder, the cost of the work involved usually will far exceed what would have been the cost of having purchased the entire set to begin with.

In addition to the above reason for purchasing rather than making your own sets is the ability of the die set manufacturer to assemble and ship the set you need in a matter of hours. And with the unlimited choice of components (which are mass produced) that he offers you, you can select one particular set out of a possible combination of hundreds of thousands. Only in rare instances, where an exceptional number of special requirements in a die set are needed, would a set be less costly to make than purchase.

DIE SET SELECTION AND ORDERING

Now, with a knowledge of the physical appearance of die sets, the tolerances to which they are manufactured, and the choices available, turn to the selection of the die set that best meets the requirements of your die. It is important in choosing a set that you include all the specifications on your order so that the set you receive exactly meets your tooling needs. With your own delivery promises to meet, an oversight may make it necessary to make do with the set you have ordered incorrectly or lose time waiting for a replacement. Here is a suggested checklist for selecting and ordering a *catalog* die set:

1. Lay out the die. Allow sufficient area for feeding devices and auxiliary stock guides that may be required. Determine also what areas must be left clear for part ejection or scrap.
2. Determine the type of set needed: rear pin, two pin, four pin, center pin, etc., according to tolerances and construction of the die.
3. Select the die area that best meets the limits set in step 1. This may be done more easily by using layout templates furnished by the die-set manufacturer.
4. Select the material from which the die set should be made—semisteel, all steel, or a combination, depending on the die's strength requirements.
5. Select thickness of punch and die holders in line with strength requirements and allowable press shut height.
6. Determine length of guide pins required based on press shut height and length of stroke. *Most manufacturers consider guide-pin length as the distance from the bottom of the die holder to the top of the pin, not the actual pin length.*
7. Determine type and length of bushing and material from which it should be made, based on accuracy, speed of operation, and required life expectancy of die.
8. Select shank diameter for the press on which the die will be run. If the press is not yet determined, either order the smallest-size shank available so that it can be fitted with an adapter to a larger size for use in any press, or have the punch holder tapped and an inserted shank installed which can be changed, if necessary, to the final size required.
9. Choose the grade of precision required—super precision, precision, or commercial—based on tolerances and number of pieces to be produced.
10. Order the die set which most closely meets all of the above requirements, making certain that all ordering information is included:

a. Catalog number
b. Precision or commercial grade
c. Length of guide pin
d. Type of bushing
e. Diameter of shank (lengths are standard) or no shank
f. How to ship

Keep in mind that special thicknesses of punch and die holder can be ordered, if required, with little extra cost and very little delay in shipping time. Of course, it may not be possible to check these points in the exact order outlined. However, if you will make it a habit to use this checklist for each die set needed, you can avoid making costly errors in ordering.

HOW TO ORDER SPECIAL DIE SETS

When it is found that no catalog die set meets the requirements for the die, a special die set must be considered. The procedure outlined above should be followed for special sets also, with some variation in the information needed. The simplest method of ordering is to send in a print or a rough sketch with all dimensions indicated clearly. When urgency is a factor, and no time is allowed to make a sketch, the manufacturer's catalog will serve as a guide to what information is necessary for ordering specials. Basically, the style of set, thicknesses of plates, pin lengths, bushing style, shank size, overall length and width of plate and/or die area clear of pins necessary, are all that are needed to order the set desired. Never hesitate to phone the die-set supplier for assistance in ordering or for technical advice and assistance. He welcomes every opportunity to be of service and may be in a position to make helpful suggestions that save time in getting the set.

BUYING DIE SETS WITH SPECIAL FEATURES OR EXTRAS

When ordering additional operations on the components being purchased, such as pockets, slots, machined edges, keyways, or drilled, tapped, or counterbored holes, etc., do not ask for closer tolerances, smaller radii, or better finishes than are needed. These unnecessarily increase the cost of the set. Below are suggested ways to keep special die-set costs to a minimum.

While the following recommendations may seem elementary to the designer and user of die sets, we bring them to your attention because each year we see literally hundreds of prints where one or more of the suggestions could have cut costs and shortened delivery time. In 90 per cent of the cases, one of these alternates could have been used.

1. On rectangular or circular plates for die holders, punch holders, strippers, bolsters, etc., if overall dimensions do not have decimal tolerances and a machined appearance is not essential, a clean-looking job at lowest cost can be obtained by flame-cutting.
2. Where cutouts are required for drop-out holes, bolt slots, clearance, or other purposes, and tolerances on dimensions are fractional, the fastest and most economical method to use is flame-cutting.
3. However, if appearance is an important factor and the fine rippled edge left by flame-cutting would not be satisfactory, edges, drop-out holes, and slots can in many cases be saw-cut to give a machined look at lower cost.
4. Large radii cut costs. Small radii in corners of machined pockets and cutouts necessitate use of a small-diameter cutter, increasing the number of cuts required and reducing cutter speed and feed. Large radii reduce cutting time by more than half.
5. Liberal hole location tolerances keep machining time down and keep the order from being held up at precision machines usually working under heavy load.
6. Pin or bushing supports in order of cost (high to low):
   a. Machined from solid
   b. Screwed and doweled
   c. Welded. (Welded supports, in addition to being lowest in cost, require much less time for fabrication and help to speed up delivery.)
7. Bolt slots in hold-down pads or flanges can be flame-cut at the same time plates are being cut to size, saving time and eliminating cost of an extra machining operation.
8. When machining of bolt slots is a must and a square slot with small corner radii is not essential, a single-radius slot is cheaper and faster.
9. Try to use catalog pins and bushings. There are comparatively few cases where selection of the right pin and bushing cannot be made directly from the catalog. Special dimensions always mean higher cost and longer delivery.

Finally, it is suggested that you always review your specifications and ask, "Will one of these alternates do the job as well?" Each "yes" is money saved.

IMPORTANCE OF PROPER LUBRICATION

It does not necessarily hold true that precisely made punches and dies mounted in precision die sets will run continuously without problems arising. Scoring or galling of the guide pins while the die is in the press can lead to serious problems, even to the possibility of the die being broken or severely damaged. One of the major causes of damage to dies is the pulling out of pins or bushings which have scored or overheated due to faulty lubrication or none. While this is a major reason for proper lubrication of sets, even the minor need to lubricate is important. A well-lubricated guide pin and bushing will last longer and remain at the original clearance built into them for longer periods.

While the problem of lubrication is easily solved in most cases, in many it is a critical procedure for which special systems have been devised. The following methods are typical:

1. Relief in top of bushing is used as an oil or grease reservoir to run lubricant over guide pins with each stroke of press.
2. Oil cups on bushings feed oil into a simple ring groove in the bushing.
3. Grease fittings, gun-fed or spring-loaded reservoir type, feed grease into simple ring grooves or double figure-eight spiral grooves.
4. Flexible reservoir lubricators, pressed over the pin, hold a supply of oil which is drawn up over the pin by the bushing with each stroke of press.
5. Specially fitted caps containing felt washers saturated with oil are fitted over bushings after they are assembled in die set to provide a combination wiping and lubricating action for guide pins. They are used in conjunction with self-lubricating types of pins.
6. Simple rings around the guide pin or thread-type spirals act as reservoirs for lubricant picked up from relief in bushings, as described in example 1. These types of pins, however, have a tendency to pick up dirt and chips. They also tend to cause the pin to act as a broach which opens the bushing diameter or scores it.
7. Guide pins with a hole through the length intersected by holes through the diameter are automatically fed with oil or grease, which runs through the holes and coats the pin and bushing.
8. An oil reservoir pin with a hole drilled in the top stores a small amount of oil and feeds pins through four feed holes intersecting from the sides. This type is usually filled intermittently but can be set up as in example 7.

High press speeds often create heat friction to the extent that lubricants thin out and lose their effectiveness, which usually results in a special lubrication problem. Use of solid-bronze or bronze-plated steel bushings greatly reduces running temperatures caused by friction and in many cases eliminates the need for special lubrication methods.

While some of these lubrication methods are standard with some users, they are special from the standpoint of die-set manufacture. They are brought to your attention so that you may consider them as a possible solution to some of the more common lubrication problems. Whatever method you choose, whether standard or special as shown here, or one you have devised, proper lubrication will add life to your die sets and help maintain die accuracy.

Lubrication is one form of maintenance necessary to long and accurate die life. Maintenance of the die itself requires even more consideration.

MAINTENANCE PROCEDURES

The vital die maintenance procedures required periodically when a die is removed from the press for sharpening should be coupled with systematic die-set inspection and maintenance in the pressroom.

In addition to having a planned schedule of lubrication (based on any of the methods discussed in the preceding section), you can make sure of getting the most out of your investment in die sets (and the costly dies and presses with which they are used) by following a simple set of procedures for inspecting and maintaining sets in the pressroom.

Although there are no hard-and-fast rules that can be established for all die sets, the procedure listed below is generally followed at leading stamping plants in the East. These companies point out that die sets used in the cutting of extremely thin material or for very critical parts would naturally require a more stringent inspection than those used for blanking heavy material or for comparatively simple parts.

At one plant specializing in long runs of close-tolerance parts for the electronics industry, the procedure followed is:

1. There is no inspection of the die set during the production run except, of course, if there is overheating, pick-up on the guide pins, or other obvious difficulties.

2. When the die is taken off the press for sharpening or maintenance, the die set is visually inspected for pick-up or score marks on either the guide pins or bushings. Pins and bushings are carefully cleaned to remove excess lubricant.
3. After approximately 1,500,000 strokes, the set is thoroughly checked for wear on both the pins and bushings. On the average, up to 0.0003 in. wear on pins and bushings is allowed before replacement with new components is scheduled. In addition, the die set is placed on a surface plate and an indicator is placed over one end of the set. An attempt is then made to rock or twist the set with a pry bar, to establish how much the die set can be deflected. Total deflection allowed on a 10 in. set is a maximum of 0.005 in., but usually less. Because the die set plays such an important part in prolonging the accuracy and lifetime of a die, it pays to take the little extra time required to do justice to die-set maintenance in the pressroom.
4. During regular or periodic regrinds, the die set is not checked for bent or distorted punch and die holders unless the die has been damaged or there is reason to believe the set may have been warped. During the overall period, if the pins and bushings are replaced, the punch and die holders themselves are checked for warpage or distortion. If necessary, these components are reground prior to assembling the new pins and bushings.

DIE INVENTORY CONTROL

Let us assume that you have designed a die, made the parts, mounted it in the press, seen that it was properly lubricated while running, periodically maintained it (including sharpening, cleaning, etc.), and now, with the run of stampings complete, have removed the die from the press and are preparing to store it against future use. After you have it on the shelf, how fast will you be able to get it back into operation again?

To show how systematic die maintenance and die inventory control help to guarantee the profits your company will make, here is a hypothetical case:

"Stamp 200,000 Part 227-C Urgent! Need Yesterday."

That is how the work order you have received might read. The die for Part 227-C was made in your toolroom two years ago. It was last taken off the press over one year ago. Can you produce the 200,000 urgently needed parts quickly and profitably? You can if. . . .

The die can be located on your shelves instantly.
It is clean and sharp.
Data relating to Part 227-C (material, thickness, temper, etc.) are readily at hand.

But all too often a rush order of this type is the signal for a prolonged hunt through storage racks, a hurried cleaning and sharpening session, and a search through incomplete files for essential data. The result: broken delivery promises . . . no profit.

Successful managers of stamping shops invariably declare that such details as the optimum speed at which a die can be run and the die's lubrication requirements are subjects for a systematic file—not for memory.

Die maintenance and die inventory control procedures are generally similar as conducted from plant to plant; differences lie in details or emphasis. For example, a manufacturing firm with its own pressroom facilities and over two thousand dies on its shelves will probably find it necessary to have a more complex method of locating dies by area and rack than a combined tool, die, and stamping shop which retains only a fraction of the dies made by it and has perhaps 200 customers' dies on hand.

The following are the main points in the die maintenance and inventory control methods of one company, a progressive Eastern tool, die, and stamping firm which retains many of the dies made for its customers and handles production and assembly of components as an important adjunct to its business.

1. A 5-by-7-in. central card file is the heart of the system. These cards identify every die in the shop. (A separate file records those returned to customers.) They show: Name of customer, type of die, lubrication requirements, operational speeds, customer's part number and description of part, maintenance, history of die (dates, hours, costs, press runs, sharpening dates), and a three-part code. For example: J-43-R has been stored on rack J and is the 43d die made for customer R.
2. Every die is thoroughly cleaned upon removal from press. At this particular company an unusual and efficient method of agitating dies, completely immersed, in a chemical solvent solution is employed. This bath provides an additional advantage in that it leaves a powdery rust-inhibiting residue over the entire die surface.

3. Upon completion of the last production run, the last part produced is attached to the die. This part serves as a reference when the die is inspected before storage and later when the die is again taken out for use.
4. Die inspection is standard procedure for every die going into storage. At this time, the die is sharpened if necessary. Now, at a moment's notice, it is in complete readiness for production.
5. In addition to the three-part code, an assigned color painted on the die set identifies the dies of major customers.

It is hoped that this outline will suggest refinements which may step up the efficiency of your own die inventory methods. There is little doubt that dies kept in harum-scarum style in various stages of unpreparedness are a hazard to expeditious service. On the other hand, dies kept clean and sharp, where they can be taken off the shelf and put on the job with no loss of time and no operational guesswork, are an excellent form of profit insurance.

This discussion has covered a wide range of problems on one small phase of die design and manufacture, most of which have involved the use of die sets. As your experience in the field of diemaking grows, you will advance far beyond this basic knowledge of the tools used in your chosen career. But these basic ideas should always be kept in mind as the foundation on which to build good diemaking practice.

INDEX